Ivan G. Graham, Ulrich Langer, Jens M. Melenk, Mourad Sini (Eds.)

Direct and Inverse Problems in Wave Propagation and Applications

Radon Series on Computational and Applied Mathematics

Volume 14

Direct and Inverse Problems in Wave Propagation and Applications

Edited by
Ivan G. Graham
Ulrich Langer
Jens M. Melenk
Mourad Sini

DE GRUYTER

Mathematics Subject Classification 2010
78A46, 78A45, 65N21, 65N38, 65N30

ISBN 978-3-11- 028223-8
e-ISBN 978-3-11-028228-3
ISSN 1865-3707

Library of Congress Cataloging-in-Publication Data
A CIP catalog record for this book has been applied for at the Library of Congress.

Bibliographic information published by the Deutsche Nationalbibliothek
The Deutsche Nationalbibliothek lists this publication in the Deutsche Nationalbibliografie; detailed bibliographic data are available in the Internet at http://dnb.dnb.de.

Typesetting: le-tex publishing services GmbH, Leipzig
Printing and binding: Hubert & Co. GmbH & Co. KG, Göttingen
♾ Printed on acid-free paper
Printed in Germany

MIX
Papier aus verantwortungsvollen Quellen
FSC www.fsc.org FSC® C016439

www.degruyter.com

Preface

This book surveys recent developments in the analysis of wave propagation and scattering problems, covering aspects of both forward and inverse problems, as well as applications in the earth sciences. We collect selected material from a workshop on "Wave Propagation and Scattering, Inverse Problems, and Applications in Energy and the Environment" held at the Radon Institute in Linz, Austria, November 21–25, 2011. The workshop was embedded in the "Radon Special Semester 2011 on Multiscale Simulation and Analysis in Energy and the Environment", which took place in Linz, October 3–December 16, 2011. Two companion volumes in the "Radon Series on Computational and Applied Mathematics" series cover closely related topics of this particular "Radon Special Semester."

Wave propagation problems are ubiquitous in environmental applications such as seismic analysis, and in acoustic and electromagnetic scattering. The design of efficient numerical methods for the forward problem, in which the scattered field is computed from known geometric configurations is very challenging due to the multiscale nature of the problems. Inverse problems where material parameters and configurations have to be determined from measurements in conjunction with the forward problem are even more challenging. This book contains review articles covering several state-of-the-art numerical methods for both forward and inverse problems.

Acoustic and electromagnetic wave propagation problems in the frequency domain in piecewise homogeneous media can be cast in the form of integral equations, which leads to the boundary element method after discretization. Several formulations are possible. Established ones and new ones are presented in the contribution by Claeys, Hiptmair & Jerez-Hanckes on multitrace boundary integral equations. The contribution by Steinbach focuses on issues of well-posedness of boundary integral formulations for Helmholtz problems. Integral equation techniques are a powerful tool to treat exterior (unbounded) domains; nevertheless, a volume-based discretization is often desired, especially for heterogeneous problems. Then, a reduction to a bounded computational domain is necessary. The key issues arising when truncating the computational domain are discussed in the contribution by Nataf on absorbing boundary conditions. The systems of equations that arise in volume-based discretizations in frequency domain are typically very large and sparse. Their efficient numerical solution is, however, very challenging. The contribution by Ernst and Gander surveys the current state-of-the-art in this area and proposes some new improvements to multigrid methods. Moving to the time domain, the chapter by Grote and Mitkova addresses the issue of local time stepping, which is an efficient way to accommodate both the need for fine spatial mesh resolution near small features and the fact that explicit discretizations in time impose conditions on the time step, the so-called CFL condition. The contribution by Elschner and Hu studies both direct and inverse problems for elastic wave diffraction gratings with a particular emphasis on global uniqueness within the class of piecewise linear profiles. The article by Ammari reviews recent techniques to estimate material and geometrical properties of small objects from electromagnetic measurements. These properties are encoded in the high-order polarization tensors appearing in the asymptotic expansions of the measurements. Finally, many inverse scattering problems have time-dependent features. A typical example is the scattering by moving objects. In this direction, the presentation by Potthast discusses, in particular, an approach that combines dynamical probing/sampling methods with data assimilation techniques.

Bath, Linz, Vienna, March 2013

Ivan G. Graham
Ulrich Langer
Jens M. Melenk
Mourad Sini

Contents

Habib Ammari

Differential electromagnetic imaging

Abstract: This chapter is an introduction to recent approaches for target imaging with electromagnetic waves. We are interested in imaging small perturbations with respect to a known situation. We provide asymptotic effects due to these perturbations. We show how to represent the shape of a target using high-order polarization tensors. Based on this, we design very effective imaging functionals, aimed at determining certain geometric and material properties of the small target from electromagnetic multistatic measurements.

Keywords: electromagnetic waves, imaging, generalized polarization tensors, shape representation, small-volume targets, shape deformation

2010 Mathematics Subject Classification: 35R30, 35B30

Habib Ammari: Department of Mathematics and Applications, Ecole Normale Supérieure, 45 Rue d'Ulm, 75005 Paris, France, habib.ammari@ens.fr

1 Introduction

Electromagnetic waves can be used in order to probe an unknown medium. They are generated and recorded by a set of antennas. The governing equations are the full Maxwell equations. In the case where the electromagnetic parameters are independent of one of the variables and the electromagnetic fields have a given exponential variation in the direction of invariance, the Maxwell equations split into two sets of independent scalar Helmholtz equations. On the other hand, Maxwell's equations reduce to the eddy currents model or the conductivity equation at low frequencies.

The objective of this chapter is to review recent approaches for target imaging with electromagnetic waves at single or multiple frequencies. The electromagnetic waves are generated by point sources on a transmitter array and measured on a receiver array. For the sake of simplicity, we consider coincident transmitter and receiver arrays. There are two interesting problems: finding small electromagnetic targets and reconstructing shape deformations of an extended target. A target is called small when its characteristic size times the operating frequency is less than one while it is called extended when this factor is much larger than one. In both situations, we are interested in imaging small perturbations with respect to known situations.

This work was supported by ERC Advanced Grant Project MULTIMOD–267184.

In this chapter, we focus on imaging small targets. We refer the interested reader to [3, 4, 16, 24] for imaging small-shape deformations.

Our approach is based on an asymptotic analysis of the measured data in terms of the size of the unknown targets or the order of magnitude of the shape deformation. The asymptotic analysis plays a key role in characterizing all the information about the small target or the shape deformations of an extended target that can be stably reconstructed from the measured data. It provides robust and accurate reconstruction of the location and of some geometric features of the small targets as well as small changes of the shape of an extended target, even with moderately noisy data.

When we are dealing with target imaging problems, a fundamental problem is to have a shape representation well suited for solving the inverse problem. Parametric representations do not do a good job. However, high-order polarization tensors provide a well-suited representation of the shape of a target. The high-order polarization tensors generalize the (classical) magnetic and electric polarization tensors associated with a small target with a given set of electromagnetic parameters. For an arbitrary shape, we can find two equivalent ellipsoids with, respectively, the same electric and magnetic polarization tensors. Using high-order polarization tensors we can not only recover finer details of the shape of a given target, but also separate its electromagnetic parameters from its volume. For extended targets, an equivalent concept can be introduced. In this chapter, we derive an optimization procedure to reconstruct a target from its high-order polarization tensors. Some care should be taken in dealing with topology problems related to Maxwell equations (Dirichlet and Neumann fields).

Since our approach is based on asymptotic expansions together with polarization tensors for shape or deformation representations, we refer to it as a *differential imaging* approach. The main potential applications of our approach include microwave breast cancer detection, airport security screening, and detection of antipersonnel mines.

The chapter is organized as follows. In Section 2, we discuss the basic theory of electromagnetic waves. Section 3 is devoted to the magnetic and electric polarization tensors. In Section 4, we provide formal derivations of the far-field behaviors of the electromagnetic fields in the presence of a small target. We also establish asymptotic expansions of the field perturbations induced by the target in the transient regime. In Sections 5–7, we introduce the main algorithms for differential electromagnetic imaging. The high-order polarization tensors are introduced in Section 8 and their use for shape representations is discussed. Section 9 is devoted to a brief discussion on the near-field imaging. The chapter ends with a list of open problems.

2 Basic theory of electromagnetic waves

2.1 The Helmholtz equation

Consider the scalar wave equation $\partial_t^2 u - \Delta u = 0$. We obtain a time-harmonic solution $u(\mathbf{r}, t) = \Re e[e^{-i\omega t} u(\mathbf{r})], t > 0$, if the space-dependent part $u(\mathbf{r})$ satisfies the Helmholtz equation

$$\Delta u + \omega^2 u = 0. \tag{1.1}$$

Here ω is the frequency.

The Helmholtz equation has a very special family of solutions called plane waves:

$$u(\mathbf{r}) = e^{i\omega\theta\cdot\mathbf{r}}, \qquad \theta \in \mathbb{S}^{d-1} := \{\theta \in \mathbb{R}^d : |\theta| = 1\}.$$

Another remarkable set of solutions is the family of spherical waves, which depend only on the radial variable $r = |\mathbf{r}|$. For example, we can verify that the following function satisfies (1.1):

$$u(\mathbf{r}) = \begin{cases} j_0(\omega r) = \dfrac{\sin \omega r}{\omega r} = \dfrac{1}{4\pi} \displaystyle\int_{\mathbb{S}^2} e^{i\omega\theta\cdot\mathbf{r}}\, ds(\theta), & d = 3, \\ J_0(\omega r) = \dfrac{1}{2\pi} \displaystyle\int_{\mathbb{S}^1} e^{i\omega\theta\cdot\mathbf{r}}\, ds(\theta), & d = 2, \end{cases}$$

where j_0 is the spherical Bessel function of the order zero and J_0 is the Bessel function of the first kind and of the order zero. Note that, for $d = 2$ or 3, $r^{(d-1)/2}$ times a spherical wave is asymptotically a plane wave for large value of r.

2.2 The Maxwell equations

Maxwell's equations describe all (classical) electromagnetic phenomena. In the absence of electric and magnetic charges and currents, the electric and magnetic fields in a conducting medium are governed by the time-dependent Maxwell equations

$$\begin{cases} -\varepsilon_0 \dfrac{\partial \mathbf{E}}{\partial t} + \nabla \times \mathbf{H} - \sigma \mathbf{E} = 0, \\ \mu_0 \dfrac{\partial \mathbf{H}}{\partial t} + \nabla \times \mathbf{E} = 0, \end{cases}$$

where ε_0, μ_0, and σ denote, respectively, the permittivity, the permeability, and the conductivity. If the medium is dielectric then $\sigma = 0$. In general, the electromagnetic parameters of a medium may be inhomogeneous, anisotropic, nonlinear, frequency dependent (dispersive), or all of the above. In this chapter, we restrict ourselves to linear, isotropic, nondispersive media.

Maxwell's equations simplify considerably in the case of harmonic time dependence. If the electromagnetic fields are of the form: $\mathbf{E}(\mathbf{r}, t) = \Re e[e^{-i\omega t}\mathbf{E}(\mathbf{r})]$ and $\mathbf{H}(\mathbf{r}, t) = \Re e[e^{-i\omega t}\mathbf{H}(\mathbf{r})]$, where ω denotes a given frequency then $\mathbf{E}$ and $\mathbf{H}$ satisfy the time-harmonic Maxwell equations

$$\begin{cases} i\omega\varepsilon_0\mathbf{E} + \nabla \times \mathbf{H} - \sigma\mathbf{E} = 0, \\ -i\omega\mu_0\mathbf{H} + \nabla \times \mathbf{E} = 0. \end{cases}$$

By taking the curl we obtain the following system of equations for $\mathbf{E}$:

$$\nabla \times \nabla \times \mathbf{E} - \omega^2\mu_0\left(\varepsilon_0 + i\frac{\sigma}{\omega}\right)\mathbf{E} = 0.$$

Looking for plane waves propagating in a dielectric medium,

$$\begin{cases} \mathbf{E}(\mathbf{r}) = \mathbf{p}e^{ik_0\theta\cdot\mathbf{r}}, \\ \mathbf{H}(\mathbf{r}) = \mathbf{q}e^{ik_0\theta\cdot\mathbf{r}}, \end{cases}$$

with $k_0 = \omega\sqrt{\varepsilon_0\mu_0}$, leads to

$$\begin{cases} \sqrt{\dfrac{\varepsilon_0}{\mu_0}}\mathbf{p} + \theta \times \mathbf{q} = 0, \\ -\sqrt{\dfrac{\mu_0}{\varepsilon_0}}\mathbf{q} + \theta \times \mathbf{p} = 0. \end{cases} \tag{1.2}$$

Therefore, the three vectors $\theta, \mathbf{p}, \mathbf{q}$ must be mutually orthogonal. Moreover, equation (1.2) describes all the plane wave solutions to Maxwell's equations.

2.3 Fundamental solutions and radiation conditions

Fundamental solution to the Helmholtz equation

A fundamental solution $g(\mathbf{r}, \mathbf{r}'; \omega)$ to the Helmholtz operator $\Delta + \omega^2$ in $\mathbb{R}^d$ is a solution (in the sense of distributions) of

$$(\Delta + \omega^2)g(\mathbf{r}, \mathbf{r}'; \omega) = \delta_{\mathbf{r}'}, \tag{1.3}$$

where $\delta_{\mathbf{r}'}$ is the Dirac mass at $\mathbf{r}'$. Solutions are not unique, since we can add to a solution any plane wave (of the form $e^{i\omega\theta\cdot\mathbf{r}}, \theta \in \mathbb{S}^{d-1}$) or any combination of such plane waves. We need to specify the behavior of the solutions at infinity. We look for solutions subject to the extra (outgoing) Sommerfeld radiation condition

$$\left|\frac{\partial g}{\partial r} - i\omega g\right| = O(r^{-(d+1)/2}) \quad \text{as} \quad r \to +\infty, \tag{1.4}$$

uniformly in $\hat{\mathbf{r}} := \mathbf{r}/r$ for $d = 2, 3$. The following lemma holds:

Lemma 1.1. *The outgoing fundamental solution g is given by*

$$g(\mathbf{r},\mathbf{r}';\omega) = \begin{cases} -\frac{i}{4}H_0^{(1)}(\omega|\mathbf{r}-\mathbf{r}'|), & d=2, \\ -\frac{e^{i\omega|\mathbf{r}-\mathbf{r}'|}}{4\pi|\mathbf{r}-\mathbf{r}'|}, & d=3, \end{cases}$$

for $\mathbf{r} \neq \mathbf{r}'$, where $H_0^{(1)}$ is the Hankel function of the first kind of order 0.

Note that $\Re e H_0^{(1)} = J_0$.

Fundamental solution to the Maxwell equations

Using g we define the fundamental solution $\mathbf{G}$ to the Maxwell equations in a homogeneous dielectric medium by

$$\mathbf{G}(\mathbf{r},\mathbf{r}';k_0) = \left[\mathbf{I} + \frac{\nabla\nabla}{k_0^2}\right] g(\mathbf{r},\mathbf{r}';k_0), \tag{1.5}$$

where $\mathbf{I}$ is the 3×3 identity matrix.

Lemma 1.2. *The function $\mathbf{G}(\mathbf{r},\mathbf{r}';k_0)$ defined by (1.5) is the solution of the Maxwell equations*

$$\nabla\times\nabla\times\mathbf{G}(\mathbf{r},\mathbf{r}';k_0) - k_0^2\mathbf{G}(\mathbf{r},\mathbf{r}';k_0) = \mathbf{I}\delta(\mathbf{r}-\mathbf{r}') \quad \text{in } \mathbb{R}^3,$$

and satisfies the outgoing radiation condition of Silver–Müller:

$$|\nabla\times\mathbf{G}(\mathbf{r},\mathbf{r}';k_0) - ik_0\hat{\mathbf{r}}\times\mathbf{G}(\mathbf{r},\mathbf{r}';k_0)| = O(r^{-2}) \quad \text{as } r\to+\infty,$$

uniformly in $\hat{\mathbf{r}}$. It also satisfies the reciprocity property

$$[\mathbf{G}(\mathbf{r}',\mathbf{r};k_0)]^T = \mathbf{G}(\mathbf{r},\mathbf{r}';k_0),$$

where T denotes the transpose.

The expression (1.5) is still valid for the Maxwell equations in a conducting medium with k replaced by $\omega\sqrt{\mu_0(\varepsilon_0 + i\sigma/\omega)}$. The fundamental solution $\mathbf{G}$ decreases as $e^{-r\sqrt{\omega\mu_0\sigma/2}}$ for $\omega\varepsilon_0/\sigma \ll 1$. The electromagnetic fields in a conducting medium tend to zero as the conductivity σ tends to infinity with a boundary layer effect, whose thickness is proportional to the skin depth, $1/\sqrt{\omega\mu_0\sigma}$.

2.4 Transmission and boundary conditions

Consider a dielectric medium with permittivity and permeability $\varepsilon(\mathbf{r})$ and $\mu(\mathbf{r})$, respectively. The tangential components $\mathbf{E}\times\nu$ and $\mathbf{H}\times\nu$ of the electric and magnetic fields **E** and **H** are continuous across a surface of discontinuity of the permittivity $\varepsilon(\mathbf{r})$ or the permeability $\mu(\mathbf{r})$. Here ν denotes the outward normal to the surface of discontinuity. Therefore, $\varepsilon(\mathbf{r})\mathbf{E} \cdot \nu$ and $\mu(\mathbf{r})\mathbf{H} \cdot \nu$ have no jump.

Now, if a conducting medium is surrounded by a dielectric medium and its conductivity goes to infinity, then **E** and **H** tend to zero in the conductor. At the interface between the conductor and the dielectric medium, the interior limits of the tangential components of **E** and **H** are zero and so is the exterior limit of the tangential component of **E** while the exterior limit of the tangential component of **H** and the normal component of **E** are nonzero. This is the model for the perfectly conducting medium.

2.5 Dirichlet and Neumann functions and the Hodge decomposition

Let B be a bounded smooth domain in $\mathbb{R}^3$. Let ν denote the outward normal to B. Introduce the functional spaces

$$\begin{aligned} H^1(B) &= \{q \in L^2(B) : \nabla q \in L^2(B)\}, \\ H_0^1(B) &= \{q \in H^1(B) : q = 0 \text{ on } \partial B\}, \\ H(\mathrm{curl}, B) &= \{\mathbf{v} \in L^2(B) : \nabla \times \mathbf{v} \in L^2(B)\}, \end{aligned}$$

and

$$H_0(\mathrm{curl}, B) = \{\mathbf{v} \in L^2(B) : \nabla \times \mathbf{v} \in L^2(B), \mathbf{v} \times \nu = 0 \text{ on } \partial B\}.$$

The functional space

$$\mathcal{D}_{\mathrm{int}}(B) := \{\mathbf{v} \in L^2(B) : \nabla \times \mathbf{v} = 0, \nabla \cdot \mathbf{v} = 0 \text{ in } B \text{ and } \mathbf{v} \times \nu = 0 \text{ on } \partial B\}$$

is formed only by the Dirichlet fields; it is finite dimensional and its dimension is equal the number of connected components of the boundary ∂B. The same holds for the exterior domain $\mathbb{R}^3 \setminus \overline{B}$ [32].

The functional space

$$\mathcal{N}_{\mathrm{int}}(B) := \{\mathbf{v} \in L^2(B) : \nabla \times \mathbf{v} = 0, \nabla \cdot \mathbf{v} = 0 \text{ in } B \text{ and } \mathbf{v} \cdot \nu = 0 \text{ on } \partial B\}$$

is formed only by the Neumann fields; it is finite dimensional and its dimension is equal the first Betti number of ∂B, which is roughly speaking the number of holes in ∂B. If ∂B is simply connected, then the first Betti number of ∂B is zero. The set of Neumann fields, $\mathcal{N}_{\mathrm{ext}}(B)$, associated with $\mathbb{R}^3 \setminus \overline{B}$ can be defined analogously.

The set of harmonic vectors on ∂B:

$$\mathcal{N}(\partial B) := \{\mathbf{v} \cdot \nu = 0 : \Delta_{\partial B} \mathbf{v} = 0\} = \{\mathbf{v} \cdot \nu = 0 : \mathrm{curl}_{\partial B} \mathbf{v} = \nabla_{\partial B} \cdot \mathbf{v} = 0\},$$

where $\Delta_{\partial B}$ is the Laplace–Beltrami operator on ∂B, $\mathrm{curl}_{\partial B}$ the scalar surface rotational, and $\nabla_{\partial B} \cdot$ the tangential divergence, is of dimension twice the first Betti number of ∂B. It is formed by the tangential traces of the Neumann fields associated with B and $\mathbb{R}^3 \setminus \overline{B}$ or equivalently by the vector surface rotational and the surface gradient of the scalar functions that are in the kernel of the scalar Beltrami–Laplace operator.

Let $TH^{-1/2}(\mathrm{div}, \partial B)$ denote the set of tangential vectors on ∂B such that $\mathbf{v}$ and $\nabla_{\partial B} \cdot \mathbf{v}$ are in $H^{-1/2}(\partial\Omega)$.

Lemma 1.3. *We have*

$$TH^{-1/2}(\mathrm{div}, \partial B) = \nabla_{\partial B} H^{3/2}(\partial B) \oplus \nabla_{\partial B} \times H^{1/2}(\partial B) \oplus \mathcal{N}(\partial B), \tag{1.6}$$

where $\nabla_{\partial B}\times$ *denotes the vector surface rotational.*

Analogously, the following decomposition of the dual of $TH^{-1/2}(\mathrm{div}, \partial B)$ with L^2 as pivot space holds.

Lemma 1.4. *We have*

$$\begin{aligned} TH^{-1/2}(\mathrm{curl}, \partial B) &:= \{\mathbf{v} \cdot \nu = 0 : \mathbf{v} \in H^{-1/2}(\partial B), \mathrm{curl}_{\partial B}\mathbf{v} \in H^{-1/2}(\partial B)\} \\ &= \nabla_{\partial B} H^{1/2}(\partial B) \oplus \nabla_{\partial B} \times H^{3/2}(\partial B) \oplus \mathcal{N}(\partial B). \end{aligned} \tag{1.7}$$

The Hodge decompositions (1.6) and (1.7) of the tangential fields on ∂B are the key to study the invertibility of the layer potentials associated with the Maxwell equations.

When dealing with the low frequency or the small-volume limits for the Maxwell equations, new difficulties compared to the Helmholtz equation arise because of the topology. We add source terms (orthogonality relations) to overcome them.

Finally, recall the following useful identities for studying layer potentials associated with the Maxwell equations

$$-\mathrm{curl}_{\partial B}(\nabla_{\partial B}\times) = \Delta_{\partial B} = (\nabla_{\partial B}\cdot)\nabla_{\partial B}.$$

2.6 Trace theorems and first Green identity

The spaces $H(\mathrm{curl}, B \text{ or } \mathbb{R}^3 \setminus \overline{B})$ and $H(\mathrm{div}, B \text{ or } \mathbb{R}^3 \setminus \overline{B}) = \{\mathbf{v} \in L^2 : \nabla \cdot \mathbf{v} \in L^2\}$ are the natural spaces associated with the Maxwell equations. It is then important to know whether we can take traces of functions in these spaces on the boundary. The following result holds:

Lemma 1.5. *The mappings* $\mathbf{v} \in H(\mathrm{curl}) \mapsto \mathbf{v} \times \nu \in H^{-1/2}(\partial B)$ *and* $\mathbf{v} \in H(\mathrm{div}) \mapsto \mathbf{v} \cdot \nu \in H^{-1/2}(\partial B)$ *are continuous.*

As a consequence, the following first Green identity holds:

Lemma 1.6. *For any* $\mathbf{u}, \mathbf{v} \in H(\mathrm{curl}, B)$, *we have*

$$\int_B \nabla \times \mathbf{u} \cdot \mathbf{v} - \mathbf{u} \cdot \nabla \times \mathbf{v} = \int_{\partial B} \mathbf{u} \times \nu \cdot \mathbf{v} - \mathbf{u} \cdot \mathbf{v} \times \nu.$$

2.7 Lippman–Schwinger representation formulas

Let D be an electromagnetic target in a dielectric medium with constant permittivity $\varepsilon_\star$ and permeability $\mu_\star$. We introduce the piecewise constant magnetic permeability

$$\mu(\mathbf{r}) := \begin{cases} \mu_\star & \text{in } D, \\ \mu_0 & \text{in } \mathbb{R}^d \backslash \overline{D}. \end{cases}$$

The piecewise constant electric permittivity ε is defined analogously by

$$\varepsilon(\mathbf{r}) := \begin{cases} \varepsilon_\star & \text{in } D, \\ \varepsilon_0 & \text{in } \mathbb{R}^d \backslash \overline{D}. \end{cases}$$

Lippman–Schwinger representation formula for the Helmholtz equation

Suppose that D is illuminated by a time-harmonic wave generated at the point source $\mathbf{r}_t$ with the operating frequency ω. In this case, the incident field is given by $g(\mathbf{r}, \mathbf{r}_t; k_0)$ and the field perturbed in the presence of the target is the solution to the following transmission problem:

$$\nabla \cdot \frac{1}{\mu(\mathbf{r})} \nabla u + \omega^2 \varepsilon(\mathbf{r}) u = \frac{1}{\mu_0} \delta_{\mathbf{r}_t}, \tag{1.8}$$

and is subject to the outgoing radiation condition, or equivalently

$$\begin{cases} \Delta u + k_0^2 u = \delta_{\mathbf{r}_t} & \text{in } \mathbb{R}^d \setminus \overline{D}, \\ \Delta u + k_\star^2 u = 0 & \text{in } D, \\ u|_+ - u|_- = 0 & \text{on } \partial D, \\ \dfrac{1}{\mu_0} \dfrac{\partial u}{\partial \nu} \Big|_+ - \dfrac{1}{\mu} \dfrac{\partial u}{\partial \nu} \Big|_- = 0 & \text{on } \partial D, \\ u \quad \text{satisfies the outgoing radiation condition.} \end{cases} \tag{1.9}$$

Here $\partial / \partial \nu$ denotes the normal derivative on the boundary ∂D and $k_\star = \omega \sqrt{\varepsilon_\star \mu_\star}$.

The following Lippman–Schwinger integral representation formula holds:

$$\begin{aligned} u(\mathbf{r}) &= g(\mathbf{r}, \mathbf{r}_t; k_0) + (\frac{\mu_0}{\mu_\star} - 1) \int_D \nabla u(\mathbf{r}') \cdot \nabla g(\mathbf{r}', \mathbf{r}_t; k_0)\, d\mathbf{r}' \\ &+ k_0^2 (\frac{\varepsilon_\star}{\varepsilon_0} - 1) \int_D u(\mathbf{r}') g(\mathbf{r}', \mathbf{r}_t; k_0)\, d\mathbf{r}', \quad \mathbf{r} \in \mathbb{R}^d. \end{aligned}$$

Lippman–Schwinger representation formulas for the full Maxwell equations
Consider the Maxwell equations

$$\begin{cases} i\omega\varepsilon(\mathbf{r})\mathbf{E} + \nabla\times\mathbf{H} = \mathbf{p}\delta_{\mathbf{r}_t}, \\ -i\omega\mu(\mathbf{r})\mathbf{H} + \nabla\times\mathbf{E} = 0, \end{cases} \tag{1.10}$$

subject to the outgoing radiation conditions, where $\mathbf{p}$ is the direction of the electric dipole and $\mathbf{r}_t$ is its location.

The following Lippman–Schwinger integral representation formulas hold:

$$\begin{aligned} \mathbf{E}(\mathbf{r}) - i\omega\mu_0\mathbf{G}(\mathbf{r},\mathbf{r}_t;k_0)\cdot\mathbf{p} = \int_D \Big[& - i\omega\,(\mu_\star - \mu_0)\,\nabla'\times\mathbf{G}(\mathbf{r},\mathbf{r}';k_0)\cdot\mathbf{H}(\mathbf{r}') \\ & + \omega^2\mu_0\,(\varepsilon_\star - \varepsilon_0)\,\mathbf{G}(\mathbf{r},\mathbf{r}';k_0)\cdot\mathbf{E}(\mathbf{r}')\Big]\, d\mathbf{r}' \end{aligned} \tag{1.11}$$

and

$$\begin{aligned} \mathbf{H}(\mathbf{r}) - \nabla\times(g(\mathbf{r},\mathbf{r}_t;k_0)\mathbf{I})\cdot\mathbf{p} = \int_D \Big[& i\omega\,(\varepsilon_\star - \varepsilon_0)\,\nabla'\times\mathbf{G}(\mathbf{r},\mathbf{r}';k_0)\cdot\mathbf{E}(\mathbf{r}') \\ & + \omega^2\varepsilon_0\,(\mu_\star - \mu_0)\,\mathbf{G}(\mathbf{r},\mathbf{r}';k_0)\cdot\mathbf{H}(\mathbf{r}')\Big]\, d\mathbf{r}', \quad \mathbf{r}\in\mathbb{R}^3. \end{aligned}$$

Here $\nabla'\times$ denotes the differentiation with respect to the variable $\mathbf{r}'$.

2.8 The Helmholtz–Kirchhoff theorems

Helmholtz–Kirchhoff theorems play a key role in understanding the resolution limit in imaging with waves. The following holds:

Theorem 1.7. *We have*

$$\begin{aligned} & \int_{r=R} \left(\frac{\partial\overline{g}}{\partial\nu}(\mathbf{r},\mathbf{r}';\omega)g(\mathbf{r},\mathbf{r}'';\omega) - \overline{g}(\mathbf{r},\mathbf{r}';\omega)\frac{\partial g}{\partial\nu}(\mathbf{r},\mathbf{r}'';\omega)\right) d\sigma(\mathbf{r}) \\ & = 2i\Im m\, g(\mathbf{r}',\mathbf{r}'';\omega), \end{aligned} \tag{1.12}$$

which yields

$$\lim_{R\to+\infty} \int_{r=R} \overline{g}(\mathbf{r},\mathbf{r}';\omega)g(\mathbf{r},\mathbf{r}'';\omega)\, d\sigma(\mathbf{r}) = -\frac{1}{\omega}\Im m\, g(\mathbf{r}',\mathbf{r}'';\omega), \tag{1.13}$$

by using the radiation outgoing condition. Analogously, for any vectors $\mathbf{p}$ *and* $\mathbf{q}$, *we have*

$$\begin{aligned} & \int_{r=R} \nabla\times(\overline{\mathbf{G}}(\mathbf{r},\mathbf{r}';k_0)\mathbf{p})\times\hat{\mathbf{r}}\cdot\mathbf{G}(\mathbf{r},\mathbf{r}'';k_0)\mathbf{q} \\ & -(\overline{\mathbf{G}}(\mathbf{r},\mathbf{r}';k_0)p)\cdot\nabla\times(\mathbf{G}\mathbf{q})(\mathbf{r},\mathbf{r}'';k_0)\times\hat{\mathbf{r}}\, d\sigma(\mathbf{r}) \\ & \qquad = 2i\mathbf{q}\cdot(\Im m\mathbf{G}(\mathbf{r}',\mathbf{r}'';k_0))\mathbf{p}, \end{aligned} \tag{1.14}$$

which yields

$$\lim_{R\to+\infty}\int_{r=R}\left(\hat{\mathbf{r}}\times\overline{\mathbf{G}}(\mathbf{r},\mathbf{r}';k_0)\mathbf{p}\times\hat{\mathbf{r}}\right)\cdot(\hat{\mathbf{r}}\times\mathbf{G}(\mathbf{r},\mathbf{r}'';k_0)\mathbf{q}\times\hat{\mathbf{r}})\,d\sigma(\mathbf{r})$$
$$=\frac{1}{k_0}\mathbf{q}\cdot(\Im m\mathbf{G}(\mathbf{r}',\mathbf{r}'';k_0))\,\mathbf{p}, \tag{1.15}$$

by once again using the outgoing radiation condition.

Identities (1.12)–(1.15) are valid in inhomogeneous media. They show that the sharper the behavior of the imaginary part of the Green function around the source, the higher the resolution.

2.9 Limiting models

The eddy currents model

Dealing with problems in connection with machines working at power frequencies, the full system of the Maxwell equations is usually simplified. In the approximation, $\omega\varepsilon_0 \ll 1$, the displacement currents can be neglected. Moreover, the Silver–Müller radiation condition is replaced by $\mathbf{E},\mathbf{H} = O(\frac{1}{r})$, which holds uniformly for $r \to +\infty$. Assume that the source current is supported outside the conductor D and is divergence free: $\nabla\cdot\mathbf{J} = 0$. The mathematical model of the eddy currents has then the following form:

$$\begin{cases}\nabla\times\mathbf{E} = i\omega\mu\mathbf{H} \quad \text{in } \mathbb{R}^3,\\ \nabla\times\mathbf{H} = -i\omega\varepsilon'\mathbf{E}+\mathbf{J} \quad \text{in } \mathbb{R}^3,\\ \nabla\cdot\mathbf{E}=0 \quad \text{in } \mathbb{R}^3\setminus\overline{D},\\ \nu\times\mathbf{E} \text{ and } \nu\times\mathbf{H} \text{ are continuous across } \partial D,\\ \nu\cdot(\mathbf{E})^- = 0 \text{ on } \partial D,\\ \nu\cdot(\mu\mathbf{H})^+ - \nu\cdot(\mu\mathbf{H})^- = 0 \text{ on } \partial D,\\ \mathbf{E} = O\left(\frac{1}{r}\right),\quad \mathbf{H} = O\left(\frac{1}{r}\right) \quad \text{as } r\to+\infty,\end{cases} \tag{1.16}$$

where ε' is defined by

$$\varepsilon'(\mathbf{r}) := \begin{cases} i\dfrac{\sigma}{\omega} & \text{in } D,\\ 0 & \text{in } \mathbb{R}^3\setminus\overline{D}.\end{cases}$$

The eddy currents model is a correction of the quasi-static model. It has the advantage to exhibit an elliptic character. In the time-dependent case, the transient eddy currents model yields parabolic elliptic interface problems; mainly, the Laplace equation in one domain coupled with the heat equation in the other domain. The topology of a conductor strongly influences the surface current density in the presence of an exciting electromagnetic field. This is not the case for a dielectric material. The topology of the surface of a dielectric material does not have any significant effect. The topology of the conductor affects the use of the eddy currents model for approximating Maxwell's equations. In fact, the eddy currents model approximates the full Maxwell system up to the second order with respect to the frequency ω and only if the limit of the source current as the frequency goes to zero is divergence free in the conductor and is orthogonal to all the Neumann fields in the conductor. Otherwise, it is a first-order approximation to the Maxwell equations [10].

The conductivity equation

The conductivity equation is an approximation of the Maxwell equations as $\omega\mu \ll 1$. From $\nabla \times \mathbf{E} = 0$ in $\mathbb{R}^3$, we can conclude that $\mathbf{E} = \nabla u$, where u is the electric potential. Then by taking the divergence of $\nabla \times \mathbf{H} = \sigma \mathbf{E} + \mathbf{J}$, we obtain $\nabla \cdot (\sigma + i\omega\varepsilon)\nabla u = -\nabla \cdot \mathbf{J}$. If $\mathbf{J}$ is applied outside a conductor Ω and $\sigma + i\omega\varepsilon$ is supposed to be negligible outside Ω then

$$\frac{\partial u}{\partial \nu} = -\mathbf{J} \cdot \nu \qquad \text{on } \partial\Omega,$$

where $-\mathbf{J} \cdot \nu$ is the current entering Ω.

2.10 The Maxwell equations with axis invariance

Suppose that the electromagnetic parameters are independent of one of the variables, say z. Suppose that the fields $\mathbf{E}$ and $\mathbf{H}$ have a given variation in z of the form $e^{i\gamma z}$ (γ is possibly zero). In this case the Maxwell Equations split into two sets of independent equations, one for the fields $(0, 0, E_z)$ and $(H_x, H_y, 0)$ and one for the fields $(E_x, E_y, 0)$ and $(0, 0, H_z)$. The first set of equations are associated with the terminology transverse electric (TE), the second set with the terminology transverse magnetic (TM). In both cases, the electromagnetic fields can be computed from one scalar function (E_z or H_z) which satisfies a two-dimensional Helmholtz equation. For example, for $\gamma = 0$, E_z and H_z solve the following Helmholtz equations:

$$\nabla \cdot \frac{1}{\varepsilon(\mathbf{r})} \nabla H_z + \omega^2 \mu(\mathbf{r}) H_z = 0,$$
$$\nabla \cdot \frac{1}{\mu(\mathbf{r})} \nabla E_z + \omega^2 \varepsilon(\mathbf{r}) E_z = 0,$$

see, for instance, [36].

2.11 The Maxwell equations versus the Helmholtz equation

There are two fundamental differences between the Maxwell and the Helmholtz equations. The first one is, as seen before, the sensitivity of the Maxwell operator to topology. The second problem is the lack of coercivity (or compactness). For the Helmholtz operator, the Laplacian is the principle part. However, for the Maxwell operator $\nabla \times \nabla \times$ does not play the role of a principle part. It lacks the property of strong ellipticity. To solve the problem (by applying the Fredholm alternative), we can use a Hodge decomposition of the electromagnetic fields together with a compact embedding of the set of functions in L^2 of a bounded domain with L^2 divergence and rotational and either L^2 tangential or normal components.

Lemma 1.8. *The embedding of* $\{\mathbf{v} \in H(\mathrm{curl}, B) \cap H(\mathrm{div}, B) : \mathbf{v} \cdot \nu \in L^2(\partial B)$ *(or* $\mathbf{v} \times \nu \in L^2(\partial B))\}$ *into* $L^2(B)$ *is compact.*

In homogeneous domains, boundary layer potentials can be used to represent the solutions and prove existence and uniqueness of solutions. Again, the Hodge decomposition (1.6) is the key tool to involve standard compact perturbation arguments.

3 Electric and magnetic polarization tensors

Let $d = 2, 3$. Let q be given by

$$q(\mathbf{r}) := \begin{cases} q_0, & \mathbf{r} \in \mathbb{R}^d \setminus \overline{B}, \\ q_\star, & \mathbf{r} \in B. \end{cases}$$

By v^q, we denote the solution of

$$\begin{cases} \nabla \cdot q(\mathbf{r}) \nabla v^q(\mathbf{r}) = 0 & \text{in } \mathbb{R}^d, \\ v^q(\mathbf{r}) - \mathbf{r} \to 0 & \text{as } r \to \infty. \end{cases} \tag{1.17}$$

The function v^q depends only on the ratio $q_\star / q_0$.

Define the polarization tensor $\mathbf{M}(q_\star / q_0, B) \in \mathbb{R}^{d \times d}$ associated with the inclusion B and the contrast $q_\star / q_0$ by

$$\mathbf{M}\left(\frac{q_\star}{q_0}, B\right) := \int_B \nabla v^q(\mathbf{r}) \, d\mathbf{r}. \tag{1.18}$$

We call $\mathbf{M}(q_\star / q_0, B)$, respectively, the magnetic polarization tensor if $q(\mathbf{r}) = \mu(\mathbf{r})$ and the electric polarization tensor if $q(\mathbf{r}) = \varepsilon(\mathbf{r})$.

Introduce the vectors φ and ψ defined in $\mathbb{R}^d \setminus \overline{B}$ and satisfying

$$\begin{cases} \Delta \varphi = 0 & \text{in } \mathbb{R}^d \setminus \overline{B}, \\ \partial_\nu \varphi = -\nu & \text{on } \partial B, \\ \lim_{r \to +\infty} \varphi(\mathbf{r}) = 0, \end{cases}$$

and

$$\begin{cases} \Delta\psi = 0 & \text{in } \mathbb{R}^d \setminus \overline{B}, \\ \psi = -\mathbf{r} + \mathbf{c} & \text{on } \partial B, \\ \lim_{r\to+\infty} \psi(\mathbf{r}) = 0, \end{cases}$$

where the constant vector $\mathbf{c}$ is to be chosen such that $\psi(\mathbf{r}) = O(r^{1-d})$ as $r \to +\infty$ or equivalently $\int_{\partial B} \partial_\nu \psi \big|_+ = 0$. Here ν is the outward unit normal to ∂B and $(\partial_\nu u)^\pm(\mathbf{r}) = \lim_{t\to 0^+} \nabla u(\mathbf{r} \pm t\nu(\mathbf{r}))$ if the limit exists.

The tensor

$$\mathbf{W}(B) := \int_{\mathbb{R}^d\setminus\overline{B}} \nabla\varphi(\mathbf{r})\,\nabla\varphi(\mathbf{r})^T\,d\mathbf{r} = \int_{\partial B} \nu\varphi^T(\mathbf{r})\,d\sigma$$

is the so-called virtual mass and the tensor

$$\mathbf{Q}(B) := \int_{\mathbb{R}^d\setminus\overline{B}} \nabla\psi(\mathbf{r})\,\nabla\psi(\mathbf{r})^T\,d\mathbf{r} = \int_{\partial B} \mathbf{r}\left(\frac{\partial\psi}{\partial\nu}\right)^T d\sigma \tag{1.19}$$

is the so-called Pólya–Szegö tensor.

The following convergence result holds as $q_\star/q_0$, respectively, tends toward infinity and zero:

$$\begin{aligned} &\int_{\partial B} \mathbf{r}\left((\partial_\nu v^q)^+\right)^T d\sigma \to |B|\mathbf{I} + \mathbf{Q}(B) \quad \text{as } \frac{q_\star}{q_0} \to +\infty, \\ &\int_{\partial B} \mathbf{r}\left((\partial_\nu v^q)^-\right)^T d\sigma \to |B|\mathbf{I} + \mathbf{W}(B) \quad \text{as } \frac{q_\star}{q_0} \to 0. \end{aligned} \tag{1.20}$$

We refer to [26] for a proof.

4 small-volume expansions

4.1 The full Maxwell equations

Suppose that there is an electromagnetic inclusion D in $\mathbb{R}^3$ of the form $\mathbf{r}_\star + \delta B$, where $B \subset \mathbb{R}^3$ is a bounded, smooth domain containing the origin and δ a small parameter. Let $(\mathbf{E}_\delta, \mathbf{H}_\delta)$ be the solution to the Maxwell equations (1.10) and let

$$\begin{aligned} \mathbf{E}_0(\mathbf{r}) &= i\omega\mu_0 \mathbf{G}(\mathbf{r}, \mathbf{r}_t; k_0) \cdot \mathbf{p}, \\ \mathbf{H}_0 &= \nabla \times (g(\mathbf{r}, \mathbf{r}_t; k_0)\mathbf{I}) \cdot \mathbf{p}, \quad \mathbf{r} \neq \mathbf{r}_t, \end{aligned}$$

be the background electromagnetic fields in the absence of the target D.

The asymptotic expansions of $\mathbf{E}_\delta$ and $\mathbf{H}_\delta$ are constructed by the method of matched asymptotic expansions for δ small. To reveal the nature of the perturbations in the electric and magnetic fields, we introduce the local variables $\xi = (\mathbf{r} - \mathbf{r}_\star)/\delta$ and set the field $\mathbf{e}_\delta(\xi) = \mathbf{E}_\delta(\mathbf{r}_\star + \delta\xi)$ and $\mathbf{h}_\delta(\xi) = \mathbf{H}_\delta(\mathbf{r}_\star + \delta\xi)$. We expect that $\mathbf{E}_\delta(\mathbf{r})$ and $\mathbf{H}_\delta(\mathbf{r})$ will differ appreciably from $\mathbf{E}_0(\mathbf{r})$ and $\mathbf{H}_0(\mathbf{r})$, respectively, for $\mathbf{r}$ near $\mathbf{r}_\star$, but they will differ little from $\mathbf{E}_0(\mathbf{r})$ and $\mathbf{H}_0(\mathbf{r})$ for $\mathbf{r}$ far from $\mathbf{r}_\star$. Therefore, in the spirit of matched asymptotic expansions, we shall represent each of the fields $\mathbf{E}_\delta$ and $\mathbf{H}_\delta$ by two different expansions, an inner expansion for $\mathbf{r}$ near $\mathbf{r}_\star$, and an outer expansion for $\mathbf{r}$ far from $\mathbf{r}_\star$. The outer expansion must begin with $\mathbf{E}_0$ and $\mathbf{H}_0$, respectively, so we write the outer expansions as

$$\begin{aligned} \mathbf{E}_\delta(\mathbf{r}) &= \mathbf{E}_0(\mathbf{r}) + \delta^{\tau_1}\mathbf{E}_1(\mathbf{r}) + \delta^{\tau_2}\mathbf{E}_2(\mathbf{r}) + \cdots, \\ \mathbf{H}_\delta(\mathbf{r}) &= \mathbf{H}_0(\mathbf{r}) + \delta^{\tau_1}\mathbf{H}_1(\mathbf{r}) + \delta^{\tau_2}\mathbf{H}_2(\mathbf{r}) + \cdots \end{aligned} \tag{1.21}$$

for $|\mathbf{r} - z| \gg O(\delta)$, where $0 < \tau_1 < \tau_2 < \cdots$, and $(\mathbf{E}_j, \mathbf{H}_j), j = 1, 2, \ldots$, are to be found.

We write the inner expansions as

$$\begin{aligned} \mathbf{E}_\delta(\mathbf{r}_\star + \delta\xi) &= \mathbf{e}_\delta(\xi) = \mathbf{e}_0(\xi) + \delta\mathbf{e}_1(\xi) + \delta^2\mathbf{e}_2(\xi) + \cdots, \quad \text{for } |\xi| = O(1), \\ \mathbf{H}_\delta(\mathbf{r}_\star + \delta\xi) &= \mathbf{h}_\delta(\xi) = \mathbf{h}_0(\xi) + \delta\mathbf{h}_1(\xi) + \delta^2\mathbf{h}_2(\xi) + \cdots, \quad \text{for } |\xi| = O(1), \end{aligned} \tag{1.22}$$

where $(\mathbf{e}_j, \mathbf{h}_j)$, $j = 0, 1, \ldots$, are to be found.

In order to determine the functions $\mathbf{E}_i(\mathbf{r}), \mathbf{H}_i(\mathbf{r}), \mathbf{e}_i(\xi)$, and $\mathbf{h}_i(\xi)$ we have to equate the inner and the outer expansions in some "overlap" domain within which the stretched variable ξ is large and $\mathbf{r} - \mathbf{r}_\star$ is small. In this domain the matching conditions are as follows:

$$\begin{aligned} \mathbf{E}_0(\mathbf{r}) + \delta^{\tau_1}\mathbf{E}_1(\mathbf{r}) + \delta^{\tau_2}\mathbf{E}_2(\mathbf{r}) + \cdots &\sim \mathbf{e}_0(\xi) + \delta\mathbf{e}_1(\xi) + \delta^2\mathbf{e}_2(\xi) + \cdots, \\ \mathbf{H}_0(\mathbf{r}) + \delta^{\tau_1}\mathbf{H}_1(\mathbf{r}) + \delta^{\tau_2}\mathbf{H}_2(\mathbf{r}) + \cdots &\sim \mathbf{h}_0(\xi) + \delta\mathbf{h}_1(\xi) + \delta^2\mathbf{h}_2(\xi) + \cdots. \end{aligned} \tag{1.23}$$

Define the stretched coefficients $\varepsilon(\xi)$ and $\mu(\xi)$ by

$$\varepsilon(\xi) = \begin{cases} \varepsilon_0, & \xi \in \mathbb{R}^3 \setminus \overline{B}, \\ \varepsilon_*, & \xi \in B, \end{cases}$$

$$\mu(\xi) = \begin{cases} \mu_0, & \xi \in \mathbb{R}^3 \setminus \overline{B}, \\ \mu_*, & \xi \in B. \end{cases}$$

If we substitute the inner expansions (1.22) into the Maxwell equations (1.10) and formally equate coefficients of δ^{-1} we get from use of the first matching condition in (1.23) that

$$\begin{cases} \nabla \times \mathbf{e}_0(\xi) = 0 & \text{in } \mathbb{R}^3, \\ \nabla \cdot (\varepsilon(\xi)\mathbf{e}_0(\xi)) = 0 & \text{in } \mathbb{R}^3, \\ \mathbf{e}_0(\xi) \to \mathbf{E}_0(\mathbf{r}_\star) & \text{as } |\xi| \to +\infty, \end{cases} \tag{1.24}$$

and

$$\begin{cases} \nabla \times \mathbf{h}_0(\xi) = 0 & \text{in } \mathbb{R}^3, \\ \nabla \cdot (\mu(\xi)\mathbf{h}_0(\xi)) = 0 & \text{in } \mathbb{R}^3, \\ \mathbf{h}_0(\xi) \to \mathbf{H}_0(\mathbf{r}_\star) & \text{as } |\xi| \to +\infty. \end{cases} \tag{1.25}$$

Therefore,

$$\begin{aligned} \mathbf{e}_0(\xi) &= \nabla v^{\varepsilon}(\xi)(\mathbf{E}_0(\mathbf{r}_\star)), \\ \mathbf{h}_0(\xi) &= \nabla v^{\mu}(\xi)(\mathbf{H}_0(\mathbf{r}_\star)), \end{aligned}$$

where the function v^q, for $q(\xi) = \varepsilon(\xi)$ or $\mu(\xi)$, is defined by (1.17).

Define the polarization tensors by (1.18). Inserting the inner expansions into the Lippman–Schwinger integral representation formula (1.11) for $\mathbf{E}_\delta$ we arrive at the following result which was proved in [28]. Theorem 1.9 shows that the leading-order effect of a small target in the far-field corresponds to the superposition of two uncoupled electric and magnetic dipoles.

Theorem 1.9. *For any* $\mathbf{r}$ *away from* $\mathbf{r}_t$ *and* $\mathbf{r}_\star$,

$$\begin{aligned} \mathbf{E}_\delta(\mathbf{r}) - \mathbf{E}_0(\mathbf{r}) = \delta^3 \Bigg[& -i\omega\, (\mu_\star - \mu_0)\, \nabla' \times \mathbf{G}(\mathbf{r}, \mathbf{r}_\star) \cdot \mathbf{M}\left(\frac{\mu_\star}{\mu_0}, B\right) \mathbf{H}_0(\mathbf{r}_\star) \\ & + k_0^2 \left(\frac{\varepsilon_\star}{\varepsilon_0} - 1\right) \mathbf{G}(\mathbf{r}, \mathbf{r}_\star) \cdot \mathbf{M}\left(\frac{\varepsilon_\star}{\varepsilon_0}, B\right) \mathbf{E}_0(\mathbf{r}_\star) \Bigg] + O(\delta^4). \end{aligned} \tag{1.26}$$

Moreover, if B is symmetric about its center, then the correction of order four in (1.26) *is zero. The remainder in this case is in fact $O(\delta^5)$.*

Note that if B is a ball, then its polarization tensor $\mathbf{M}(q_\star/q_0, B)$ has the following explicit form:

$$\mathbf{M}\left(\frac{q_\star}{q_0}, B\right) = \frac{3q_0}{2q_0 + q_\star} |B| \mathbf{I}, \quad q(\mathbf{r}) = \varepsilon(\mathbf{r}) \text{ or } \mu(\mathbf{r}),$$

and therefore (1.26) becomes

$$\begin{aligned} \mathbf{E}_\delta(\mathbf{r}) - \mathbf{E}_0(\mathbf{r}) = \delta^3 \Bigg[& -i\omega\, \mu_0 \frac{3(\mu_\star - \mu_0)}{2\mu_0 + \mu_\star} |B|\, \nabla' \times \mathbf{G}(\mathbf{r}, \mathbf{r}_\star) \cdot \mathbf{H}_0(\mathbf{r}_\star) \\ & + k_0^2 \frac{3(\varepsilon_\star - \varepsilon_0)}{2\varepsilon_0 + \varepsilon_\star} |B|\, \mathbf{G}(\mathbf{r}, \mathbf{r}_\star) \cdot \mathbf{E}_0(\mathbf{r}_\star) \Bigg] + O(\delta^5). \end{aligned} \tag{1.27}$$

Formula (1.27) would easily extend to the case of triaxial ellipsoidal inclusions, using their standard tensors which, for B, read in the ellipsoidal coordinate system attached to it as

$$\mathbf{M}\left(\frac{q_\star}{q_0}, B\right) = \begin{bmatrix} m_1 & 0 & 0 \\ 0 & m_2 & 0 \\ 0 & 0 & m_3 \end{bmatrix} |B|,$$

where the coefficients $m_l, l = 1,2,3$, are equal to $q_0/(q_0 + L_l(q_\star - q_0))$, the L_l being the well-known depolarization factors of the ellipsoid proportional to harmonic elliptic integrals of degree 1.

We also note that from (1.20) it follows that

$$\frac{\varepsilon_\star}{\varepsilon_0}\mathbf{M}\left(\frac{\varepsilon_\star}{\varepsilon_0}, B\right) \to \mathbf{Q}(B) \quad \text{as} \quad \left|\frac{\varepsilon_\star}{\varepsilon_0}\right| \to +\infty$$

and

$$\mathbf{M}\left(\frac{\mu_\star}{\mu_0}, B\right) \to \mathbf{W}(B) \quad \text{as} \quad \frac{\mu_\star}{\mu_0} \to 0,$$

which (formally) extends the validity of (1.26) to the limiting perfectly electric conductor case.

Theorem 1.10. *If the target $D = \mathbf{r}_\star + \delta B$ is a small perfect conductor, then for any $\mathbf{r}$ away from $\mathbf{r}_t$ and $\mathbf{r}_\star$,*

$$\begin{aligned}\mathbf{E}_\delta(\mathbf{r}) - \mathbf{E}_0(\mathbf{r}) = \delta^3 \Big[& i\omega\,\mu_0 \nabla' \times \mathbf{G}(\mathbf{r}, \mathbf{r}_\star) \cdot \mathbf{W}(B)\mathbf{H}_0(\mathbf{r}_\star) \\ & - k_0^2 \mathbf{G}(\mathbf{r}, \mathbf{r}_\star) \cdot \mathbf{Q}(B)\mathbf{E}_0(\mathbf{r}_\star)\Big] + O(\delta^4).\end{aligned} \tag{1.28}$$

In order to prove (1.28) we shall use instead of (1.11) integral representations of the solutions to Maxwell's equations associated with the conductor problem. If target D is a perfect conductor, then we have

$$\begin{aligned}\mathbf{E}_\delta(\mathbf{r}) - \mathbf{E}_0(\mathbf{r}) = {} & i\omega\mu_0 \int_{\partial D} g(\mathbf{r}, \mathbf{r}'; k_0)\mathbf{j}_\delta(\mathbf{r}')\, d\sigma(\mathbf{r}') \\ & + \frac{i}{\omega\varepsilon_0}\nabla \int_{\partial D} g(\mathbf{r}, \mathbf{r}'; k_0)\nabla_{\partial D} \cdot \mathbf{j}_\delta(\mathbf{r}')\, d\sigma(\mathbf{r}') \\ & - \nabla \times \int_{\partial D} g(\mathbf{r}, \mathbf{r}'; k_0)(\mathbf{E}_0 \times \nu)(\mathbf{r}')\, d\sigma(\mathbf{r}'), \qquad \mathbf{r} \in \mathbb{R}^3 \setminus \overline{D},\end{aligned}$$

where

$$\mathbf{j}_\delta = (\mathbf{H}_\delta - \mathbf{H}_0) \times \nu \qquad \text{on } \partial D.$$

We can prove that

$$\mathbf{j}_\delta(\mathbf{r}') \approx \left(\mathbf{h}_0\left(\frac{\mathbf{r}' - \mathbf{r}_\star}{\delta}\right) - \mathbf{H}_0(\mathbf{r}_\star)\right) \times \nu, \qquad \mathbf{r}' \in \partial D,$$

where $\mathbf{h}_0$ is defined by

$$\begin{cases} \nabla \times \mathbf{h}_0 = 0 & \text{in } \mathbb{R}^3 \setminus \overline{B}, \\ \nabla \cdot \mathbf{h}_0 = 0 & \text{in } \mathbb{R}^3 \setminus \overline{B}, \\ \mathbf{h}_0 \cdot \nu = 0 & \text{on } \partial B, \\ \mathbf{h}_0 \to \mathbf{H}_0(\mathbf{r}_\star) & \text{at } +\infty. \end{cases} \tag{1.29}$$

Note that

$$\nabla_{\partial D} \cdot \mathbf{j}_\delta(\mathbf{r}') \approx 0,$$

since

$$\nabla_{\partial D} \cdot (\mathbf{h}_0 \times \nu) = (\nabla \times \mathbf{h}_0) \cdot \nu = 0.$$

Indeed, if B is not simply connected, then $\mathbf{h}_0$ is defined modulo $\mathcal{N}_{\text{ext}}(B)$. But for any $\mathbf{v} \in \mathcal{N}_{\text{ext}}(B)$,

$$\int_{\partial B} \mathbf{v} \times \nu \, d\sigma = 0.$$

Therefore, the topology of B does not play any role at the leading-order term. The dependence of the asymptotic expansion of the electric field on the topology of the conductor appears for the first time at the second-order term.

If ∂B is simply connected then we can involve the symmetric roles that the electromagnetic fields play to write an asymptotic expansion for $\mathbf{H}_\delta$ similar to (1.28).

4.2 The eddy currents model

Suppose that

$$\mathbf{J}(\mathbf{r}) = \nabla \times (\mathbf{p}\delta(\mathbf{r} - \mathbf{r}_t)),$$

where $\mathbf{r}_t$ is the position and the unit vector $\mathbf{p}$ the direction of the magnetic dipole $\mathbf{J}$. The electric and magnetic fields $\mathbf{E}_0$ and $\mathbf{H}_0$ due to $\mathbf{J}(\mathbf{r})$ are given by

$$\mathbf{E}_0(\mathbf{r}) = i\omega\mu_0 \nabla \times \left(\frac{\mathbf{p}}{4\pi|\mathbf{r} - \mathbf{r}_t|}\right), \quad \mathbf{H}_0(\mathbf{r}) = \nabla \times \nabla \times \left(\frac{\mathbf{p}}{4\pi|\mathbf{r} - \mathbf{r}_t|}\right). \tag{1.30}$$

Now let $\mathbf{r}$ be the position of the receiver and let $\mathbf{q}$ be a unit vector. Assume that the inclusion is nonmagnetic, i.e., $\mu_\star = \mu_0$. Writing $\varepsilon_\delta = \Re e \varepsilon_\delta + i\frac{\sigma}{\omega}\chi(D)$, and considering the regime where δ is of the same order as the skin depth $1/\sqrt{\omega\mu_0\sigma}$:

$$\omega\mu_0\sigma\delta^2 = \nu = O(1),$$

we arrive at the following expansion for the eddy currents model [11]:

$$\begin{aligned}(\mathbf{H}_\delta - \mathbf{H}_0)(\mathbf{r}) \cdot \mathbf{q} \approx i\nu\delta^3 &\left(D_\mathbf{r}^2 \left(\frac{1}{4\pi|\mathbf{r} - \mathbf{r}_\star|}\right)\mathbf{q}\right) \\ &\cdot \mathbf{N}(\nu, B)\left(D_\mathbf{r}^2 \left(\frac{1}{4\pi|\mathbf{r}_\star - \mathbf{r}_t|}\right)\mathbf{p}\right),\end{aligned} \tag{1.31}$$

where $D_\mathbf{r}^2$ denotes the second derivative with respect to the first variable and $\mathbf{N}$ is a complex symmetric tensor depending only on B (which is supposed to be a sphere) and ν. Formula (1.31) can be generalized for arbitrary shapes. An asymptotic formula in the magnetic case can be derived as well [11].

4.3 The Helmholtz equation

Let u_δ be the outgoing solution to

$$\nabla \cdot \frac{1}{\mu_\delta(\mathbf{r})} \nabla u_\delta + \omega^2 \varepsilon_\delta(\mathbf{r}) u_\delta = \frac{1}{\mu_0} \delta_{\mathbf{r}_t}. \tag{1.32}$$

The following holds.

Theorem 1.11. *For any* $\mathbf{r}$ *away from* $\mathbf{r}_t$ *and* $\mathbf{r}_\star$,

$$\begin{aligned} u_\delta(\mathbf{r}, \mathbf{r}_t) - g(\mathbf{r}, \mathbf{r}_t; k_0) = \delta^d \Bigg[&\left(\frac{\mu_\star}{\mu_0} - 1 \right) \nabla g(\mathbf{r}, \mathbf{r}_\star; k_0) \cdot \mathbf{M}\left(\frac{\mu_\star}{\mu_0}, B \right) \nabla g(\mathbf{r}_\star, \mathbf{r}_t; k_0) \\ &+ k_0^2 \left(\frac{\varepsilon_\star}{\varepsilon_0} - 1 \right) |B| g(\mathbf{r}, \mathbf{r}_\star; k_0) g(\mathbf{r}_\star, \mathbf{r}_t; k_0) \Bigg] + O(\delta^{d+1}). \end{aligned} \tag{1.33}$$

Moreover, (1.33) *holds for any* ω *such that* $\omega\delta \ll 1$ *and the remainder* $O(\delta^{d+1})$ *is uniform with respect to the magnetic contrast* $\mu_\star/\mu_0$.

4.4 The conductivity equation

By taking $k_0 = 0$ in (1.33), the limit of g as k_0 goes to zero, and then replacing $\mu_\star/\mu_0$ by $(\sigma_0 + i\omega\varepsilon_0)/(\sigma_\star + i\omega\varepsilon_\star)$, we find a small-volume expansion for the conductivity problem.

4.5 Asymptotic formulas in the time domain

For $\rho > 0$, define the operator P_ρ on tempered distributions by

$$P_\rho[\psi](\mathbf{r}, s) = \int\limits_{|\omega| \le \rho} e^{-i\omega s} \hat{\psi}(\mathbf{r}, \omega)\, d\omega, \tag{1.34}$$

where $\hat{\psi}(\mathbf{r}, \omega)$ denotes the Fourier transform of $\psi(\mathbf{r}, s)$ in the s-variable. Clearly, the operator P_ρ truncates the high-frequency component of ψ.

Let $d = 3$ and let

$$\mathcal{G}(\mathbf{r}, \mathbf{r}_t, s) := \frac{\delta_{s = |\mathbf{r} - \mathbf{r}_t|/c}}{4\pi |\mathbf{r} - \mathbf{r}_t|} \quad \text{for } \mathbf{r} \neq \mathbf{r}_t, \tag{1.35}$$

where $c = 1/\sqrt{\varepsilon_0 \mu_0}$.

The function $\mathcal{G}$ is the outgoing (or the retarded) Green function to the wave equation

$$\left(\frac{1}{c^2} \partial_s^2 - \Delta \right) \mathcal{G}(\mathbf{r}, \mathbf{r}_t, s) = \delta_{\mathbf{r}_t} \delta_{s=0} \quad \text{in } \mathbb{R}^3 \times]0, +\infty[.$$

Moreover, $\mathcal{G}$ satisfies the initial conditions: $\mathcal{G}(\mathbf{r},\mathbf{r}_t,0) = \partial_s\mathcal{G}(\mathbf{r},\mathbf{r}_t,0) = 0$ for $\mathbf{r} \neq \mathbf{r}_t$. Consider now for the wave equation in the whole three-dimensional space with appropriate initial conditions:

$$\begin{cases} \varepsilon_\delta \partial_s^2 u_\delta - \nabla \cdot \dfrac{1}{\mu_\delta} \nabla u_\delta = \dfrac{1}{\mu_0} \delta_{\mathbf{r}_t} \delta_{s=0} & \text{in } \mathbb{R}^3 \times]0,+\infty[, \\ u_\delta(\mathbf{r},0) = 0, \quad \partial_s u_\delta(\mathbf{r},0) = 0 & \text{for } \mathbf{r} \in \mathbb{R}^3, \mathbf{r} \neq \mathbf{r}_t. \end{cases} \tag{1.36}$$

Theorem 1.12. *For $\rho\delta \ll 1$, the following expansion holds:*

$$\begin{aligned} &P_\rho[u_\delta](\mathbf{r},s) - P_\rho[\mathcal{G}](\mathbf{r},\mathbf{r}_t,s) \\ &\approx -\delta^3 \Bigg[\left(\frac{\mu_\star}{\mu_0} - 1\right) \int_{\mathbb{R}} \nabla P_\rho[\mathcal{G}](\mathbf{r},\mathbf{r}_\star,s-\tau) \cdot \mathbf{M}\left(\frac{\mu_\star}{\mu_0}, B\right) \nabla P_\rho[\mathcal{G}](\mathbf{r}_\star,\mathbf{r}_t,\tau)\, d\tau \\ &\quad - c^{-2} \left(\frac{\varepsilon_\star}{\varepsilon_0} - 1\right) |B| P_\rho[\mathcal{G}](\mathbf{r},\mathbf{r}_\star,s-\tau) P_\rho[\partial_s^2 \mathcal{G}](\mathbf{r}_\star,\mathbf{r}_t,\tau)\, d\tau \Bigg], \end{aligned} \tag{1.37}$$

for $\mathbf{r}$ away from $\mathbf{r}_\star$.

Formula (1.37) shows that the perturbation due to the target is in the time domain a wavefront emitted by a dipolar source located at $\mathbf{r}_\star$.

Taking the Fourier transform of (1.37) in the time variable yields expansions (1.33) given for the perturbations resulting from the presence of a small target for solutions to the Helmholtz equation at any frequency ω such that $\omega\delta \ll 1$.

Analogously to (1.37), we can prove that for the time-dependent Maxwell equations, the following small-volume expansion holds:

$$\begin{aligned} &P_\rho[\mathbf{E}_\delta](\mathbf{r},s) - P_\rho[\mathbf{E}_0](\mathbf{r},s) \\ &\approx \delta^3 \int_{\mathbb{R}} \Bigg[(\mu_\star - \mu_0) \nabla' \times P_\rho[\mathcal{G}](\mathbf{r},\mathbf{r}_\star,s-\tau) \cdot \mathbf{M}\left(\frac{\mu_\star}{\mu_0}, B\right) P_\rho[\partial_s \mathbf{H}_0](\mathbf{r}_\star,\tau)\, d\tau \\ &\quad - c^{-2} \left(\frac{\varepsilon_\star}{\varepsilon_0} - 1\right) P_\rho[\mathcal{G}](\mathbf{r},\mathbf{r}_\star,s-\tau) \cdot \mathbf{M}\left(\frac{\varepsilon_\star}{\varepsilon_0}, B\right) P_\rho[\partial_s^2 \mathbf{E}_0](\mathbf{r}_\star,\tau) \Bigg]\, d\tau, \end{aligned} \tag{1.38}$$

where $\mathcal{G}$ is the inverse Fourier transform of $\mathbf{G}$.

5 Imaging in the frequency domain

Suppose that we have transmitter and receiver arrays $\{\mathbf{r}_1,\dots,\mathbf{r}_N\}$ and $\{\mathbf{r}'_1,\dots,\mathbf{r}'_{N'}\}$ of, respectively, N and N' elements, used to detect the target. In the presence of the target the scattered field induced on the ith receiving element, $\mathbf{r}'_i$, from the scattering of an incident wave generated at $\mathbf{r}_j$ can be expressed as follows:

$$A_{ij} := u_\delta(\mathbf{r}'_i, \mathbf{r}_j) - g(\mathbf{r}'_i, \mathbf{r}_j; \omega), \quad i = 1, \ldots, N', \quad j = 1, \ldots, N. \tag{1.39}$$

The problem we now consider is to image the target D from the MSR matrix $\mathbf{A}$ of entries A_{ij}.

For the sake of simplicity we consider the Helmholtz equation. Indeed, we suppose that the distance between the sensor array and the target is large. The emitted waves are then approximately plane waves. All the imaging methods presented below apply for the full Maxwell equations.

5.1 MUSIC-type imaging at a single frequency

Consider m well-separated targets $D_s = \delta B_s + \mathbf{r}_\star^{(s)}$, $s = 1, \ldots, m$. The permittivities and permeabilities of D_s are denoted by ε_s and μ_s, respectively. The background permittivity and permeability, ε_0 and μ_0, are equal to 1. Suppose that all the domains B_s are balls. Let $(\theta_1, \ldots, \theta_N)$ be N unit vectors in $\mathbb{R}^d$. For arbitrary $\theta, \theta' \in \{\theta_1, \ldots, \theta_N\}$, one assumes that one is in possession of u_δ in the direction θ' when the target D is illuminated with the plane wave $e^{-i\omega\theta\cdot\mathbf{r}}$. Therefore, taking the far-field behavior of the Green function g shows that one is in possession of

$$\sum_{s=1}^{m} |D_s| \left(\frac{d(\mu_s - 1)}{\mu_s + d - 1} \theta \cdot \theta' + (\varepsilon_s - 1) \right) e^{i\omega(\theta - \theta')\cdot\mathbf{r}_\star^{(s)}},$$

for $\theta, \theta' \in \{\theta_1, \ldots, \theta_N\}$. Define the response matrix $\mathbf{A} = (A_{ll'})_{l,l'=1}^N \in \mathbb{C}^{N\times N}$ by

$$A_{ll'} = \sum_{s=1}^{m} |D_s| \left(\frac{d(\mu_s - 1)}{\mu_s + d - 1} \theta_l \cdot \theta_{l'} + (\varepsilon_s - 1) \right) e^{i\omega(\theta_l - \theta_{l'})\cdot\mathbf{r}_\star^{(s)}}, \quad l, l' = 1, \ldots, N.$$

Introduce

$$\mathbf{g}(\mathbf{r}) = \left((1, \theta_1)^T e^{i\omega\theta_1\cdot\mathbf{r}}, \ldots, (1, \theta_N)^T e^{i\omega\theta_N\cdot\mathbf{r}} \right)^T.$$

One has the following characterization of the location of the targets in terms of the range of the matrix $\mathbf{A}$.

Lemma 1.13. *There exists $n_0 \in \mathbb{N}, n_0 > (d+1)m$, such that for any $N \geq n_0$ the following statement holds:*

$$g^j(\mathbf{r}) \in \mathit{Range}(\mathbf{A}) \textit{ if and only if } \mathbf{r} \in \{\mathbf{r}_\star^{(1)}, \ldots, \mathbf{r}_\star^{(m)}\} \quad \textit{for } j = 1, \ldots, d+1,$$

where $g^j(\mathbf{r})$ is the jth column of $\mathbf{g}(\mathbf{r})$.

The MUSIC algorithm can now be used to determine the location of the targets. Let $\mathbf{P}_{\text{noise}} = \mathbf{I} - \mathbf{P}$, where $\mathbf{P}$ is the orthogonal projection onto the range of $\mathbf{A}$. The imaging functional

$$W_{\text{MU}}(\mathbf{r}) := \frac{1}{\sum_{j=1}^{d+1} \|\mathbf{P}_{\text{noise}} g^j(\mathbf{r})\|}$$

has large peaks only at the locations of the targets, see Figure 1.2.

The significant singular vectors of $\mathbf{A}$ can be computed by the singular value decomposition. The number of significant singular values determines the number of targets. If, for example, $\mu_s \neq 1$ and $\varepsilon_s \neq 1$ for all $s = 1, \ldots, m$, then there are exactly $(d+1)m$ significant singular values of $\mathbf{A}$ and the rest are zero or close to zero, see Figure 1.1. Note that for each target, there are d significant singular values associat-

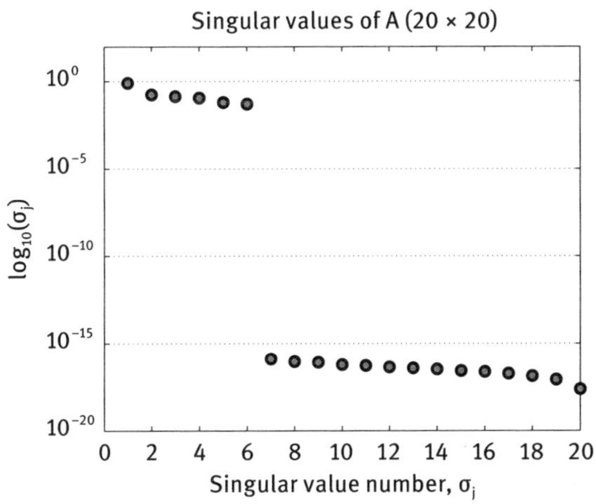

Figure 1.1: Singular value decomposition of the response matrix $\mathbf{A}$ corresponding to two well-separated inclusions of general shape for $N = 20$, using a standard log scale. Six singular values emerge from the 14 others in the noise subspace.

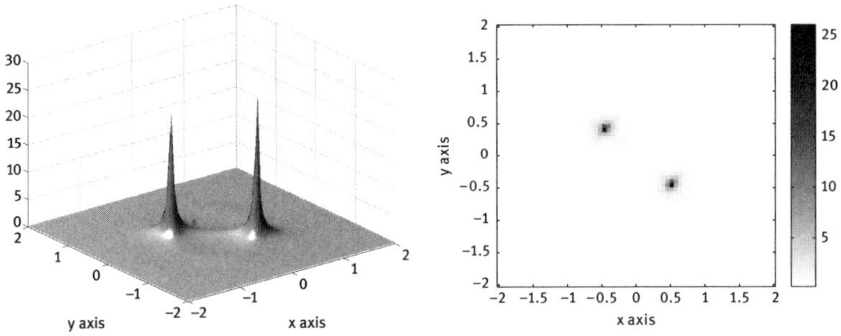

Figure 1.2: MUSIC-type reconstruction from the singular value decomposition of $\mathbf{A}$ represented in Figure 1.1.

ed with the magnetic contrast and only one associated with the electric contrast. The significant singular values of **A** can be used to estimate $\frac{d(\mu_s-1)}{\mu_s+d-1}|D_s|$ and $(\varepsilon_s-1)|D_s|$.

5.2 Backpropagation type imaging at a single frequency

A backpropagation imaging functional at a single frequency ω is given by

$$W_{\mathrm{BP}}(\mathbf{r}) := \frac{1}{N}\sum_{l=1}^{N} e^{i\omega(\theta_l+\theta')\cdot\mathbf{r}} I[\theta_l, \theta'; \omega], \tag{1.40}$$

where $I[\theta_l, \theta'; \omega]$ is the data in the direction θ_l for $\theta_l \in \{\theta_1, \ldots, \theta_N\}$ when the target is illuminated with the plane wave $e^{-i\omega\theta'\cdot\mathbf{r}}$.

Suppose that $(\theta_1, \ldots, \theta_N)$ are equidistant points on the unit sphere. For sufficiently large N, since

$$\frac{1}{N}\sum_{l=1}^{N} e^{i\omega\theta_l\cdot\mathbf{r}} \approx \begin{cases} j_0(\omega r) & \text{for } d = 3, \\ J_0(\omega r) & \text{for } d = 2, \end{cases}$$

where j_0 is the spherical Bessel function of order zero and J_0 is the Bessel function of the first kind and of order zero, it follows that

$$W_{\mathrm{BP}}(\mathbf{r}) \propto \begin{cases} j_0(\omega|\mathbf{r}-\mathbf{r}_\star^{(s)}|) & \text{for } d = 3, \\ J_0(\omega|\mathbf{r}-\mathbf{r}_\star^{(s)}|) & \text{for } d = 2, \end{cases}$$

for $\mathbf{r}$ near $\mathbf{r}_\star^{(s)}$.

Note that if many plane waves (say with directions $\theta_{l'}$) are used to image the target, then we shall replace (1.40) by

$$W_{\mathrm{BP}}(\mathbf{r}) := \frac{1}{N}\sum_{l,l'=1}^{N} e^{i\omega(\theta_l+\theta_{l'})\cdot\mathbf{r}} I[\theta_l, \theta_{l'}; \omega] \propto \begin{cases} j_0(\omega|2\mathbf{r}-\mathbf{r}_\star^{(s)}|) & \text{for } d = 3, \\ J_0(2\omega|\mathbf{r}-\mathbf{r}_\star^{(s)}|) & \text{for } d = 2, \end{cases}$$

for $\mathbf{r}$ near $\mathbf{r}_\star^{(s)}$, and therefore increase the resolution by factor 2. Indeed, in this case, an analogy between the backpropagation and MUSIC-type imaging can be established. Suppose that $\mu_s = 1$ for $s = 1, \ldots, m$. One can see that

$$W_{\mathrm{MU}}(\mathbf{r}) \propto \frac{1}{|D_s|(\varepsilon_s - 1) - W_{\mathrm{BP}}(\mathbf{r})}$$

for $\mathbf{r}$ near $\mathbf{r}_\star^{(s)}$.

Figure 1.3 shows the detection result by backpropagation where a linear array of 46 receivers placed parallel to the y-axis and spaced by half the wavelength is used.

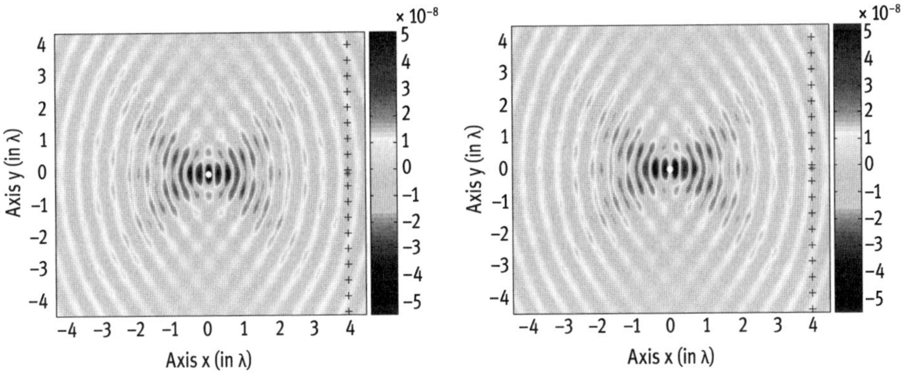

Figure 1.3: (See Color Plates for a color version.) Real and imaginary part of the backpropagation. Here "*" and "+," respectively, show the transceiver and the receiver positions.

5.3 Imaging with a broad range of frequencies

Set $m = 1$. The Kirchhoff imaging functional is given by

$$W_{\mathrm{KI}}(\mathbf{r}) = \frac{1}{J} \sum_{\omega_j, j=1,\dots,J} \frac{1}{\omega_j^2} e^{i\omega_j \mathbf{r}\cdot(\theta+\theta')} I[\theta, \theta'; \omega_j],$$

where J is the number of frequencies.

Since

$$W_{\mathrm{KI}}(\mathbf{r}) \propto \delta_{(\mathbf{r}-\mathbf{r}_\star)\cdot(\theta'+\theta)=0},$$

it follows that in order to determine the location $\mathbf{r}_\star$ of the target, one needs d different measurement directions θ.

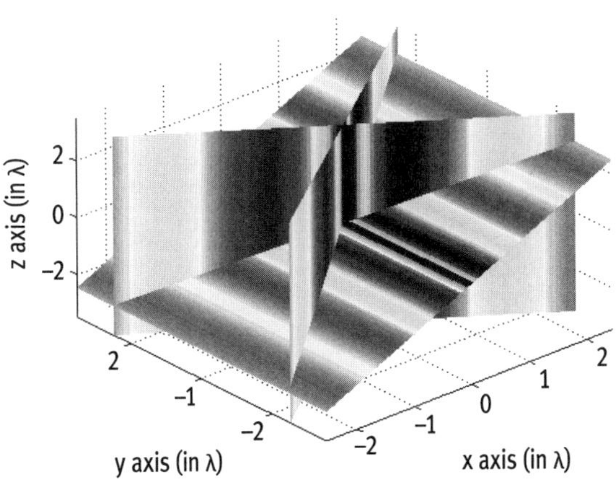

Figure 1.4: (See Color Plates for a color version.) Intersection of the three planes where the real parts of the Kirchhoff functionals attain their maximum for three different receivers.

Figure 1.4 shows the results of the Kirchhoff imaging functionals for three different receiver locations. The position of the target is obtained as the intersection of the three planes where each of the Kirchhoff functional attains its maximum.

It is worth noticing that MUSIC uses one frequency, many incident and measurement directions; the backpropagation uses one frequency, many measurement directions but one direction of incidence, and Kirchhoff uses d measurement directions, one direction of incidence but broad range of frequencies.

6 Imaging in the time domain

6.1 Time-domain imaging with full view measurements

The main idea of time reversal is to measure the perturbation of the wave on a closed surface surrounding the target, and to retransmit it through the background medium in a time-reversed chronology. Then the perturbation will travel back to the location of the target. Suppose that we are able to measure the perturbation and its normal derivative at any point $\mathbf{r}$ on a sphere S englobing the target D. The time-reversal operation is described by the transform $s \mapsto s_0 - s$. Both the perturbation and its normal derivative on S are time reversed and emitted from S. Then a time-reversed perturbation, denoted by w_{tr}, propagates inside the volume surrounded by S. Using the asymptotic formula (1.37), it was proved in [12] that the time-reversed perturbation w_{tr} due to D can be approximated by the superposition of incoming and outgoing polarized waves, centered on the location $\mathbf{r}_\star$ of the target. In fact,

$$w_{\mathrm{tr}}(\mathbf{r}, s) \propto \delta^3 \mathbf{p} \cdot \nabla \left[P_\rho[\mathcal{G}](\mathbf{r}, \mathbf{r}_\star, s_0 - T - s) - P_\rho[\mathcal{G}](\mathbf{r}, s - s_0 + T) \right], \tag{1.41}$$

for some constant (polarization) vector $\mathbf{p}$.

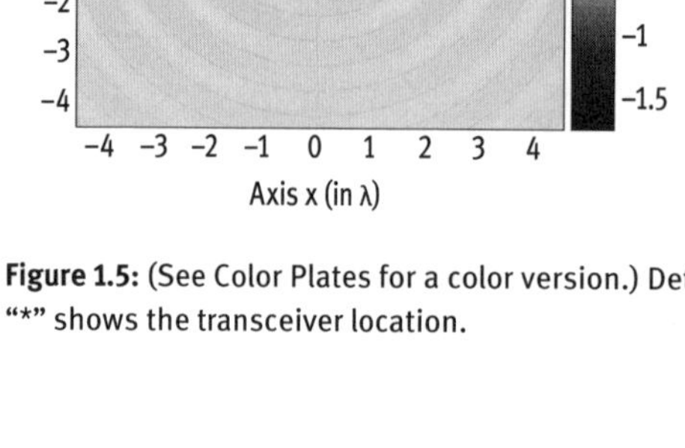

Figure 1.5: (See Color Plates for a color version.) Detection result using the time-reversal technique, "*" shows the transceiver location.

By taking Fourier transform of (1.41) over the time variable s, we obtain that

$$\hat{w}_{\mathrm{tr}}(\mathbf{r}, \omega) \propto \delta^3 \mathbf{p} \cdot \nabla \left(\frac{\sin(\omega(|\mathbf{r} - \mathbf{r}_\star|))}{|\mathbf{r} - \mathbf{r}_\star|} \right), \tag{1.42}$$

where ω is the wave number. This yields that the antiderivative of time-reversal perturbation $\hat{w}_{\mathrm{tr}}$ focuses on the location $\mathbf{r}_\star$ of the target with the focal spot size limited to one-half the wavelength which is in agreement with the Rayleigh resolution limit.

Figure 1.5 shows the performance of the time-reversal technique for detecting the target.

6.2 Time-domain imaging in a cavity with limited-view data

While physical time reversal is a powerful tool for focalizing a wave emitted by a point source, it does not yield to a better (computational) imaging than Kirchhoff-type algorithms. However, in a cavity, it can be modified to perform imaging of small targets with the same resolution using limited-view as using full-view data. For doing so, we shall make use of the geometrical control method [2]. This seems very hard to achieve with Kirchhoff imaging.

Note that the link between time reversal and the geometric control was first observed by Bardos [30]. Moreover, he mathematically explained how the ergodicity of the inaccessible part of the boundary of a cavity can lead to a very good refocusing of a localized initial signal measured at one single receiver.

A model problem

Let Ω be a bounded domain. The basic model to be considered is the following wave equation:

$$\frac{1}{c^2} \frac{\partial^2 u}{\partial s^2}(\mathbf{r}, s) - \Delta u(\mathbf{r}, s) = 0, \quad \mathbf{r} \in \Omega, \quad s \in]0, T[, \tag{1.43}$$

for some final observation time T, with the Dirichlet boundary conditions

$$u(\mathbf{r}, s) = 0 \quad \text{on } \partial\Omega \times]0, T[, \tag{1.44}$$

the initial conditions

$$u(\mathbf{r}, s)|_{s=0} = 0 \quad \text{in } \Omega, \tag{1.45}$$

and

$$\partial_s u(\mathbf{r}, s)|_{s=0} = \delta_{\mathbf{r}_\star} \text{ in } \Omega. \tag{1.46}$$

Here c is constant. We suppose that T is large enough so that

$$T > \frac{\operatorname{diam}(\Omega)}{c}. \tag{1.47}$$

The problem is to image the source point $\mathbf{r}_\star$ from partial measurements of $\partial u/\partial \nu$ on $\Gamma\times]0,T[$, where Γ is part of the boundary $\partial\Omega$. In this section, we construct Kirchhoff-, backpropagation-, and MUSIC-type algorithms for imaging point sources from limited-view data. Our approach is based on averaging of the limited-view data, using weights constructed by the geometrical control method. It is quite robust with respect to perturbations of the nonaccessible part of the boundary. If one can construct accurately the geometric control then one can perform imaging with the same resolution using partial data as using complete data.

Geometrical control

Suppose that T and Γ are such that they geometrically control Ω, which roughly means that every geometrical optic ray, starting at any point $\mathbf{r} \in \Omega$, at time $t = 0$, hits Γ before time T at a nondiffractive point. Let $\beta \in C_0^\infty(\Omega)$ be a cutoff function such that $\beta(\mathbf{r}) \equiv 1$ in a subdomain Ω' of Ω, which contains the source point $\mathbf{r}_\star$.

For a given function w which will be specified later, we construct by the geometrical control method a function $v(\mathbf{r}, s)$ satisfying

$$\frac{1}{c^2}\frac{\partial^2 v}{\partial s^2} - \Delta v = 0 \quad \text{in } \Omega\times]0,T[, \tag{1.48}$$

with the initial condition

$$v(\mathbf{r},0) = c^2\beta(\mathbf{r})w(\mathbf{r}), \quad \partial_s v(\mathbf{r},0) = 0, \tag{1.49}$$

the boundary condition $v = 0$ on $\partial\Omega \setminus \overline{\Gamma}$, and the final conditions

$$v\Big|_{s=T} = \frac{\partial v}{\partial s}\Big|_{s=T} = 0 \quad \text{in } \Omega. \tag{1.50}$$

Let

$$g_w(\mathbf{r}, s) := v(\mathbf{r}, s) \quad \text{on } \Gamma\times]0,T[. \tag{1.51}$$

Multiplying (1.43) by v and integrating over $\Omega \times [0,T]$ lead to the following key identity:

$$\int_0^T\int_\Gamma \frac{\partial u}{\partial \nu}(\mathbf{r}', s) g_w(\mathbf{r}', s)\, d\sigma(\mathbf{r}')\, ds = w(\mathbf{r}_\star). \tag{1.52}$$

The existence of a solution to (1.48)–(1.50) under the geometric control condition was established in the seminal paper [31]. Its numerical computation is based on the Hilbert uniqueness method (HUM) of Lions [34].

Kirchhoff algorithm

Let $\mathbf{r}_t \in \mathbb{R}^d \setminus \overline{\Omega}$, $d = 2, 3$, and $\omega \in \mathbb{R}$. Set

$$w(\mathbf{r}) = e^{i\omega|\mathbf{r}-\mathbf{r}_t|}, \quad \mathbf{r} \in \Omega.$$

Then, for a given search point $\mathbf{r}$ in Ω, we have from (1.52)

$$\int_{\mathbb{R}} e^{i\omega|\mathbf{r}-\mathbf{r}_t|} \int_0^T \int_\Gamma \frac{\partial u}{\partial \nu}(\mathbf{r}, s) g_w(\mathbf{r}, s)\, d\sigma(\mathbf{r})\, ds\, d\omega = \int_{\mathbb{R}} e^{-i\omega(|\mathbf{r}-\mathbf{r}_t|-|\mathbf{r}_\star-\mathbf{r}_t|)}\, d\omega$$
$$= \delta_{|\mathbf{r}-\mathbf{r}_t|-|\mathbf{r}_\star-\mathbf{r}_t|=0}.$$

Taking a (virtual) planar array of receivers $\mathbf{r}_t$ outside Ω yields then a Kirchhoff-type algorithm for finding $\mathbf{r}_\star$.

In fact, let $\omega_j, j = 1, \ldots, J$, be a set of frequencies and let $\mathbf{r}_t^{(1)}, \ldots, \mathbf{r}_t^{(N)}$, be a set of virtual receivers. To find the location $\mathbf{r}_\star$ one maximizes over $\mathbf{r}$ the following imaging functional:

$$W_{\mathrm{KI}}(\mathbf{r}) := \frac{1}{J} \Re e \sum_{\omega_j} \sum_{\mathbf{r}_t^{(n)}} e^{-i\omega|\mathbf{r}-\mathbf{r}_t^{(n)}|} \int_0^T \int_\Gamma \frac{\partial u}{\partial \nu}(\mathbf{r}', s) g_{w_{j,n}}(\mathbf{r}', s)\, d\sigma(\mathbf{r}')\, ds,$$

where $w_{j,n}(\mathbf{r}') = e^{i\omega_j|\mathbf{r}'-\mathbf{r}_t^{(n)}|}$.

Backpropagation algorithm

If one takes w to be a plane wave:

$$w(\mathbf{r}) = e^{i\omega\theta\cdot\mathbf{r}}, \qquad \theta \in \mathbb{S}^{d-1},$$

then one computes for a given search point $\mathbf{r} \in \Omega$,

$$\int_{\mathbb{S}^{d-1}} e^{-i\omega\theta\cdot\mathbf{r}} \int_0^T \int_\Gamma \frac{\partial u}{\partial \nu}(\mathbf{r}', s) g_w(\mathbf{r}', s)\, d\sigma(\mathbf{r}')\, ds\, d\sigma(\theta) = \int_{\mathbb{S}^{d-1}} e^{i\omega\theta\cdot(\mathbf{r}_\star-\mathbf{r})}\, d\sigma(\theta).$$

This is a backpropagation algorithm. Let $\theta_1, \ldots, \theta_N$, be a discretization of the unit sphere $\mathbb{S}^{d-1}$. One plots at each point $\mathbf{r}$ in the search domain the following imaging functional:

$$W_{\mathrm{BP}}(\mathbf{r}) := \frac{1}{N} \Re e \sum_{\theta_n} e^{-i\omega\theta_n\cdot\mathbf{r}} \int_0^T \int_\Gamma \frac{\partial u}{\partial \nu}(\mathbf{r}', s) g_{w_n}(\mathbf{r}', s)\, d\sigma(\mathbf{r}')\, ds,$$

where $w_n(\mathbf{r}) = e^{i\omega\theta_n\cdot\mathbf{r}}$. The resulting plot will have a large peak at $\mathbf{r}_\star$. Note that the higher the frequency ω is, the better is the resolution. However, high-frequency oscillations cause numerical instabilities. There is a trade-off between resolution and stability.

MUSIC-type algorithm

Take

$$w(\mathbf{r}) = e^{i\omega(\theta+\theta')\cdot\mathbf{r}}, \quad \theta, \theta' \in \mathbb{S}^{d-1}.$$

It follows from (1.52) that

$$\int_0^T \int_\Gamma \frac{\partial u}{\partial \nu}(\mathbf{r}', s) g_w(\mathbf{r}', s)\, d\sigma(\mathbf{r}')\, ds = e^{i\omega(\theta+\theta')\cdot \mathbf{r}_\star}.$$

Therefore, one can design a MUSIC-type algorithm for locating $\mathbf{r}_\star$. For doing so, let $\theta_1, \dots, \theta_N$ be N unit vectors in $\mathbb{R}^d$. Define the matrix $\mathbf{A} = (A_{nn'})_{n,n'=1}^N$ by

$$A_{nn'} := \int_0^T \int_\Gamma \frac{\partial u}{\partial \nu}(\mathbf{r}', s) g_{w_{n,n'}}(\mathbf{r}', s)\, d\sigma(\mathbf{r}')\, ds,$$

with

$$w_{n,n'}(\mathbf{r}') = e^{i\omega(\theta_n+\theta_{n'})\cdot \mathbf{r}}.$$

Let $\mathbf{P}$ be the orthogonal projection onto the range of $\mathbf{A}$. Given any point $\mathbf{r}$ in the search domain, form the vector

$$\mathbf{g}(\mathbf{r}) := \left(e^{i\omega\theta_1\cdot \mathbf{r}}, \dots, e^{i\omega\theta_N\cdot \mathbf{r}}\right)^T.$$

Then plot the MUSIC imaging functional:

$$W_{\mathrm{MU}}(\mathbf{r}) := \frac{1}{\|(\mathbf{I}-\mathbf{P})\mathbf{g}(\mathbf{r})\|}.$$

The resulting plot will have a large peak at $\mathbf{r}_\star$. Again, the higher the frequency ω is, the better is the resolution.

Reconstruction results

We present here some results obtained by algorithms presented below. For each algorithm we consider three different geometries and both the full view and the partial view cases Figures 1.6–1.8.

In real experiments, one does not necessarily know the uncontrolled part of the boundary with good accuracy. A major concern for real applications of the method is thus its robustness with respect to perturbations of the boundary. We give the results, with three perturbations, for the Kirchhoff (Figure 1.9) and the backpropagation (Figure 1.10). The estimation of the source position is quite robust with respect to perturbations of the nonaccessible part of the boundary and deteriorates as we increase the boundary uncertainty.

6.3 Time-domain imaging in dissipative media

In dissipative media, time-reversal techniques do not apply because of the lack of reversibility of the wave equation in attenuating media. Our strategy is to first com-

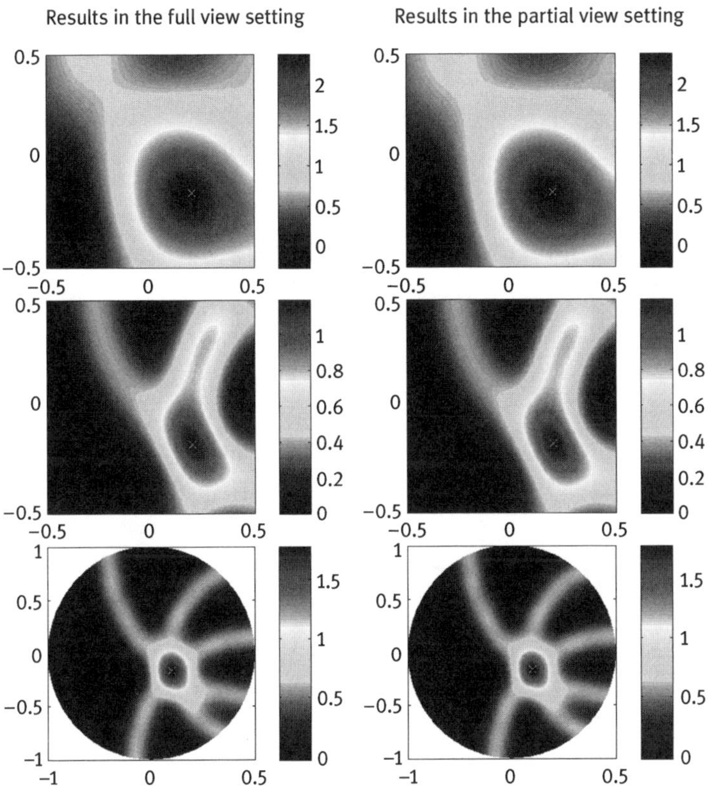

Figure 1.6: (See Color Plates for a color version.) Kirchhoff results. The (black/white) × denotes the (numerical/theoretical) center of the target.

pensate for the effect of attenuation and then apply classical time-reversal techniques to reconstruct the point source $\mathbf{r}_\star$ from attenuated data, see [6].

Let $g_a(\mathbf{r}, \mathbf{r}_\star, s)$ be the Green function of

$$\frac{\partial^2 g_a}{\partial s^2}(\mathbf{r}, \mathbf{r}_\star, s) - \Delta g_a(\mathbf{r}, \mathbf{r}_\star, s) - L(s) * g_a(\mathbf{r}, \mathbf{r}_\star, s) = \delta_{s=0}\delta_{\mathbf{r}_\star},$$

where L is defined by

$$L(s) = \frac{1}{2\pi}\int_{\mathbb{R}} \left(K^2(\omega) - \omega^2\right) e^{i\omega s} d\omega. \tag{1.53}$$

A common model for $K(\omega)$, known as the thermoviscous model, is given by

$$K(\omega) = \frac{\omega}{\sqrt{1 - ia\omega}} \approx \omega + \frac{i}{2}a\omega^2.$$

Recall that the Fourier transforms of g and g_a satisfy

$$\left(\Delta + \omega^2\right)\hat{g}(\mathbf{r}, \mathbf{r}_\star, \omega) = \delta_{\mathbf{r}_\star} \quad \text{and} \quad \left(\Delta + K(\omega)^2\right)\hat{g}_a(\mathbf{r}, \mathbf{r}_\star, \omega) = \delta_{\mathbf{r}_\star},$$

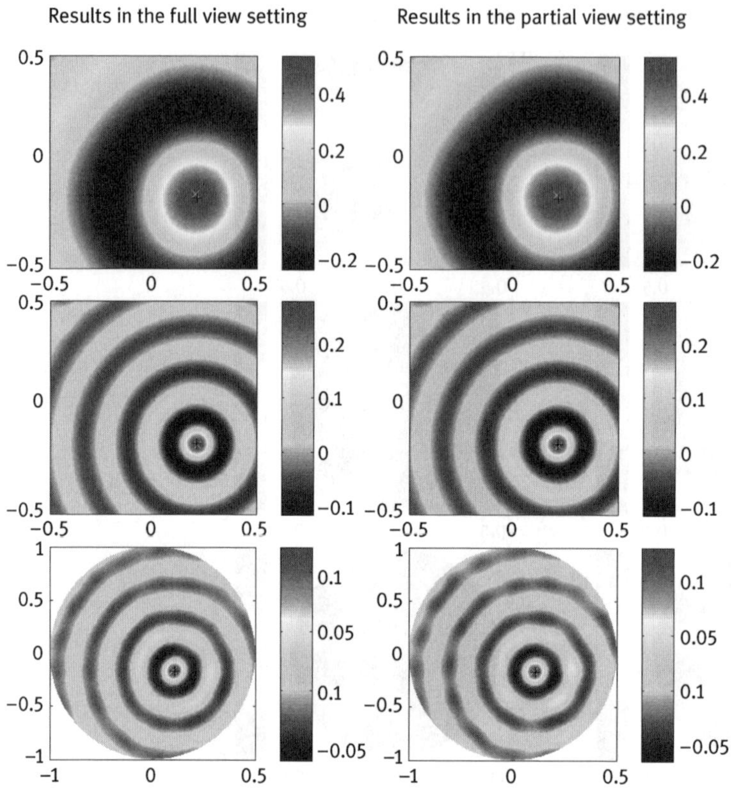

Figure 1.7: (See Color Plates for a color version.) Backpropagation results. The (black/white) × denotes the (numerical/theoretical) center of the source.

which implies that

$$\hat{g}(\mathbf{r}, \mathbf{r}_\star, K(\omega)) = \hat{g}_a(\mathbf{r}, \mathbf{r}_\star, \omega).$$

The issue is then to estimate g from g_a using the relationship $g_a = \mathcal{L}[g]$, where $\mathcal{L}$ is defined by

$$\mathcal{L}[\phi](s) = \frac{1}{2\pi} \int_{\mathbb{R}} e^{-i\omega s} \int_0^\infty \phi(t) e^{iK(\omega)t}\, dt\, d\omega.$$

The main difficulty is that $\mathcal{L}$ is not well conditioned. We compare two approaches. The first one uses a regularized inverse of $\mathcal{L}$ via a singular value decomposition (SVD). The second one is based on the asymptotic behavior of $\mathcal{L}$ as the attenuation coefficient a tends to zero.

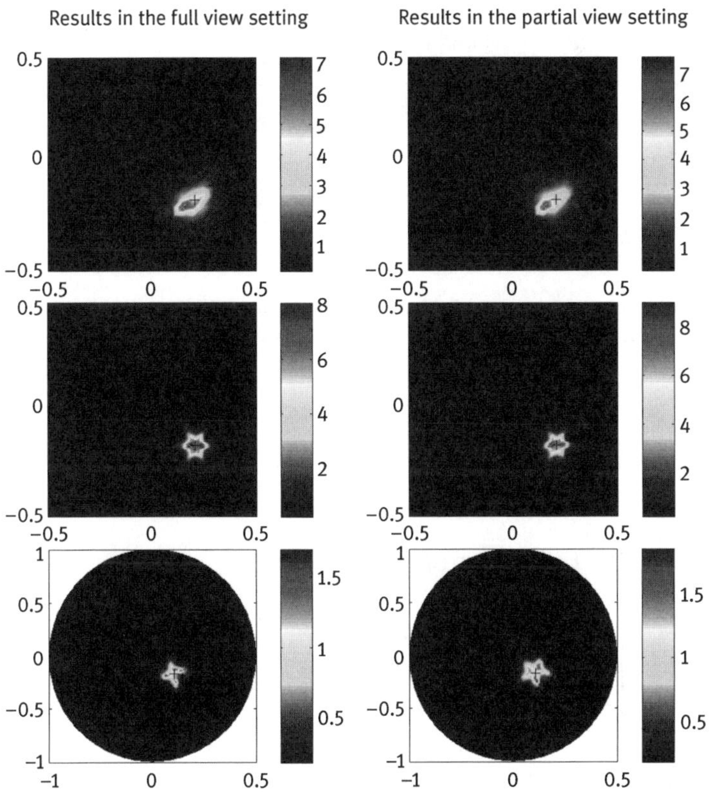

Figure 1.8: (See Color Plates for a color version.) MUSIC results. The (black/white) $\times$ denotes the (numerical/theoretical) center of the source.

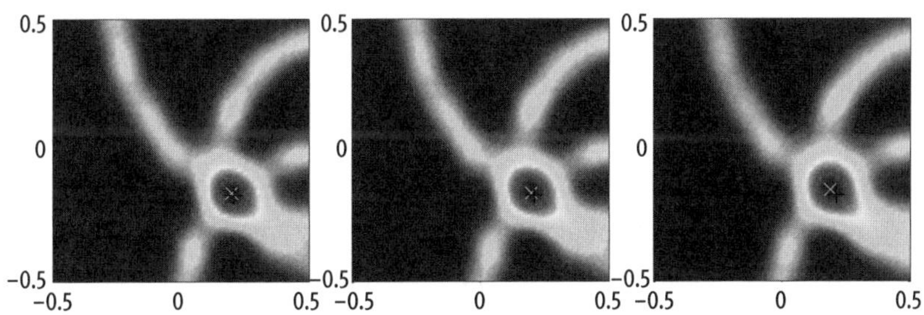

Figure 1.9: (See Color Plates for a color version.) Kirchhoff results with perturbed boundary (from left-to-right, 0.01, 0.025, and 0.05). The (black/white) $\times$ denotes the (numerical/theoretical) center of the source.

A singular value decomposition (SVD) approach

A regularized inverse of the operator $\mathcal{L}$ can be obtained by a standard SVD approach:

$$\mathcal{L}[\phi] = \sum_l \sigma_l \langle \phi, \tilde{\psi}_l \rangle \psi_l,$$

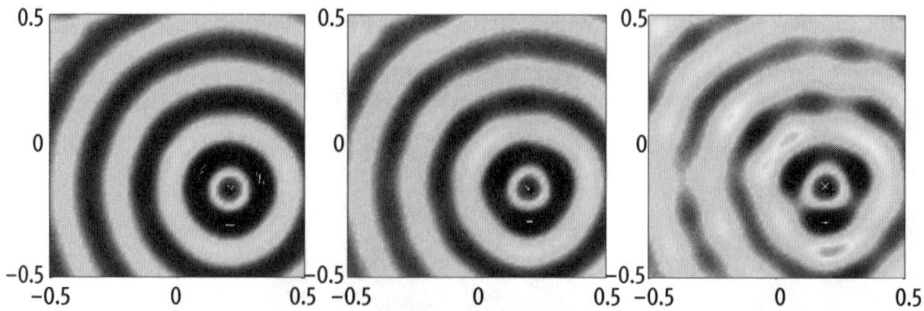

Figure 1.10: (See Color Plates for a color version.) Backpropagation results with perturbed boundary (from left-to-right, 0.01, 0.025, and 0.05). The (black/white) $\times$ denotes the (numerical/theoretical) center of the source.

where $(\tilde{\psi}_l)$ and (ψ_l) are two orthonormal bases of $L^2(0,T)$ and σ_l are positives eigenvalues such that

$$\begin{cases} \mathcal{L}^*[\phi] = \sum_l \sigma_l \langle \phi, \psi_l \rangle \tilde{\psi}_l, \\ \mathcal{L}^*\mathcal{L}[\phi] = \sum_l \sigma_l^2 \langle \phi, \tilde{\psi}_l \rangle \tilde{\psi}_l, \\ \mathcal{L}\mathcal{L}^*[\phi] = \sum_l \sigma_l^2 \langle \phi, \psi_l \rangle \psi_l. \end{cases}$$

A η-approximation inverse of $\mathcal{L}$ is then given by

$$\mathcal{L}_{1,\eta}^{-1}[\phi] = \sum_l \frac{\sigma_l}{\sigma_l^2 + \eta^2} \langle \phi, \psi_l \rangle \tilde{\psi}_l,$$

where $\eta > 0$. This algorithm corrects a part of the attenuation effect but is unstable when η tends to zero.

Asymptotics of $\mathcal{L}$

In physical situations, the coefficient of attenuation a is very small. We will take into account this phenomenon and introduce an approximation of $\mathcal{L}$ and $\mathcal{L}^{-1}$ as a goes to zero:

$$\mathcal{L}_k[\phi] = \mathcal{L}[\phi] + o(a^{k+1}) \quad \text{and} \quad \mathcal{L}_{2,k}^{-1}[\phi] = \mathcal{L}^{-1}[\phi] + o(a^{k+1}),$$

where k represents an order of approximation.

The operator $\mathcal{L}$ is approximated as follows:

$$\mathcal{L}[\phi](s) \simeq \frac{1}{2\pi} \int_0^\infty \phi(t) \int_{\mathbb{R}} e^{-\frac{1}{2}a\omega^2 t} e^{i\omega(t-s)} \, d\omega \, dt.$$

Since

$$\frac{1}{\sqrt{2\pi}}\int_{\mathbb{R}} e^{-\frac{1}{2}a\omega^2 t} e^{i\omega(t-s)}\,d\omega = \frac{1}{\sqrt{a\,t}}e^{-\frac{1}{2}\frac{(s-t)^2}{a\,t}},$$

it follows that

$$\mathcal{L}[\phi] \simeq \frac{1}{\sqrt{2\pi}}\int_0^{+\infty} \phi(t)\frac{1}{\sqrt{a\,t}}e^{-\frac{1}{2}\frac{(s-t)^2}{a\,t}}\,dt.$$

We then investigate the asymptotic behavior of $\tilde{\mathcal{L}}$ defined by

$$\tilde{\mathcal{L}}[\phi] = \frac{1}{\sqrt{2\pi}}\int_0^{+\infty} \phi(t)\frac{1}{\sqrt{a\,t}}e^{-\frac{1}{2}\frac{(s-t)^2}{a\,t}}\,dt. \tag{1.54}$$

Since the phase in (1.54) is quadratic and a is small, by the stationary phase theorem we can prove that

$$\tilde{\mathcal{L}}[\phi](s) = \sum_{i=0}^{k} \frac{a^i}{2^i i!}D_i[\phi](s) + o(a^k), \tag{1.55}$$

where the differential operators D_i satisfy $D_i[\phi](s) = (t^i\phi(t))^{(2i)}(s)$. We can also deduce the following approximation of order k of $\tilde{\mathcal{L}}^{-1}$

$$\tilde{\mathcal{L}}_l^{-1}[\psi] = \sum_{j=0}^{l} a^j \psi_{l,j}, \tag{1.56}$$

where $\psi_{l,j}$ are defined recursively by

$$\psi_{l,0} = \psi \quad \text{and} \quad \psi_{l,j} = -\sum_{i=1}^{j}\frac{1}{i!}D_i[\psi_{l,j-i}],\ \text{for all } j \le l.$$

In [6], numerical reconstructions using (1.56) seem to be as good as those obtained by the SVD regularization approach. Moreover, this new algorithm has better stability properties. We refer to [5] reconstructing extended sources using a modified time-reversal technique.

7 Numerical examples of MUSIC reconstructions for the full Maxwell equations

Here we present some numerical examples of MUSIC reconstruction for the full Maxwell equations from [18].

The eigenvalue structure of the multistatic response (MSR) matrix is employed within the framework of the MUSIC method in order to retrieve either dielectric inclusions or magnetic inclusions, or combinations thereof, with the limiting perfectly electric conductor or perfectly magnetic conductor encompassed as well.

In practice this means carrying out two steps: first, the calculation of singular values (and the corresponding eigenvectors), the number of nonzero values depending upon the number and the electromagnetic nature of the inclusions and upon the transmitters' and/or receivers' geometrical arrangement and polarization; second, the orthogonal projection of a properly built vector propagator onto the null space of the MSR matrix, coincidence with an inclusion being then associated to a peak of the inverse norm of the projection.

Simple forms of the indicator can be used if the electromagnetic nature of the inclusions is known from prior information and is either purely dielectric or purely magnetic (e.g., all inclusions are known to be nonmagnetic dielectric ones), while a slightly more complex form should be used in the other situations.

It is worth emphasizing that the number of nonzero singular values allows us to characterize the electromagnetic nature of the inclusion.

In the case of a single triaxial ellipsoidal inclusion with arbitrary semiaxes and location with respect to the transmitter and receiver arrays, the singular vectors and values can be exhibited in closed form.

The configuration of interest is sketched in Figure 1.11. We consider two coincident transmitter and receiver arrays centered within the plane $z = 5$ and covering a square area of side 10, both consisting of $N = 13^2$ points.

At each transmitter, an electric dipole with three orthogonal polarizations is radiated. But, only two polarizations of the induced magnetic field are received on the array of receivers.

The operating frequency $\omega = 4$, the corresponding wavelength in the embedding free space, with as usual unit permittivity and permeability value ($\varepsilon_0 = \mu_0 = 1$), is $\lambda = \pi/2$.

Two identical prolate spheroidal inclusions D_1 and D_2 (needle-like, the length of the longest semiaxis being three times the ones of the other two semiaxes) with size parameter $\delta = 0.1$ and orthogonal orientations (the longest semiaxis of D_1 is orientated along z, the longest one of D_2 along x) are to be retrieved within a cubical box $\Omega = [-1, 1]^3 \subset \mathbb{R}^3$ assumed to be situated below the arrays. The dielectric permittivities of the targets are set to $\varepsilon_1 = 5$, $\varepsilon_2 = 2$, and their magnetic permeabilities to $\mu_1 = 5$, $\mu_2 = 2$.

Within the above setting, the retrieval involves first the calculation of the SVD of the MSR matrix $\mathbf{A} \in \mathbb{C}^{169\times 169}$ and then the plot of the MUSIC imaging functional W_{MU} which is calculated within Ω.

The first 25 singular values of the matrix $\mathbf{A} \in \mathbb{C}^{169\times 169}$ are displayed in Figure 1.12 (left); in the case of noisy data the full set is shown in Figure 1.12 (right). So, if

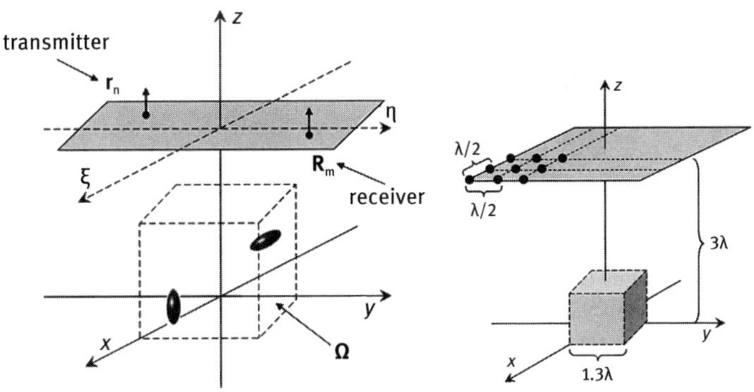

Figure 1.11: Sketch of the configuration under numerical study.

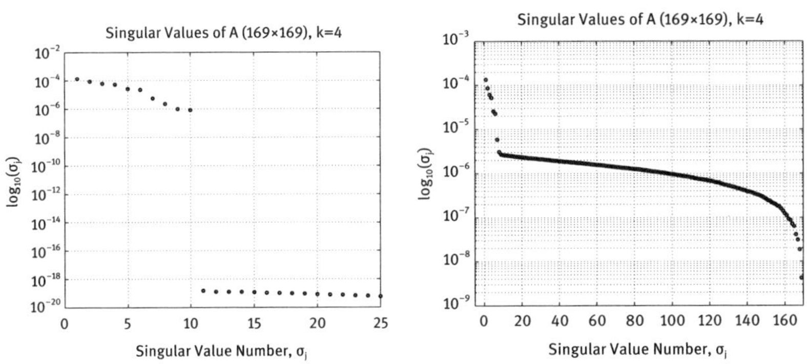

Figure 1.12: (Dielectric and permeability contrasts): distribution of the first 25 singular values of **A** or 13^2 singly polarized dipolar transmitters and receivers in the absence of noise (left) and distribution of the full set in the case of noisy data with 20 dB signal-to-noise ratio (right).

ten of them (six associated to the dielectric features, four to the magnetic features) emerge in the absence of noise, this is far less clear cut in the case of a good amount of noise on the data, since the two or three singular values of lesser magnitude are confounded with those outside the signal subspace.

Maps of the MUSIC imaging functional in the cross-sectional planes $z = \{-0.1, 0.7\}$, $x = \{-0.6, 0.6\}$, and $y = \{-0.6, 0.6\}$ are displayed in Figures 1.13 and corresponding 3D plots are displayed in Figure 1.14.

The singular values of the matrix A and corresponding 3D plots of W_{MU} are displayed in Figures 1.15–1.18 in the case of noisy data (white Gaussian noise) with 20 dB signal-to-noise ratio. Note that if the quantity of data (the dimension of the multistatic response matrix) is not enough to detect multiple inclusions, then, from the numerical results above, it is clear that our inversion procedure of MUSIC type may miss some of these inclusions or wrongly determine their nature.

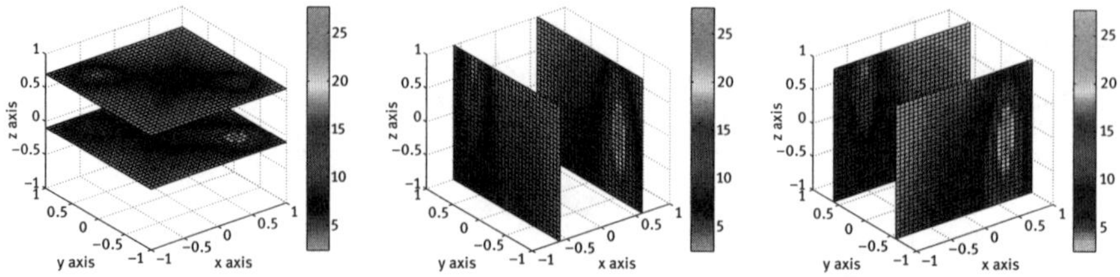

Figure 1.13: (Dielectric and permeability contrasts): color maps of W_{MU} in cross-sectional planes $z = \{-0.1, 0.7\}$, $x = \{-0.6, 0.6\}$, and $y = \{-0.6, 0.6\}$ in the case of noisy data with 20 dB signal-to-noise ratio. (See Color Plates for a color version.)

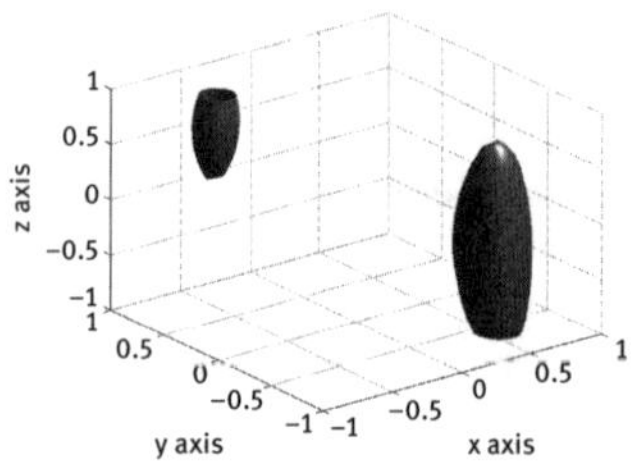

Figure 1.14: (Dielectric and permeability contrasts): 3D plots of W_{MU} in the case of noisy data with 20 dB signal-to-noise ratio (refer to Figure 1.13).

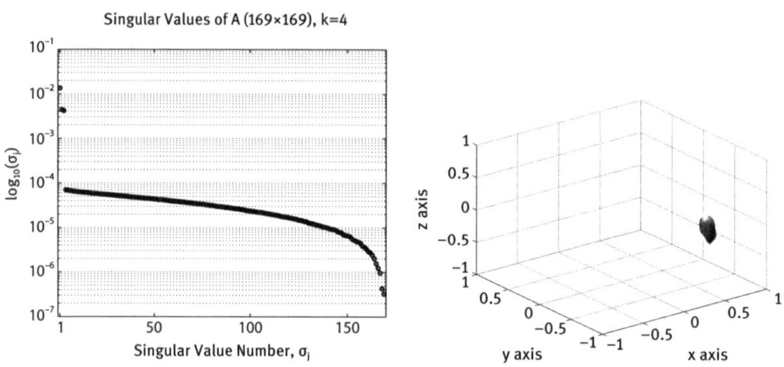

Figure 1.15: (Dielectric contrasts only): distribution of the singular values of $\mathbf{A}$ in the case of a single dielectric inclusion D_1 for 13^2 singly polarized dipolar transmitters and receivers with 20 dB signal-to-noise ratio (left); corresponding 3D plots of W_{MU} (right).

We refer to the recent work [14] (see also [15]) for a topological derivative-based imaging functional and for an investigation of the stability of the imaging algorithms discussed here with respect to medium and measurement noises as well as their resolution.

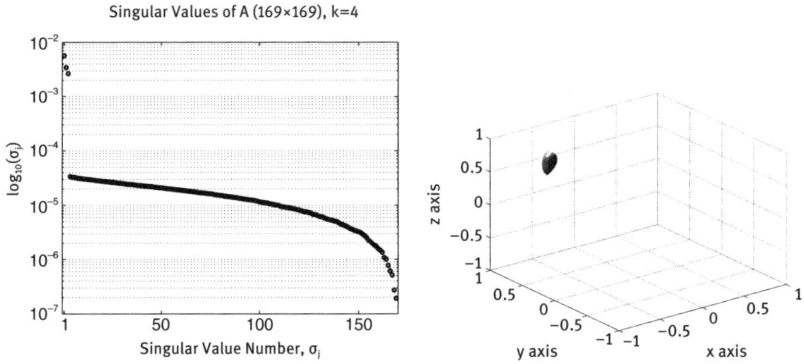

Figure 1.16: (Dielectric contrasts only): distribution of the singular values of $\mathbf{A}$ in the case of a single dielectric inclusion D_2 for 13^2 singly polarized dipolar transmitters and receivers with 20 dB signal-to-noise ratio (left); corresponding 3D plots of W_{MU} (right).

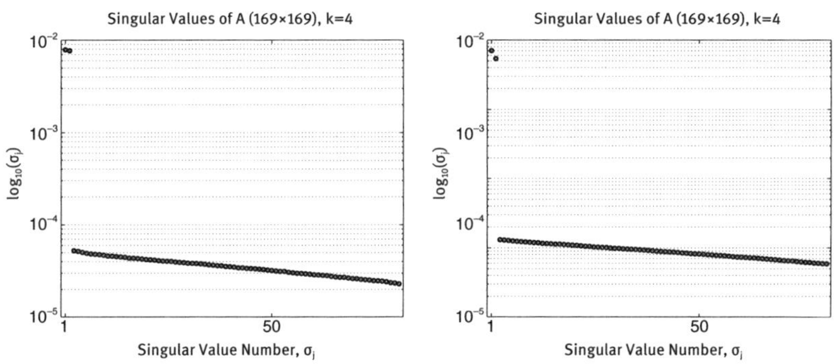

Figure 1.17: (Permeability contrasts only): distribution of the first 80 singular values of $\mathbf{A}$ in the case of a single permeable inclusion D_1 (left), D_2 (right) for 13^2 singly polarized dipolar transmitters and receivers in the case of noisy data with 20 dB signal-to-noise ratio.

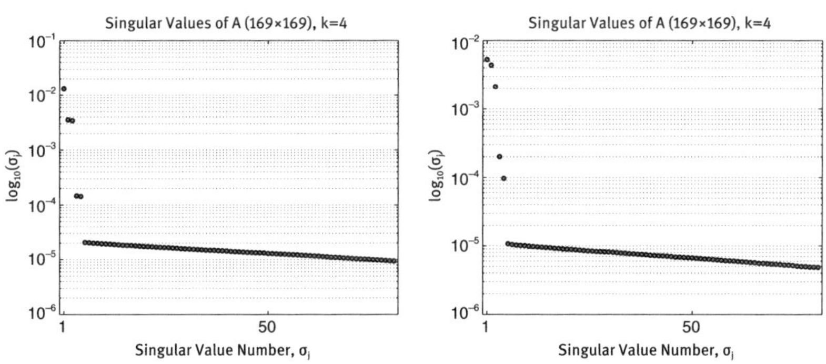

Figure 1.18: (Dielectric and permeability contrasts): distribution of the first 80 singular values of $\mathbf{A}$ in the case of a single dielectric and permeable inclusion D_1 (left), D_2 (right) for 13^2 singly polarized dipolar transmitters and receivers in the absence of noise (left) and distribution of the full set in the case of noisy data with 20 dB signal-to-noise ratio (right).

8 Shape representations

A fundamental problem in imaging is to represent inclusion shapes.

8.1 High-order polarization tensors

In this section we show that the generalized polarization tensors (GPTs) provide a natural basis to represent shapes.

For simplicity let $d = 2$. Consider the following conductivity transmission problem:

$$\begin{cases} \nabla \cdot \left(\chi(\mathbb{R}^2 \setminus \overline{B}) + k\chi(B)\right) \nabla u = 0 & \text{in } \mathbb{R}^2, \\ u(\mathbf{r}) - H(\mathbf{r}) = O(r^{-1}) & \text{as } r \to \infty, \end{cases} \tag{1.57}$$

where $\chi(B)$ denotes the characteristic function of the domain B, $k \neq 1$ its material parameter, and H is a given harmonic function in $\mathbb{R}^2$.

The coefficient $\chi(\mathbb{R}^2 \setminus \overline{B}) + k\chi(B)$ represents the material parameter distribution. The function ∇H is the background electric field and u is the voltage potential in the presence of the target B. Then the perturbation, $u - H$, is given by the multipolar expansion

$$(u - H)(\mathbf{r}) = \sum_{|\alpha|,|\beta|=1}^{+\infty} \frac{(-1)^{|\alpha|}}{\alpha!\beta!} \partial^\alpha \Gamma(\mathbf{r}) \mathbf{M}_{\alpha\beta}(k, B) \partial^\beta H(\mathbf{r}_\star) \quad \text{as } r \to +\infty, \tag{1.58}$$

where $\mathbf{r}_\star$ is the center of mass of D and Γ is the fundamental solution to the Laplacian, i.e.,

$$\Gamma(\mathbf{r}) = \frac{1}{2\pi} \ln r. \tag{1.59}$$

Here $\alpha = (\alpha_1, \alpha_2)$ and $\beta = (\beta_1, \beta_2)$ are multi-indices and $|\alpha| = \alpha_1 + \alpha_2$.

The quantity $\mathbf{M}_{\alpha\beta}$ is the GPT. Formula (1.58) shows that through the GPTs we have complete information about the far-field expansion of the perturbation $u - H$. When $|\alpha| = |\beta| = 1$, if we denote by $\mathbf{M}_{\alpha\beta} = m_{ij}$, then $\mathbf{M} = (m_{ij})_{i,j=1}^2$ is the polarization tensor (PT).

The PT associated with an unknown inclusion can be detected from boundary measurements. The detected PT in turn yields the "equivalent ellipse" of a single inclusion. In other words, in terms of the PT associated with an inclusion and a material parameter (or a cluster of inclusions and a set of material parameters) we are able to recover an equivalent ellipse with the same PT. On the other hand, the full set of GPTs uniquely determines the inclusion (and its material parameter).

Theorem 1.14. *If all harmonic combinations of GPTs of two domains are the same, i.e.,*

$$\sum_{\alpha,\beta} a_\alpha b_\beta \mathbf{M}_{\alpha\beta}(k_1, B_1) = \sum_{\alpha,\beta} a_\alpha b_\beta \mathbf{M}_{\alpha\beta}(k_2, B_2)$$

for all pairs (a_α), (b_β) *of harmonic coefficients, then* $B_1 = B_2$ *and* $k_1 = k_2$.

Therefore, it is natural to ask the question whether we can recover more shape details than the equivalent ellipse using a finite number of GPTs.

Because of the canonical one-to-one correspondence between the class of PTs and the class of ellipses, one can find easily the equivalent ellipse if one knows the PT. However, there is no (and it is unlikely to have one) known class of geometric shapes which has such a property for higher order polarization tensors. Therefore, only an optimization approach seems possible in order to recover fine shape details using GPTs.

Let B_{true} be an unknown domain. Let $\mathbf{M}_{\alpha\beta}(k, B_{\mathrm{true}})$ denote the GPTs associated with B_{true} and the material parameter k. Suppose that $\mathbf{M}_{\alpha\beta}(k, B_{\mathrm{true}})$ are known for all $|\alpha| + |\beta| \leq J$ for some number J. Suppose also that the material parameter k is known. Our recursive optimization procedure is to minimize over B

$$\frac{1}{2} \sum_{|\alpha|+|\beta|\leq J} w^{(n)}_{|\alpha|+|\beta|} \left| \sum_{\alpha,\beta} a_\alpha b_\beta \mathbf{M}_{\alpha\beta}(k, B) - \sum_{\alpha,\beta} a_\alpha b_\beta \mathbf{M}_{\alpha\beta}(k, B_{\mathrm{true}}) \right|^2 . \tag{1.60}$$

Here the coefficients a_α and b_β are such that $H = \sum a_\alpha \mathbf{r}^\alpha$ and $F = \sum b_\beta \mathbf{r}^\beta$ are homogeneous harmonic polynomials and $w^{(n)}_{|\alpha|+|\beta|}$ are binary weights. We assign either 0 or 1 to $w^{(n)}_{|\alpha|+|\beta|}$ depending which GPTs are used at step n. Moreover, at each step n we use as an initial guess the result of step $n - 1$. In the first step we get an equivalent ellipse with the same PT as well as the location of the inclusion. This recursive approach was introduced in [25].

In order to compute the shape derivative of the discrepancy functional, we shall derive an asymptotic expansion of the perturbations of the GPTs that are due to a small deformation of the boundary of the domain.

Suppose that ∂B is of class C^2 and let, for δ small, B_δ be a δ-perturbation of B, i.e., there is a function $h \in C^1(\partial B)$ such that

$$\partial B_\delta := \{\tilde{\mathbf{r}} = \mathbf{r} + \varepsilon h(\mathbf{r})\nu(\mathbf{r}) \mid \mathbf{r} \in \partial B\}, \tag{1.61}$$

where ν is the outward unit normal vector field on ∂B. Let v be the solution to the following dual transmission problem:

$$\begin{cases} \Delta v = 0 & \text{in } B \cup (\mathbb{R}^2 \setminus \overline{B}), \\ kv|_+ - v|_- = 0 & \text{on } \partial B, \\ \left.\dfrac{\partial v}{\partial \nu}\right|_+ - \left.\dfrac{\partial v}{\partial \nu}\right|_- = 0 & \text{on } \partial B, \\ (v - F)(\mathbf{r}) = O(r^{-1}) & \text{as } r \to \infty. \end{cases} \tag{1.62}$$

Theorem 1.15. *Suppose that a_α and b_β are constants such that $H = \sum_\alpha a_\alpha \mathbf{r}^\alpha$ and $F = \sum_\beta b_\beta \mathbf{r}^\beta$ are harmonic polynomials. Then*

$$\begin{aligned}&\sum_{\alpha,\beta} a_\alpha b_\beta \mathbf{M}_{\alpha\beta}(k, B_\delta) - \sum_{\alpha,\beta} a_\alpha b_\beta \mathbf{M}_{\alpha\beta}(k, B)\\ &= \delta(k-1)\int_{\partial B} h(\mathbf{r})\left[\frac{\partial v}{\partial \nu}\bigg|_- \frac{\partial u}{\partial \nu}\bigg|_- + \frac{1}{k}\frac{\partial u}{\partial \tau}\bigg|_- \frac{\partial v}{\partial \tau}\bigg|_-\right](\mathbf{r})\, d\sigma(\mathbf{r}) + O(\delta^2),\end{aligned} \tag{1.63}$$

where $\partial/\partial\tau$ is the tangential derivative and u and v, respectively, satisfies (1.57) *and* (1.62).

The examples in Figure 1.19 show the reconstruction of a kite-shaped inclusion with various relative errors. They demonstrate the validity of the recursive optimization approach. Since the high-frequency oscillations of the boundary of an inclusion are only contained in its high-order GPTs, the recursive method yields a stable way to reconstruct such information.

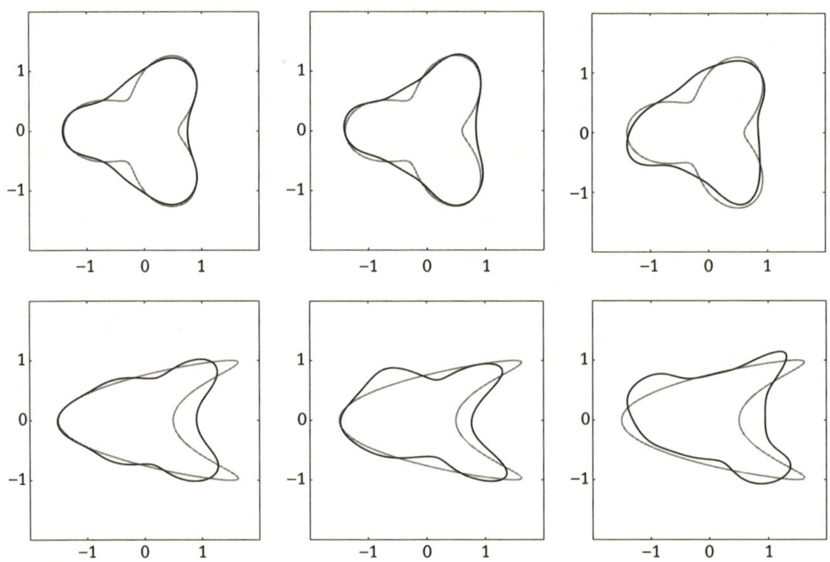

Figure 1.19: Gray curve is the target domain (B_{true}) and the black curve is the reconstructed shape for $J = 6$ starting from the equivalent ellipse. The first, second, and third column is, respectively, from the data with 0%, 10%, and 20% noise.

8.2 Frequency dependent high-order polarization tensors

In order to image targets with waves, we introduce a new notion of the GPTs depending on the frequency. For a multi-index α with $|\alpha| \geq 0$, let v_α be the solution to

$$\begin{cases} \Delta v_\alpha + (k_0\delta)^2 v_\alpha = 0 & \text{in } \mathbb{R}^d \setminus \overline{B}, \\ \Delta v_\alpha + (k\delta)^2 v_\alpha = 0 & \text{in } B, \\ v_\alpha|_+ - v_\alpha|_- = -\mathbf{r}^\alpha & \text{on } \partial B, \\ \left.\dfrac{1}{\mu_0}\dfrac{\partial v}{\partial \nu}\right|_+ - \left.\dfrac{1}{\mu}\dfrac{\partial v_\alpha}{\partial \nu}\right|_- = -\dfrac{1}{\mu_0}\dfrac{\partial \mathbf{r}^\alpha}{\partial \nu} & \text{on } \partial B, \\ v_\alpha \text{ satisfies the outgoing radiation condition.} \end{cases} \tag{1.64}$$

Then v_α can be represented using the single layer potentials as

$$v_\alpha(\mathbf{r}) = \begin{cases} S_B^{k_0\delta}[\psi_\alpha](\mathbf{r}), & \mathbf{r} \in \mathbb{R}^d \setminus \overline{B}, \\ S_B^{k\delta}[\phi_\alpha](\mathbf{r}), & \mathbf{r} \in B, \end{cases} \tag{1.65}$$

for some potentials ψ_α and ϕ_α belonging to $L^2(\partial B)$. In fact, they are unique solutions to the system of integral equations

$$\begin{cases} S_B^{k\delta}[\phi_\alpha] - S_B^{k_0\delta}[\psi_\alpha] = \mathbf{r}^\alpha \\ \dfrac{1}{\mu_\star}\dfrac{\partial}{\partial \nu}S_B^{k\delta}[\phi_\alpha] - \dfrac{1}{\mu_0}\dfrac{\partial}{\partial \nu}S_B^{k_0\delta}[\psi_\alpha] = \dfrac{1}{\mu_0}\dfrac{\partial \mathbf{r}^\alpha}{\partial \nu} \end{cases} \quad \text{on } \partial B. \tag{1.66}$$

The single layer potential $S_B^{k_0\delta}$ is defined by

$$S_B^{k_0\delta}[\phi](\mathbf{r}) := \int_{\partial B} g(\mathbf{r},\mathbf{r}';k_0\delta)(\mathbf{r},\mathbf{r}')\phi(\mathbf{r}')d\sigma(\mathbf{r}'), \quad \mathbf{r} \in \mathbb{R}^d \setminus \overline{B}.$$

Similarly, we define $S_B^{k\delta}$.

With the solution $(\phi_\alpha, \psi_\alpha)$ to (1.66), we define $\mathbf{M}_{\alpha\beta}^{(\omega)} = \mathbf{M}_{\alpha\beta}^{(\omega)}(B, \frac{\mu_0}{\mu_\star}, k_0\delta, k\delta)$ for multi-indices α and β by

$$\mathbf{M}_{\alpha\beta}^{(\omega)} = \int_{\partial B} \psi_\alpha(\mathbf{r})\mathbf{r}^\beta \, d\sigma(\mathbf{r}). \tag{1.67}$$

We call $\mathbf{M}_{\alpha\beta}^{(\omega)}$ the frequency-dependent polarization tensor (FDPT). The FDPTs are first introduced in [19].

The following proposition shows the limiting behavior of $\mathbf{M}_{\alpha\beta}^{(\omega)}$ as $\delta \to 0$ and makes the connection between $\mathbf{M}_{\alpha\beta}^{(\omega)}$ and the GPTs $\mathbf{M}_{\alpha\beta}$.

Proposition 1.16. *The FDPT $\mathbf{M}_{\alpha\beta}^{(\omega)}$ has the following asymptotic behavior as $\delta \to 0$: If $|\alpha| \geq 1$ and $|\beta| \geq 1$, then*

$$\mathbf{M}_{\alpha\beta}^{(\omega)}\left(B, \frac{\mu_0}{\mu_\star}, k_0\delta, k\delta\right) \to \mathbf{M}_{\alpha\beta}\left(B, \frac{\mu_0}{\mu_\star}\right) \quad \text{as } \delta \to 0. \tag{1.68}$$

Multipolar asymptotic expansions, as $\delta \to 0$, of the perturbation $u_\delta(\mathbf{r}, \mathbf{r}_t) - g(\mathbf{r}, \mathbf{r}_t; k_0)$ may be described most conveniently using the notion of the FDPTs.

Theorem 1.17. *The following multipolar expansion of the perturbation due to the presence of $D = \delta B + \mathbf{r}_\star$ holds:*

$$u_\delta(\mathbf{r}, \mathbf{r}_t) - g(\mathbf{r}, \mathbf{r}_t; k_0) \approx \delta^{d-2} \sum_{p=0}^{n+1} \delta^p \sum_{|\alpha|+|\beta|=p} \frac{1}{\alpha!\beta!} \mathbf{M}_{\alpha\beta}^{(\omega)} \partial^\alpha g(\mathbf{r}, \mathbf{r}_\star; k_0) \partial^\beta g(\mathbf{r}_\star, \mathbf{r}_t; k_0) + O(\delta^{n+d}). \tag{1.69}$$

Now we turn to imaging. It is possible to represent the shape of the target using the GPTs. However, because of (1.69) it seems more natural to use the FDTs. Indeed, the FDTs can be easily reconstructed from the MSR measurements by solving a linear system [21].

Recall that we have coincident transmitter and receiver arrays $\{\mathbf{r}_t^{(1)}, \dots, \mathbf{r}_t^{(N)}\}$ of N elements, used to detect the target located at $\mathbf{r}_\star$. In the presence of the target the scattered field induced on the ith receiving element from the scattering of an incident wave generated at $\mathbf{r}_t^{(j)}$ can be approximated using the multipolar expansion (1.69). The following proposition shows the structure of the MSR matrix.

Theorem 1.18. *We have*

$$u_\delta\left(\mathbf{r}_t^{(i)}, \mathbf{r}_t^{(j)}\right) - g\left(\mathbf{r}_t^{(i)}, \mathbf{r}_t^{(j)}; k_0\right) \approx \mathcal{G}\left(\mathbf{r}_t^{(i)}, \mathbf{r}_\star\right) \mathcal{W} \mathcal{G}\left(\mathbf{r}_t^{(j)}, \mathbf{r}_\star\right)^T, \tag{1.70}$$

where $\mathcal{G}(\mathbf{r}_t^{(i)}, \mathbf{r}_\star)$ is a row vector of size $\frac{1}{2}(N+1)(N+2)$ in dimension two and $\frac{1}{6}(N+1)(N+2)(N+3)$ in the three-dimensional case, which is given by

$$\mathcal{G}(\mathbf{r}_t^{(i)}, \mathbf{r}_\star) = \left(\frac{1}{\alpha!} \partial^\alpha g(\mathbf{r}_t^{(i)}, \mathbf{r}_\star; k_0)\right)_{|\alpha| \le N}, \tag{1.71}$$

and $\mathcal{W}$ is defined by

$$\mathcal{W} = (\mathcal{W}_{\alpha\beta})_{|\alpha|, |\beta| \le N} = \left(\delta^{d-2+|\alpha|+|\beta|} \mathbf{M}_{\alpha\beta}^{(\omega)}\right)_{|\alpha|, |\beta| \le N}. \tag{1.72}$$

In view of (1.71), the set of singular vectors consists of the Green function and its high derivatives on the array. Then, in order to locate the target, exactly the same MUSIC-type estimator, W_{MU}, can be used. This estimator peaks at the location of the target. However, the significant singular values are perturbed, even those associated to the dipolar approximation. The difference between those based on a point approximation and those based on a multipolar approximation (high-order approximation) are measured in terms of the difference between the polarization tensor $\mathbf{M}$ and the new quantities $\mathbf{M}_{\alpha\beta}^{(\omega)}$ for $|\alpha|, |\beta| = 0, 1$. Indeed, when δ is increasing, new significant singular values can emerge. Those are related to higher-order multipolar terms. They can be expressed in terms of $\mathbf{M}_{\alpha\beta}^{(\omega)}$ for $|\alpha|$ or $|\beta| \geq 2$. These new singular values, which are intermediate between the larger ones and zero, contain some information on the target and give better approximation of its shape and electromagnetic parameters.

Once the location of the target is recovered, the matrix $\mathcal{W}$ in (1.72) can be recovered from the MSR matrix $\mathbf{A}$. In fact, if we denote by $\mathbf{r}_\star$ the recovered location of the target, then we can find $\mathcal{W}$ by solving

$$A_{ij} \approx \mathcal{G}\left(\mathbf{r}_t^{(i)}, \mathbf{r}_\star\right) \mathcal{W} \mathcal{G}\left(\mathbf{r}_t^{(j)}, \mathbf{r}_\star\right)^T, \tag{1.73}$$

where $\mathcal{G}(\mathbf{r}_t^{(i)}, \mathbf{r}_\star)$ is given by (1.71). This is a system of $n \times n$ linear equations with $[\frac{1}{2}(N+1)(N+2)]^2$ (the number of entries of $\mathcal{W}$) unknowns in two dimensions. Hence it can be solved in general if we take n sufficiently large. We solve (1.73) in the minimum residual sense.

We first obtain the equivalent ellipse representing the shape of the target using $\mathbf{M}_{\alpha\beta}^{(\omega)}$ for $|\alpha| = |\beta| = 1$ and estimate its permittivity. To separate the permeability $\mu_\star$

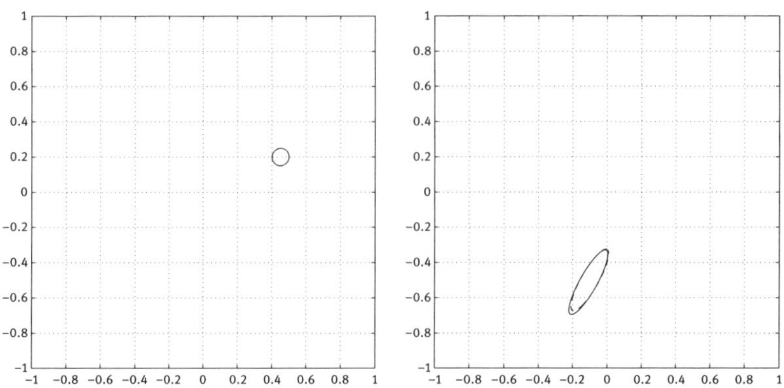

	Actual data	Computed data
z	0.4500, 0.2000	0.4508, 0.2005
δ	0.0500	0.0488, 0.0498
θ	-	−32.4909
μ	64	80.2815
ε	25	25.9793

	Actual data	Computed data
z	– 0.1000, – 0.5000	– 0.1086, – 0.5126
δ	0.2000, 0.0500	0.2089, 0.0500
θ	60	60.1418
μ	121	102.2785
ε	36	32.1234

Figure 1.20: Reconstructed ellipses for disk or ellipse shaped targets marked with dashed-dotted curves. The solid ellipses are reconstructed equivalent ellipses. We take $(\mu_0, \varepsilon_0) = (4, 1), (9, 6)$ in the first and second plots, respectively.

from $|D|$ we use the information provided by $\mathcal{W}_{(2,0),(0,0)}$. It is impossible to separate $\mu_\star$ from $|D|$ by only using $(\mathcal{W}_{\alpha\beta})_{|\alpha|=|\beta|=1}$.

Once the electromagnetic parameters $\varepsilon_\star$, $\mu_\star$, and the equivalent ellipse are reconstructed we can reconstruct fine details of shape of the target using higher-order FDPTs. For doing so, we minimize the discrepancy between the computed and the measured FDPTs.

We now present computational results for reconstruction of electromagnetic parameters and the equivalent ellipses of targets of various shapes. We emphasize that the electromagnetic parameters, μ and ε, are unknown in advance. First, Figure 1.20 shows the reconstructed equivalent ellipses and $\mu_\star$ and $\varepsilon_\star$ for a single target of the elliptic shape.

Figure 1.21 summarizes the computed results for a target of the general shape. It demonstrates that the reconstruction algorithm works pretty well even for targets of general shapes.

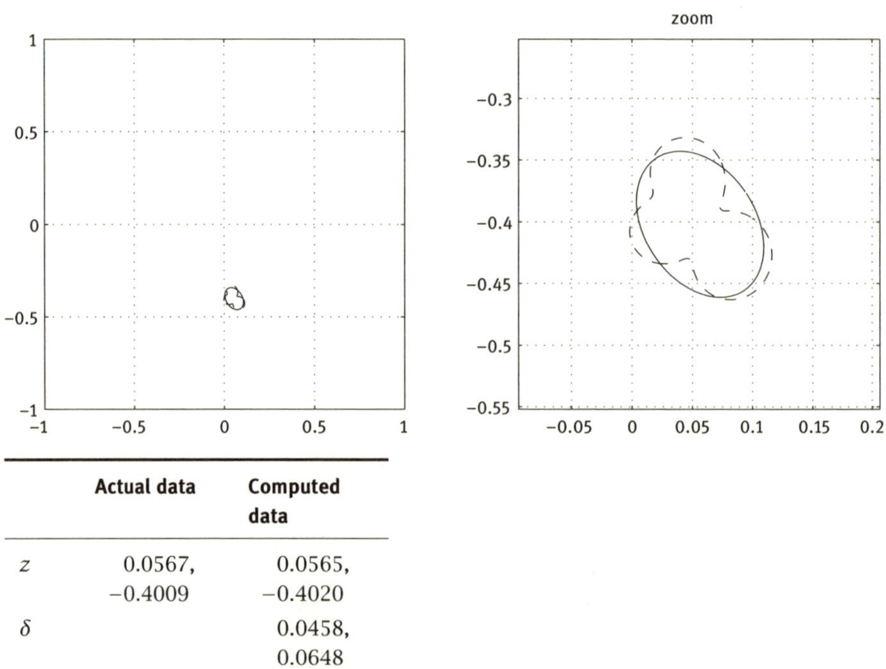

	Actual data	Computed data
z	0.0567, −0.4009	0.0565, −0.4020
δ		0.0458, 0.0648
θ		35.2369
μ	36	48.3027
ε	25	25.0300

Figure 1.21: Reconstructed ellipses for a nonellipse shaped target marked with dashed-dotted curves. The solid ellipse are reconstructed equivalent ellipse. We take $(\mu_0, \varepsilon_0) = (4, 1)$.

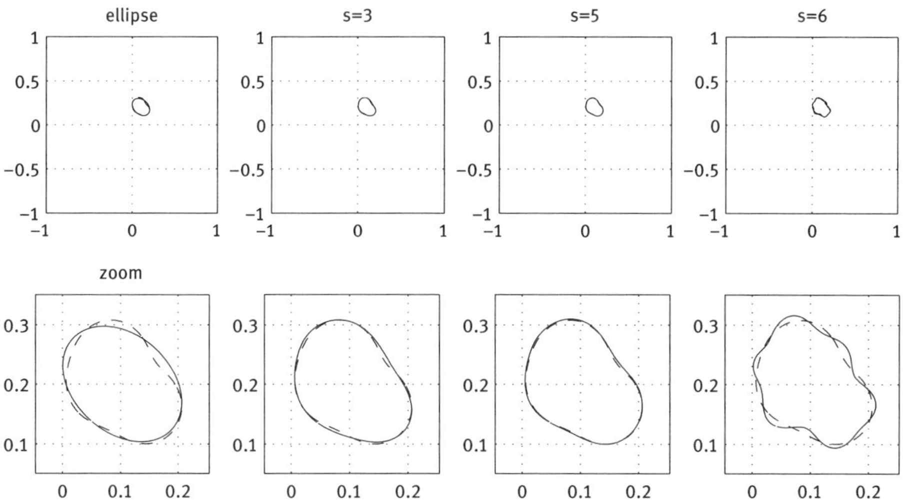

Figure 1.22: Computed shape for a general-shaped target marked with dashed–dotted curves. The solid lines are computed increasing the order s of h. Bottom figures are 6.67 times zoom-ups of the corresponding results.

For reconstructing the shape deformation, we assume that the true target is a perturbation of a disk of radius r_0 and look for h (a parametrization of the computed target) of the form

$$h(\theta) := r_0 + \sum_{\ell=1}^{s} (a_\ell^c \cos(\ell\theta) + a_\ell^s \sin(\ell\theta)),$$

where a_ℓ^c and a_ℓ^s are constants to be found by minimizing the discrepancy functional between the computed and the measured FDPTs. In the following examples, we refine the domain by increasing the order s of h while the order J of the FDPT $\mathbf{M}_{\alpha\beta}^{(\omega)}$ is set to be 7, $|\alpha| + |\beta| \leq J = 7$.

9 Far-field imaging versus near-field imaging

Let u_δ be the outgoing solution to (1.32). The following theorem gives an asymptotic expansion of $u_\delta(\mathbf{r}) - g(\mathbf{r}_\star, \mathbf{r}_t; k_0)$ for $\mathbf{r}$ near $\mathbf{r}_\star$. It shows that the perturbation is of order the size of the inclusion and not the volume. It also shows that the high-frequency information about the target shape can be described by the corrector function defined by (1.17). Recall that the PT is nothing else than the leading-order information in the far-field expansion of such a function.

Theorem 1.19. *The following inner asymptotic formula holds:*

$$u_\delta(\mathbf{r},\mathbf{r}_t) - g(\mathbf{r}_\star,\mathbf{r}_t;k_0) \approx \delta v^{\frac{1}{\mu}}(\frac{\mathbf{r}-\mathbf{r}_\star}{\delta}) \cdot \nabla g(\mathbf{r}_\star,\mathbf{r}_t;k_0) \quad \textit{for } \mathbf{r} \textit{ near } \mathbf{r}_\star \tag{1.74}$$

where $v^{\frac{1}{\mu}}$ *is the solution to* (1.17) *with* $q(\mathbf{r}) = 1/\mu(\mathbf{r})$.

Note that v is independent of the frequency. This could be important in applications to filter out small random fluctuations in the medium using multiple frequencies.

We also note that the inner asymptotic expansions (1.74) and (1.77) uniquely characterize the shape and the material parameter of the target. In fact, the following holds.

Lemma 1.20. *Suppose for two Lipschitz domains B and B' and two magnetic contrasts k and k' are such that $v(B,k) = v(B',k')$ in a domain englobing B and B', then $B = B'$ and $k = k'$.*

Let f be a binary level set representation of the scaled domain B:

$$f(\mathbf{r}) = \begin{cases} 1, & \mathbf{r} \in B, \\ -1, & \mathbf{r} \in \mathbb{R}^3 \setminus \overline{B}. \end{cases} \tag{1.75}$$

Suppose that the wavefield in a window Ω containing the target can be acquired. In view of (1.74), the shape and the material parameter of D can be approximately reconstructed by minimizing over k and f the following functional:

$$\begin{aligned} L(f,k) = {}& \frac{1}{2}\left\| u_\delta(\mathbf{r},\mathbf{r}_t) - g(\mathbf{r}_\star,\mathbf{r}_t;k_0) - \delta v^{\frac{1}{\mu}}\left(\frac{\mathbf{r}-\mathbf{r}_\star}{\delta}\right) \cdot \nabla g(\mathbf{r}_\star,\mathbf{r}_t;k_0)\right\|^2_{L^2(\Omega)} \\ & + \beta \int_\Omega |\nabla h(\mathbf{r})|\, d\mathbf{r}, \end{aligned} \tag{1.76}$$

subject to (1.17), where $2h(\mathbf{r}) = k(f(\frac{\mathbf{r}-\mathbf{r}_\star}{\delta})+1) - (f(\frac{\mathbf{r}-\mathbf{r}_\star}{\delta})-1)$, and f given by (1.75). Here, $\int_\Omega |\nabla h|\, d\mathbf{r}$ is the total variation of the shear modulus and $|\nabla h|$ is understood as a measure:

$$\int_\Omega |\nabla h| = \sup\left\{ \int_\Omega h \nabla \cdot \mathbf{v}\, d\mathbf{r},\ \mathbf{v} \in C_0^1(\Omega) \text{ and } |\mathbf{v}| \leq 1 \text{ in } \Omega \right\}.$$

This regularization indirectly controls both the length of the level curves and the jumps in the coefficients.

The local character of the method is due to the decay of $v^{\frac{1}{\mu}}(\xi) - \xi$ at infinity. This is one of the main features of the near-field imaging. In the presence of noise, because of a trade-off between accuracy and stability, one has to choose carefully the size of Ω. The size of Ω should not be too small in order to preserve some stability and not too big so that one can gain some accuracy, see Figure 1.23.

In the time-dependent regime, the following inner expansion holds.

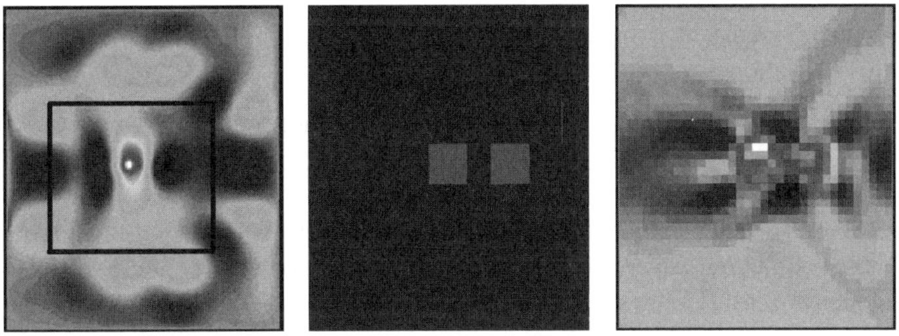

Figure 1.23: (See Color Plates for a color version.) Reconstruction using the data on the whole domain on the left, a zoom on the target in the middle, and on the right the reconstruction limited on the subregion defined by the boxed region on the left.

Theorem 1.21. *The following inner approximation holds:*

$$P_\rho[u_\delta](\mathbf{r},,\mathbf{r}_t,s)-P_\rho[g](\mathbf{r},\mathbf{r}_t;s)\approx\delta\tilde{v}^{\frac{1}{\mu}}\left(\frac{\mathbf{r}-\mathbf{r}_\star}{\delta}\right)\cdot\nabla P_\rho[g](\mathbf{r},\mathbf{r}_\star,s)\quad \textit{for } \mathbf{r} \textit{ near } \mathbf{r}_\star, \tag{1.77}$$

where $\tilde{v}^{\frac{1}{\mu}}(\xi)=v^{\frac{1}{\mu}}(\xi)-\xi$.

Therefore, we shall minimize the discrepancy functional given by

$$\begin{aligned} L(f,k) = \frac{1}{2\Delta T}\int_{T-\frac{\Delta T}{2}}^{T+\frac{\Delta T}{2}} \Big\| P_\rho[u_\delta](\mathbf{r},,\mathbf{r}_t,s) - P_\rho[g](\mathbf{r},\mathbf{r}_t;s) \\ -\delta v^{\frac{1}{\mu}}\left(\frac{\mathbf{r}-\mathbf{r}_\star}{\delta}\right)\cdot\nabla P_\rho[g](\mathbf{r},\mathbf{r}_\star,s)\Big\|^2_{L^2(\Omega)} dt + \beta\int_\Omega |\nabla h(\mathbf{r})|\, d\mathbf{r}, \end{aligned} \tag{1.78}$$

subject again to (1.17), where $T=|\mathbf{r}_t-\mathbf{r}_\star|$ is the arrival time and $\Delta T=O(\delta/\sqrt{k})$.

10 Open problems

To complete this chapter, we list some open problems.

1. Identities (1.12)–(1.15) are not valid in negative refractive index materials. Show whether by placing a negative material around a point source we can focus far-field measurements at the source beyond the resolution limit. See [40] for an excellent review on negative refractive index materials.
2. Prove that (1.26) holds for any ω such that $\omega\delta \ll 1$. Is the remainder $O(\delta^4)$ is uniform with respect to the electric and magnetic contrasts, $\varepsilon_\star/\varepsilon_0$ and $\mu_\star/\mu_0$?

3. In chiral media, electromagnetic fields satisfy the Drude–Born–Fedorov constitutive relations:

$$\begin{cases} \nabla \times \mathbf{H}(\mathbf{r}) = \beta(\mathbf{r})\mathbf{H}(\mathbf{r}) - i\omega\varepsilon(\mathbf{r})\mathbf{E}(\mathbf{r}), \\ \nabla \times \mathbf{E}(\mathbf{r}) = \beta(\mathbf{r})\mathbf{H}(\mathbf{r}) + i\omega\mu(\mathbf{r})\mathbf{H}(\mathbf{r}), \end{cases}$$

   where $\beta(\mathbf{r})$ is the chirality admittance. Derive a small-volume expansion for the effect of a chiral target and prove that the polarization tensors are in this case coupled. See [27] for the derivation of the radiation conditions and existence and uniqueness of solutions in chiral media.
4. What is the ellipse or ellipsoid that gives the same polarization tensor as two nearly touching inclusions? What happens when the distance between the inclusions goes to zero and the contrasts degenerate (go to zero or $+\infty$)?
5. In applications, the position and the orientation of the emitted (electric or magnetic dipoles) may not be known exactly. Does this affect the imaging procedures? Which imaging procedure is the most robust against such noise?
6. Establish a resolution analysis for the eddy currents imaging. Since no wavelength does exit any more it would be interesting to see whether the resolution limit comes from the analytic continuation rather than the filtering effect associated to wave propagation.
7. Develop a MUSIC-type algorithm which uses one direction of incidence but many frequencies. In this direction, see [7] and [39].
8. Suppose that the target signature is weak enough to be hidden in the noise subspace of the MSR matrix. By making multifrequency measurements, can we detect a target?
9. Adapt the multistatic imaging (in the time domain) to the situation where the target is moving. In this direction, see the recent works [8] and [9].
10. As explained in Section 6.3 time reversal does not work if the medium is dissipative because of the lack of reversibility of the attenuated wave equation. If the attenuation is not constant but randomly distributed, how to compensate its effect on image reconstruction? In this direction, see [5].
11. Measure the frequency content of the GPTs and the FDPTs. In this direction, see [17].
12. Find a basis to represent shapes from near-field measurements.

References

[1] H. Ammari, *An Introduction to Mathematics of Emerging Biomedical Imaging*, Mathématiques & Applications, Vol. 62, Springer-Verlag, Berlin, 2008.

[2] H. Ammari, M. Asch, V. Jugnon, and H. Kang, Transient imaging with limited-view data, SIAM J. Imag. Sci., 4 (2011), 1097–1121.

[3] H. Ammari, E. Beretta, E. Francini, H. Kang, and M. Lim, Optimization algorithm for reconstructing interface changes of a conductivity inclusion from modal measurements, Math. Comp., 79 (2010), 1757–1777.
[4] H. Ammari, E. Beretta, E. Francini, H. Kang, and M. Lim, Reconstruction of small interface changes of an inclusion from modal measurements II: The elastic case, J. Math. Pures Appl., 94 (2010), 322–339.
[5] H. Ammari, E. Bretin, J. Garnier, and A. Wahab, Time reversal in attenuating acoustic media, Contemp. Math., 548 (2011), 151–163.
[6] H. Ammari, E. Bretin, V. Jugnon, and A. Wahab, Photo-acoustic imaging for attenuating acoustic media, in *Mathematical Modeling in Biomedical Imaging II*, Lecture Notes in Mathematics, Vol. 2035, 57–84, Springer-Verlag, Berlin, 2011.
[7] H. Ammari, T. Boulier, and J. Garnier, Modeling active electrolocation in weakly electric fish, SIAM J. Imag. Sci., 5 (2013), 285–321.
[8] H. Ammari, T. Boulier, J. Garnier, W. Jing, H. Kang, and H. Wang, Target identification using dictionary matching of generalized polarization tensors, submitted.
[9] H. Ammari, T. Boulier, J. Garnier, H. Kang, and H. Wang, Tracking of a mobile target using generalized polarization tensors, to appear in SIAM J. Imag. Sci.
[10] H. Ammari, A. Buffa, and J. C. Nédélec, A justification of eddy currents model for the Maxwell equations, SIAM J. Appl. Math., 60 (2000), 1805–1823.
[11] H. Ammari, J. Chen, Z. Chen, J. Garnier, and D. Volkov, Target detection and characterization from electromagnetic induction data, to appear in J. Math. Pures Appl.
[12] H. Ammari, P. Garapon, L. Guadarrama Bustos, and H. Kang, Transient anomaly imaging by the acoustic radiation force, J. Diff. Equat., 249 (2010), 1579–1595.
[13] H. Ammari, J. Garnier, W. Jing, H. Kang, M. Lim, K. Sølna, and H. Wang, *Mathematical and Statistical Methods for Multistatic Imaging*, Lecture Notes in Mathematics, Springer-Verlag, submitted.
[14] H. Ammari, J. Garnier, V. Jugnon, and H. Kang, Direct reconstruction methods in ultrasound imaging of small anomalies, in *Mathematical Modeling in Biomedical Imaging II*, Lecture Notes in Mathematics, Vol. 2035, 31–55, Springer-Verlag, Berlin, 2011.
[15] H. Ammari, J. Garnier, V. Jugnon and H. Kang, Stability and resolution analysis for a topological derivative based imaging functional, SIAM J. Control Optim., 50 (2012), 48–76.
[16] H. Ammari, J. Garnier, H. Kang, M. Lim, and K. Sølna, Multistatic imaging of extended targets, SIAM J. Imag. Sci., 5 (2012), 564–600.
[17] H. Ammari, J. Garnier, H. Kang, M. Lim, and S. Yu, Generalized polarization tensors for shape description, Numer. Math., to appear.
[18] H. Ammari, E. Iakovleva, D. Lesselier, and G. Perrusson, A MUSIC-type electromagnetic imaging of a collection of small three-dimensional inclusions, SIAM J. Sci. Comput., 29 (2007), 674–709.
[19] H. Ammari and H. Kang, *Reconstruction of Small Inhomogeneities from Boundary Measurements*, Lecture Notes in Mathematics, Vol. 1846, Springer-Verlag, Berlin, 2004.
[20] H. Ammari and H. Kang, *Polarization and Moment Tensors: with Applications to Inverse Problems and Effective Medium Theory*, Applied Mathematical Sciences, Vol. 162, Springer-Verlag, New York, 2007.
[21] H. Ammari, H. Kang, E. Kim, and J. Y. Lee, The generalized polarization tensors for resolved imaging. Part II: Shape and electromagnetic parameters reconstruction of an electromagnetic inclusion from multistatic measurements, Math. of Comp., 81 (2012), 839–860.
[22] H. Ammari, H. Kang, H. Lee, and W. K. Park, Asymptotic imaging of perfectly conducting cracks, SIAM J. Sci. Comput., 32 (2010), 894–922.

[23] H. Ammari, H. Kang, E. Kim, K. Louati, and M. Vogelius, A MUSIC-type algorithm for detecting internal corrosion from electrostatic boundary measurements, Numer. Math., 108 (2008), 501–528.

[24] H. Ammari, H. Kang, M. Lim, and H. Zribi, Conductivity interface problems. Part I: small perturbations of an interface, Trans. Amer. Math. Soc., 362 (2010), 2435–2449.

[25] H. Ammari, H. Kang, M. Lim, and H. Zribi, The generalized polarization tensors for resolved imaging. Part I: Shape reconstruction of a conductivity inclusion, Math. of Comp., 81 (2012), 367–386.

[26] H. Ammari and A. Khelifi, Electromagnetic scattering by small dielectric inhomogeneities, J. Math. Pures Appl., 82 (2003), 749–842.

[27] H. Ammari and J. C. Nédélec, Time-harmonic electromagnetic fields in chiral media, in *Modern Mathematical Methods in Diffraction Theory and its Applications in Engineering*, 174–202, Methoden und Verfahen der Mathematischen Physik, 42, Lang, Frankfurt am Main, 1997.

[28] H. Ammari, M. Vogelius, and D. Volkov, Asymptotic formulas for perturbations in the electromagnetic fields due to the presence of inhomogeneities of small diameter II. The full Maxwell equations, J. Math. Pures Appl., 80 (2001), 769–814.

[29] B. A. Auld and J. C. Moulder, Review of advances in quantitative eddy current nondestructive evaluation, J. Nondest. Eval., 18 (1999), 3–36.

[30] C. Bardos, A mathematical and deterministic analysis of the time-reversal mirror, in *Inside Out: Inverse Problems*, MSRI Publications, Vol. 47, 2003.

[31] C. Bardos, G. Lebeau, and J. Rauch, Sharp sufficient conditions for the observation, control, and stabilization of waves from the boundary, SIAM J. Control Optim., 30 (1992), 1024–1065.

[32] J. Cantarella, D. DeTurck, and H. Gluck, Vector calculus and the topology of domains in 3-space, The American Mathematical Monthly, 109 (2002), 409–442.

[33] A. Friedman and M. S. Vogelius, Identification of small inhomogeneities of extreme conductivity by boundary measurements: a theorem on continuous dependence, Arch. Rat. Mech. Anal., 105 (1989), 299–326.

[34] R. Glowinski, J. L. Lions, and J. He, *Exact and Approximate Controllability for Distributed Parameter Systems. A Numerical Approach*, Encyclopedia of Mathematics and its Applications, Vol. 117, Cambridge University Press, Cambridge, 2008.

[35] G. W. Milton, *The Theory of Composites*, Cambridge Monographs on Applied and Computational Mathematics, Cambridge University Press, 2001.

[36] J. C. Nédélec, *Acoustic and Electromagnetic Equations. Integral Representations for Harmonic Problems*, Applied Mathematical Sciences, Vol. 144, Springer-Verlag, New-York, 2001.

[37] S. J. Norton and I. J. Won, Identification of buried unexploded ordnance from broadband electromagnetic induction data, IEEE Trans. Geoscience Remote Sensing, 39 (2001), 2253–2261.

[38] G. Pólya and G. Szegö, *Isoperimetric Inequalities in Mathematical Physics*, Annals of Mathematical Studies Number 27, Princeton University Press, Princeton, NJ, 1951.

[39] B. Scholz, Towards virtual electrical breast biopsy: space-frequency MUSIC for trans-admittance data, IEEE Trans. Med. Imaging, 21 (2002), 588–595.

[40] V. Veselago, L. Braginsky, V. Shklover, and C. Hafner, Negative refractive index materials, J. Comput. Theor. Nanoscience, 3 (2006), 1–30.

Xavier Claeys, Ralf Hiptmair and Carlos Jerez-Hanckes

Multitrace boundary integral equations

Abstract: We consider the scattering of acoustic or electromagnetic waves at a penetrable object composed of different homogeneous materials. This problem can be recast as a first-kind boundary integral equation posed on the interface trace spaces through what we call a single trace boundary integral equation formulation (STF). Its Ritz–Galerkin discretization by means of low-order piecewise polynomial boundary elements on fine interface triangulations leads to ill-conditioned linear systems of equations, which defy an efficient iterative solution.

Powerful preconditioners for discrete boundary integral equations are provided by the policy of operator preconditioning provided that the underlying trace spaces support a duality pairing with L^2 as pivot space. This condition is not met by the STF. As a remedy we have proposed two variants of new multitrace boundary integral equations (MTF); whereas the STF features unique Cauchy traces on material domain interfaces as unknowns, the multitrace approach tears apart the traces so that local traces are recovered. Local trace spaces are in duality with respect to the L^2-pairing, and, thus, operator preconditioning becomes available for MTF.

Keywords: Helmholtz equation, Maxwell's equation, transmission problems, boundary integral equations, PMCHWT, operator preconditioning, boundary elements, multitrace formulations

2010 Mathematics Subject Classification: 74J20,65N38,65N55

Xavier Claeys: ISAE/INSA Toulouse, France, xavier.claeys@isae.fr
Ralf Hiptmair: SAM, ETH Zurich, Switzerland, hiptmair@sam.math.ethz.ch
Carlos Jerez-Hanckes: Escuela de Ingeniería, Pontificia Universidad Católica de Chile, Santiago, Chile, cjerez@ing.puc.cl

1 Introduction

Boundary integral equations connecting traces of solutions of linear elliptic boundary value problems were a key discovery of classical analysis and potential theory. They have also become a pivotal tool for numerical simulation, since they provide the foundation of the boundary element method in all its variants.

The research on which this article is based was funded by Thales Systemes Aeroportes SA through the project "Preconditioned Boundary Element Methods for Electromagnetic Scattering at Dielectric Objects".
The authors would like to thank Prof. M. Ganesh for valuable hints on a draft version of this chapter.

No doubt, the study of boundary integral equations is a mature field of applied mathematics. Thus, the reader might be surprised to learn that the field has witnessed recent new developments. At their cradle stood transmission problems arising in the scattering of waves at composite objects. A composite object is finite and composed of a few simple parts with constant material properties inside each of them, see Section 1.1 for details. It goes without saying that this kind of geometry is very common in technical designs.

The use of boundary element methods to tackle such transmission problems numerically is well established and they have become the methods of choice for a range of industrial applications. Yet, for scattering at composite objects scalability of numerical simulations based on traditional boundary element approaches emerged as a challenge; despite the use of sophisticated matrix compression techniques like multipole [15], adaptive cross approximation [2], and panel clustering [48, Chapter 7], computational effort still increases much faster than the number of degrees of freedom. The sole culprit is sluggish convergence of iterative solvers, when applied to high resolution boundary element models. Unfortunately, it seems to be the very structure of the customary first-kind boundary integral equations that denies us the remedy of effective preconditioners. In Section 4 we are going to explain this in detail.

This takes us to the heart of this chapter: after having understood the basic difficulties inherent in the standard first-kind boundary integral equations, called *single-trace formulation* (STF) hereafter, we will try to convince the reader that those can be overcome by switching to *multitrace formulations* (MTF). They come in two different varieties, the global MTF to which Section 5 is devoted, and the local MTF treated in Section 6.

Of course, the problem has been attacked from other directions, notably by means of domain decomposition techniques like "Boundary element tearing and interconnecting" (BETI) [37, 52]. Schemes with a multitrace flavor have also been proposed in computational electromagnetics, see [42–44] and [50, Sect. 3–4]. As disclosed in [33], they can be viewed as general local MTF, but they still await thorough investigations and will not be covered in this chapter.

In this chapter, we aim to motivate and explain our novel multitrace boundary integral equations. The emphasis will be rather on their derivation than on comprehensive theoretical analysis. Simplicity will often trump generality and for rigorous proofs and technical details the reader is referred to the original publications

- [19] as concerns the global MTF for acoustic scattering,
- [18], where the global MTF for electromagnetic scattering was introduced,
- [32], which proposed the local MTF.

Discretization will be addressed, because it is a particular difficulty haunting the low-order boundary element Ritz–Galerkin discretization of the STF that has triggered the pursuit of new approaches. Yet, we are not going to delve into a full-fledged numerical analysis of a Galerkin discretization of the new MTF. Nor will we report numerical results; these can be found in the seminal publications cited above.

Eventually, we would like to remind the reader that this chapter reports very recent developments that are unfolding with numerous issues still unresolved. Here we can only provide a snapshot of current research.

List of notations

Ω_i	material subdomains $\subset \mathbb{R}^d$, Ω_0 unbounded, see Figure 2.1
N	number of subdomains
Ω_*	union of closures of bounded subdomains
Ω_0	(unbounded) complement of Ω_*
$\boldsymbol{n}, \boldsymbol{n}_i$	exterior unit normal vector fields
Γ_{ij}	interface $\partial\Omega_i \cap \partial\Omega_j$, often assumed to have nonvanishing $d-1$-dimensional measure
Σ	skeleton, union of all interfaces
κ_i, $\kappa(\boldsymbol{x})$	(local) wave numbers
α_i, $\alpha(\boldsymbol{x})$	(locally constant) coefficients for PDE operator, see (2.4)
D	first-order partial differential operator grad or **curl**
D^*	L^2-adjoint of D, either $-$ div or **curl**
L	second-order partial differential operator $\mathsf{D}^*(\alpha(\boldsymbol{x})\mathsf{D}\cdot) - \kappa^2(\boldsymbol{x})\mathbf{u}$
$\mathcal{H}(\mathsf{D}, \Omega)$	Sobolev spaces, see (2.5)
$\mathbf{u}_{\text{inc}}$	incoming wave, see (2.7)
$\mathcal{T}(\mathsf{D}, \partial\Omega)$	trace space for Sobolev space $\mathcal{H}(\mathsf{D}, \Omega)$
T_D	(Dirichlet) trace operator
$[\cdot,\cdot]$, $[\![\cdot,\cdot]\!]$	bilinear pairings, usually of L^2-type
$\langle\cdot,\cdot\rangle$	duality pairing between a vector space and its dual
T_N	Neumann trace operator, see (2.12)
$\mathcal{T}(\partial\Omega)$	Cauchy trace space, product of Dirichlet and Neumann trace spaces
$\mathfrak{u}, \mathfrak{v}, \mathfrak{w}, \ldots$	elements of Cauchy trace spaces
$\mathbb{A}, \mathbb{B}, \ldots$	linear operators on Cauchy trace spaces
$\mathbb{T}, \mathbb{T}^+$	Cauchy trace operator, see (2.15)
$\mathbb{X}$	exterior-to-interior trace transfer operator, see (2.17)
SL, DL	single and double layer potentials, see Theorem 2.9
$\mathbb{G}$	total potential, see (2.22)
$\mathbb{P}$	Calderón projector, see Definition 2.13
$\mathcal{CD}(\partial\Omega)$	space of Cauchy data for L, see Definition 2.14
$\mathbb{A}, \mathbb{A}_j[i]$	local compound boundary integral operator, see (2.26)
$\mathcal{MT}(\Sigma)$	multitrace space, see (2.29)
$\vec{\mathfrak{u}}, \vec{\mathfrak{v}}, \vec{\mathfrak{w}}, \ldots$	elements of $\mathcal{MT}(\Sigma)$

$\mathbb{L}_i$	localization operators, $\mathbb{L}_i : \mathcal{MT}(\Sigma) \to \mathcal{T}(\partial\Omega_i)$
$\mathcal{ST}(\Sigma)$	single-trace space, see (2.33)
$\mathbb{S}'_\Sigma$	single-trace boundary integral operator (weak form), see (2.48)
$\Sigma_h, \Gamma_{j,h}$	boundary meshes

1.1 Geometry

The notion of a composite scatterer inspires the geometric setting adopted throughout this chapter. It is supposed to occupy the bounded domain $\Omega_* \subset \mathbb{R}^d$, $d = 2,3$, it features a Lipschitz boundary, and it is composed of so-called subdomains $\Omega_i \subset \mathbb{R}^d$, $i = 1, \ldots, N$, that represent open and connected curvilinear Lipschitz polygons ($d = 2$) or polyhedra ($d = 3$). We demand that they do not intersect ($\Omega_i \cap \Omega_j = \emptyset$ for $i \neq j$), and that they form a partition of Ω_* in that

$$\overline{\Omega}_* = \bigcup_{i=1}^{N} \overline{\Omega}_i.$$

The unbounded complement of $\overline{\Omega}_*$ should be connected and will provide another subdomain Ω_0, which means $\Omega_0 := \mathbb{R}^d \setminus \overline{\Omega}_*$.

The generic situation that we have in mind is depicted in Figure 2.1. The subdomains will usually be adjacent, which engenders “material junction points,” that is, points on which at least three subdomains abut (marked in Figure 2.1). In this case, some Ω_i will inevitably have nonsmooth boundaries. For each $i = 1, \ldots, N$, the boundary $\partial\Omega_i$ is orientable and can be endowed with a unit normal vector field $\boldsymbol{n}_i$ pointing into the exterior of Ω_i.

We write Γ_{ij} for the common interface of Ω_i and Ω_j, $\Gamma_{ij} := \partial\Omega_i \cap \partial\Omega_j$, $i \neq j$. Two subdomains Ω_i and Ω_j are adjacent, if Γ_{ij} is a $d-1$-dimensional Lipschitz manifold (with boundary). In this case we equip Γ_{ij} with an intrinsic orientation that amounts to prescribing a specific transversal direction. Note that, when referring to “all interfaces Γ_{ij}” we only include those that have a positive $d-1$-dimensional measure. We refer to them as “genuine interfaces” in the sequel. The union of all the interfaces Γ_{ij} forms the so-called *skeleton*

$$\Sigma := \bigcup_{ij} \Gamma_{ij} = \bigcup_{i=0}^{N} \partial\Omega_i.$$

For $N > 1$ the skeleton Σ will usually not be orientable, nor be a manifold.

1.2 Transmission problems

The application that we have in mind for the new multitrace integral equations are time-harmonic wave scattering problems. The first instance is acoustic scattering with

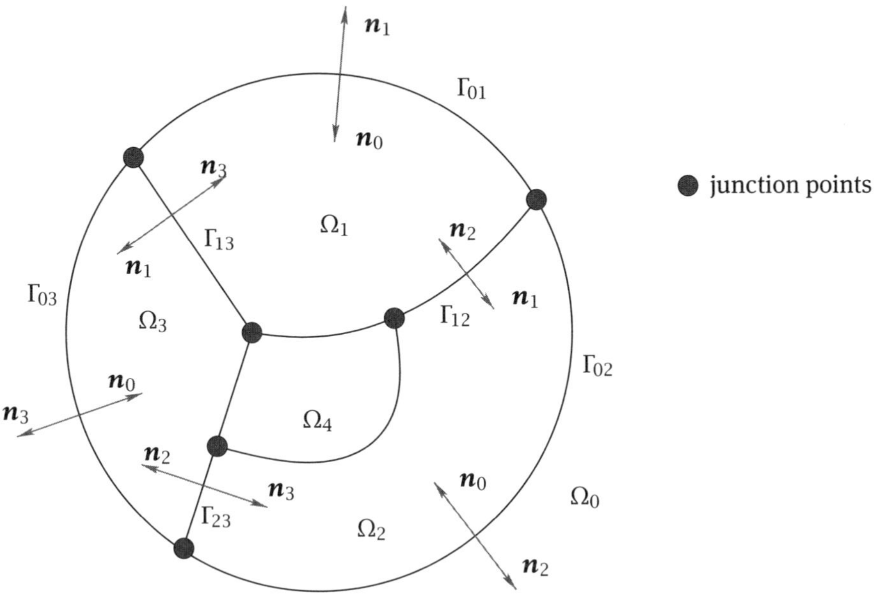

Figure 2.1: Typical geometry of a 2D "composite object," $N = 4$, induced orientations of some interfaces indicated by normal directions.

penetrable materials, modeled by a transmission problem for the Helmholtz equation [20, Chapter 2], [41, Chapter 2],

$$\begin{aligned} &-\operatorname{div}(\alpha_i \operatorname{grad} u) - \kappa_i^2 \beta_i u = 0 \quad \text{in } \Omega_i, \quad i = 0, \dots, N, \\ &+ \text{transmission conditions across } \Gamma_{ij} \text{ for adjacent subdomains,} \\ &+ \text{Sommerfeld radiation conditions at } \infty. \end{aligned} \tag{2.1}$$

Here, the $\kappa_i \geq 0$ are the local wave numbers, the coefficients α_i are positive definite $d \times d$ tensors, and $\beta_i > 0$, $i = 0, \dots, N$. The transmission and radiation conditions will be specified in detail in Section 2.1. A related scattering problem can be stated for $d = 3$ and electromagnetic waves based on the time-harmonic Maxwell equations [20, Chapter 6], [41, Chapter 5],

$$\begin{aligned} &\mathbf{curl}(\alpha_i \,\mathbf{curl}\,\mathbf{u}) - \kappa_i^2 \beta_i \mathbf{u} = 0 \quad \text{in } \Omega_i, \quad i = 0, \dots, N, \\ &+ \text{transmission conditions across } \Gamma_{ij} \text{ for adjacent subdomains,} \\ &+ \text{Silver–Müller radiation conditions at } \infty, \end{aligned} \tag{2.2}$$

where $\mathbf{u}$ stands for the electric or magnetic field. In the former case, α_i is called the magnetic susceptibility and β_i the electric permittivity. Both are positive definite 3×3 tensors.

For the sake of simplicity, in the remainder of this chapter we will assume $\beta_i \equiv 1$ in (2.1) and β_i to be the identity matrix in (2.2). In both cases α_i is supposed to be a positive number, which describes isotropic materials. Dropping the subscript from α_i and κ_i and writing $\alpha(\boldsymbol{x})$ and $\kappa(\boldsymbol{x})$ indicates that we consider the corresponding piecewise constant functions on $\mathbb{R}^d$.

The transmission problems (2.1) and (2.2) have much in common. Therefore we opt for a unified treatment in an abstract framework. We are aware that cognitive ease suggests a presentation for a concrete transmission problem. On the other hand, abstraction unveils important general patterns, which, eventually, promotes insight. In our opinion this benefit is worth the extra effort it takes to grasp abstraction. In this spirit, let us introduce the general second-order partial differential operator

$$\mathsf{L}\mathbf{u} := \mathsf{D}^*(\alpha(\boldsymbol{x})\mathsf{D}\mathbf{u}) - \kappa^2(\boldsymbol{x})\mathbf{u}, \tag{2.3}$$

where D is a suitable first-order differential operator, for instance, $\mathsf{D} = \mathrm{grad}$ for the Helmholtz equation (2.1), and $\mathsf{D} = \mathbf{curl}$ for Maxwell's equation. The other first-order differential operator occurring in (2.3) is the formal L^2-adjoint D^* of D. Then solutions of both (2.1) and (2.2) satisfy[1]

$$\mathsf{L}\mathbf{u} = 0 \quad \text{locally in } \Omega_i, \quad i = 0, \dots, N. \tag{2.4}$$

For the variational formulation of the transmission problems associated with (2.4) we have to establish a function space framework. To that end, for the remainder of this section let Ω be a generic domain in $\mathbb{R}^d$, $d = 2,3$. The natural domains of definition of both D and D^* are the Sobolev spaces[2]

$$\begin{aligned} \mathcal{H}(\mathsf{D},\Omega) &:= \{\mathbf{u} \in L^2(\Omega) : \mathsf{D}\mathbf{u} \in L^2(\Omega)\}, \\ \mathcal{H}(\mathsf{D}^*,\Omega) &:= \{\mathbf{u} \in L^2(\Omega) : \mathsf{D}^*\mathbf{u} \in L^2(\Omega)\}. \end{aligned} \tag{2.5}$$

They become Hilbert spaces when equipped with the natural graph norm. In case Ω is unbounded, a subscript "loc" will tag the Fréchet spaces of functions that belong to the corresponding function space on each compact subset. Moreover, in order to deal with the radiation conditions at ∞ for $\Omega = \Omega_0$, the space $\mathcal{H}(\mathsf{D},\Omega)$ has to be modified by switching to weaker weighted norms. This yields larger Hilbert spaces $\mathcal{H}_{\mathrm{rad}}(\mathsf{D},\Omega)$ that contain all "physically meaningful outgoing solutions". Details can be found in [48, Section 2.9.2.4] and [41, Sections 2.6.5 & 5.4].

A function $\mathbf{u}$ qualifies as a solution of $\mathsf{L}\mathbf{u} = 0$ on Ω only if it belongs to $\mathcal{H}_{\mathrm{loc}}(\mathsf{D},\Omega)$. In addition, the differential equation immediately implies $\alpha\mathsf{D}\mathbf{u} \in \mathcal{H}_{\mathrm{loc}}(\mathsf{D}^*,\Omega)$ and we conclude that $\mathbf{u}$ is an element of the function space

$$\mathcal{H}_{\mathrm{loc}}(\mathsf{L},\Omega) := \{\mathbf{v} \in \mathcal{H}_{\mathrm{loc}}(\mathsf{D},\Omega) : \alpha\mathsf{D}\mathbf{v} \in \mathcal{H}_{\mathrm{loc}}(\mathsf{D}^*,\Omega)\}. \tag{2.6}$$

1 Bold roman typeface will mark functions or vector fields defined on a domain $\subset \mathbb{R}^d$.

2 We use calligraphic font for Hilbert spaces of functions, unless this clashes with firmly established notational conventions for Sobolev spaces.

Sources will be introduced into our scattering models through a given *incident wave* $\mathbf{u}_{\mathrm{inc}} \in \mathcal{H}_{\mathrm{loc}}(\mathsf{L}, \mathbb{R}^d)$ that satisfies

$$\mathsf{L}_0\mathbf{u}_{\mathrm{inc}} := \mathsf{D}^*(\alpha_0 \mathsf{D}\mathbf{u}) - \kappa_0^2 \mathbf{u} = 0 \quad \text{everywhere in } \mathbb{R}^d. \tag{2.7}$$

Plane waves provide the most common specimens for $\mathbf{u}_{\mathrm{inc}}$.

Definition 2.1. A function $\mathbf{u} \in \mathcal{H}_{\mathrm{loc}}(\mathsf{L}, \Omega) \cap \mathcal{H}_{\mathrm{rad}}(\mathsf{D}, \Omega)$ is called a *radiating solution* (on Ω), if it solves $\mathsf{L}\mathbf{u} = 0$ in the sense of distributions.

The function spaces introduced above supply a suitable framework for the variational formulation of the transmission problems that are in the focus of this chapter: seek $\mathbf{u} \in \mathcal{H}_{\mathrm{loc}}(\mathsf{D}, \mathbb{R}^d)$ such that for all compactly supported $\mathbf{v} \in \mathcal{H}(\mathsf{D}, \mathbb{R}^d)$

$$[\alpha \mathsf{D}\mathbf{u}, \mathsf{D}\mathbf{v}]_{\mathbb{R}^d} - \left[\kappa^2 \mathbf{u}, \mathbf{v}\right]_{\mathbb{R}^d} = 0, \tag{2.8}$$

where $[\mathbf{w}, \mathbf{q}]_{\mathbb{R}^d} := \int_{\mathbb{R}^d} \mathbf{w}(\boldsymbol{x}) \cdot \mathbf{q}(\boldsymbol{x})\, d\boldsymbol{x}$, and the *scattered field* $\mathbf{u} - \mathbf{u}_{\mathrm{inc}}$ satisfies

$$(\mathbf{u} - \mathbf{u}_{\mathrm{inc}})_{|\Omega_0} \in \mathcal{H}_{\mathrm{rad}}(\mathsf{D}, \Omega_0). \tag{2.9}$$

Obviously, solutions of (2.8) are radiating solutions in each bounded Ω_i. By Rellich's lemma, analytic continuation techniques and Fredholm arguments existence and uniqueness of solutions can be established [1, 28].

Theorem 2.2. *For* $\mathsf{D} = \mathrm{grad}$ *("Helmholtz case") and* $\mathsf{D} = \mathbf{curl}$ *("Maxwell case") the variational problems* (2.8) *possess a unique solution.*

Remark 2.3 (Complex and vanishing wave numbers). Lossy materials, in which propagating waves are damped, can be modeled by wave numbers κ_i with positive imaginary part. All methods discussed in this chapter can deal with $\mathrm{Im}\, \kappa_i > 0$ without modifications.

In the Helmholtz case $\mathsf{D} = \mathrm{grad}$ all considerations of this chapter carry over to the case of local pure diffusion $\kappa_i = 0$ for some or all i's. If $\kappa_0 = 0$ the radiation conditions have to be replaced with suitable decay conditions [39, Chapter 8].

In the Maxwell case $\mathsf{D} = \mathbf{curl}$ we cannot accommodate $\kappa_i = 0$ for any i due to the notorious low-frequency instability of the representation formulas underlying our methods [22, 24] for a discussion. △

2 Boundary integral operators

Many boundary value problems for linear second-order differential operators with constant coefficients can be recast in the form of boundary integral equations, and this can be achieved through a number of standard steps. This is a venerable insight of applied mathematics dating back to the nineteenth century and monographs offer comprehensive coverage of the topic, among them [34, 39, 48]. This section is meant to review and summarize relevant formulas and results, introducing key notions and notations along the way.

In this section the exposition will mainly address a generic domain $\Omega \subset \mathbb{R}^d$ with compact Lipschitz boundary $\partial\Omega$ and exterior unit normal vectorfield $\boldsymbol{n} \in L^\infty(\partial\Omega)$. Throughout, α and κ are positive numbers and stand for constant coefficients in the partial differential operator $\mathsf{L} := \mathsf{D}^*\alpha\mathsf{D} - \kappa^2\mathsf{Id}$, cf. (2.3).

2.1 Trace spaces and operators

Writing $\mathcal{H}_0(\mathsf{D},\Omega)$ for the closure of smooth compactly supported functions in $\mathcal{H}(\mathsf{D},\Omega)$, it turns out that the quotient space $\mathcal{H}(\mathsf{D},\Omega)/\mathcal{H}_0(\mathsf{D},\Omega)$ is isomorphic to an intrinsically defined function space $\mathcal{T}(\mathsf{D},\partial\Omega)$ on the boundary $\partial\Omega$, a so-called trace space, see the Concretizations 2.5 and 2.6 below for examples. The attribute "intrinsic" indicates that we can also establish an isomorphism

$$\mathcal{T}(\mathsf{D},\partial\Omega) \cong \mathcal{H}(\mathsf{D},\mathbb{R}^d \setminus \overline{\Omega})/\mathcal{H}_0(\mathsf{D},\mathbb{R}^d \setminus \overline{\Omega}).$$

The canonical projection $\mathcal{H}(\mathsf{D},\Omega) \mapsto \mathcal{T}(\mathsf{D},\partial\Omega)$ gives rise to a surjective and continuous trace operator

$$\mathsf{T}_\mathsf{D} : \mathcal{H}(\mathsf{D},\Omega) \mapsto \mathcal{T}(\mathsf{D},\partial\Omega).$$

It owes its name to the fact that T_D applied to smooth functions in $\mathcal{H}(\mathsf{D},\Omega)$ on a domain Ω with piecewise smooth boundary agrees with a particular pointwise restriction almost everywhere on $\partial\Omega$. This relationship is addressed by so-called trace theorems.

Traces occur in the crucial integration by parts formulas ("Green's theorems")

$$[\mathsf{D}\mathbf{u},\mathbf{v}]_\Omega - [\mathbf{u},\mathsf{D}^*\mathbf{v}]_\Omega = [\mathsf{T}_\mathsf{D}\,\mathbf{u},\mathsf{T}_{\mathsf{D}^*}\,\mathbf{v}]_{\partial\Omega} \quad \forall \mathbf{u} \in \mathcal{H}(\mathsf{D},\Omega),\ \mathbf{v} \in \mathcal{H}(\mathsf{D}^*,\Omega), \tag{2.10}$$

with $[\cdot,\cdot]_\Omega$ and $[\cdot,\cdot]_{\partial\Omega}$ denoting (extensions of) the bilinear L^2-type pairings on Ω and $\partial\Omega$, respectively.

Notations and conventions. Single and double square brackets designate *bilinear* forms $[\cdot,\cdot], [\![\cdot,\cdot]\!] : V \times W \mapsto \mathbb{C}$ on function spaces, so-called pairings. Usually they will carry an additional subscript for clear identification. Angle brackets $\langle\cdot,\cdot\rangle$ are reserved for the bilinear duality pairing $\langle\cdot,\cdot\rangle : V' \times V \to \mathbb{C}$ between a normed space V and its dual V'. Round brackets $(\cdot,\cdot)_V$ are used for (sesquilinear) inner products in Hilbert spaces. Recall from [48, Theorem 2.1.44] that a pairing $[\cdot,\cdot] : V \times W \to \mathbb{C}$ induces an isomorphism $V \mapsto W'$, if

$$\begin{aligned} \exists c > 0 : \quad & \sup_{w \in W\setminus\{0\}} \frac{|[v,w]|}{\|w\|_W} \geq c\,\|v\|_V \quad \forall v \in V, \\ & \sup_{v \in V\setminus\{0\}} \frac{|[v,w]|}{\|v\|_V} > 0 \quad \forall w \in W. \end{aligned} \tag{2.11}$$

We phrase this as "W is dual to V with respect to the pairing $[\cdot,\cdot]$". If $[\cdot,\cdot]$ acts on $V \times V$, then V is called "self-dual with respect to the pairing $[\cdot,\cdot]$".

We would like to highlight an important consequence of (2.10).

Theorem 2.4. *The spaces $\mathcal{T}(\mathsf{D},\partial\Omega)$ and $\mathcal{T}(\mathsf{D}^*,\partial\Omega)$ are dual to each other with respect to the pairing $[\cdot,\cdot]_{\partial\Omega}$.*

Two trace operators are induced by the second-order partial differential operator L:

$$\begin{aligned} \mathsf{T}_D &: \mathcal{H}(\mathsf{D},\Omega) \to \mathcal{T}(\mathsf{D},\partial\Omega), \quad && \mathsf{T}_D := \mathsf{T}_{\mathsf{D}}, \\ \mathsf{T}_N &: \mathcal{H}(\mathsf{L},\Omega) \to \mathcal{T}(\mathsf{D}^*,\partial\Omega), \quad && \mathsf{T}_N := \mathsf{T}_{\mathsf{D}^*}(\alpha\mathsf{D}). \end{aligned} \tag{2.12}$$

We are going to refer to the surjective operators T_D and T_N as *Dirichlet trace* and *Neumann trace*, respectively. For the associated trace spaces we adopt the notation $\mathcal{T}_D(\partial\Omega) = \mathcal{T}(\mathsf{D},\partial\Omega)$ and $\mathcal{T}_N(\partial\Omega) := \mathcal{T}(\mathsf{D}^*,\partial\Omega)$ and the parlance "Dirichlet trace space" and "Neumann trace space". These can be merged into the Cauchy trace space

$$\mathcal{T}(\partial\Omega) := \mathcal{T}_D(\partial\Omega) \times \mathcal{T}_N(\partial\Omega), \tag{2.13}$$

which is *self-dual* with respect to the pairing[3]

$$[\![\mathfrak{u},\mathfrak{v}]\!]_{\mathcal{T}(\partial\Omega)} := [u,\varphi]_{\partial\Omega} + [v,\nu]_{\partial\Omega}, \quad \mathfrak{u} := \begin{pmatrix} u \\ \nu \end{pmatrix}, \ \mathfrak{v} := \begin{pmatrix} v \\ \varphi \end{pmatrix} \in \mathcal{T}(\partial\Omega). \tag{2.14}$$

A related compact notation is the Cauchy trace operator[4]

$$\mathbb{T} : \mathcal{H}(\mathsf{L},\Omega) \to \mathcal{T}(\partial\Omega), \quad \mathbb{T}\,\mathbf{u} := \begin{pmatrix} \mathsf{T}_D\,\mathbf{u} \\ \mathsf{T}_N\,\mathbf{u} \end{pmatrix}. \tag{2.15}$$

Of course, traces can also be taken from the exterior of Ω, with the complement $\mathbb{R}^d \setminus \overline{\Omega}$ now playing the role of Ω. The corresponding trace operators are tagged with a superscript "+," for instance

$$\mathbb{T}^+ : \mathcal{H}_{\mathrm{loc}}(\mathsf{L},\mathbb{R}^d \setminus \overline{\Omega}) \to \mathcal{T}(\partial\Omega), \quad \mathbb{T}^+\,\mathbf{u}^+ := \begin{pmatrix} \mathsf{T}_D^+\,\mathbf{u}^+ \\ \mathsf{T}_N^+\,\mathbf{u}^+ \end{pmatrix}. \tag{2.16}$$

Note that both Neumann traces T_N and T_N^+ rely on the same coefficient α, but assume opposite orientation of $\partial\Omega$! Thus, even for $\mathbf{u} \in \mathcal{H}_{\mathrm{loc}}(\mathsf{L},\mathbb{R}^d)$ exterior and interior traces agree only up to a change of sign of the Neumann component

$$\mathbb{T}^+\,\mathbf{u} = \mathbb{X}\,\mathbb{T}\,\mathbf{u} \quad \text{where} \quad \mathbb{X} : \begin{cases} \mathcal{T}(\partial\Omega) & \to \quad \mathcal{T}(\partial\Omega) \\ \begin{pmatrix} u \\ \varphi \end{pmatrix} & \mapsto \quad \begin{pmatrix} u \\ -\varphi \end{pmatrix} \end{cases}. \tag{2.17}$$

---

3 Fraktur font is used to designate functions in the Cauchy trace space, whereas Roman typeface is reserved for Dirichlet traces, and Greek symbols for Neumann traces.

4 We use blackboard bold typeface in order to distinguish operators acting on or mapping into Cauchy trace spaces.

Concretization 2.5 (Helmholtz case). Our terminology "Dirichlet trace" and "Neumann trace" is borrowed from the standard parlance used for the Helmholtz problem, that is, for $\mathsf{D} = \operatorname{grad}$ and $\mathcal{H}(\mathsf{D}, \Omega) = H^1(\Omega)$. In this special case, for smooth functions, the trace operators boil down to the pointwise restrictions

$$\mathsf{T}_D\, u = u_{|\partial\Omega}\,, \quad \mathsf{T}_N\, u = (\alpha \operatorname{grad} u)_{|\partial\Omega} \cdot \boldsymbol{n}. \tag{2.18}$$

The trace spaces are the well-known Sobolev spaces $H^{\frac{1}{2}}(\partial\Omega)$ and $H^{-\frac{1}{2}}(\partial\Omega)$, see [39, Chapter 3]. △

Concretization 2.6 (Maxwell case). In the Maxwell case $\mathsf{D} = \mathbf{curl}$ and $\mathcal{H}(\mathsf{D}, \Omega) = \boldsymbol{H}(\mathbf{curl}, \Omega)$ we are led to consider the tangential trace operators

$$\mathsf{T}_D\, \mathbf{u} = \boldsymbol{n} \times (\mathbf{u}_{|\partial\Omega} \times \boldsymbol{n}), \quad \mathsf{T}_N\, \mathbf{u} = (\alpha\, \mathbf{curl}\, \mathbf{u})_{|\partial\Omega} \times \boldsymbol{n}.$$

In [7, 10, 11] the related trace space have been characterized and we recall the notations $\mathcal{T}_D(\partial\Omega) = \boldsymbol{H}^{-\frac{1}{2}}(\mathbf{curl}_\Gamma, \Omega)$ and $\mathcal{T}_N(\partial\Omega) = \boldsymbol{H}^{-\frac{1}{2}}(\operatorname{div}_\Gamma, \Omega)$. △

Remark 2.7. Both, in (2.18) and (2.6), the Neumann traces will change signs, when the orientation of the normal vector $\boldsymbol{n}$ is flipped; the induced orientation of $\partial\Omega$ matters for the Neumann trace! This pattern has a simple explanation, when viewing (2.4) as an equation for differential forms [29, 36]. Then it turns out that multiplication with α corresponds to the action of a Hodge operator, converting straight forms into twisted forms. △

The formula (2.10) can be applied to a solution $\mathbf{u}$ of (2.4) and with $\mathbf{v} := \alpha \mathsf{D}\overline{\mathbf{u}}$, where the overbar stands for complex conjugation. Thus, by virtue of the relation (2.10), we obtain

$$[\mathsf{T}_D\, \mathbf{u}, \mathsf{T}_N\, \overline{\mathbf{u}}]_{\partial\Omega} = \int_\Omega \alpha\, |\mathsf{D}\mathbf{u}|^2 - \kappa^2\, |\mathbf{u}|^2 \,\mathrm{d}\boldsymbol{x}. \tag{2.19}$$

Notation. By a subscript index we indicate that a trace operator is applied to a particular subdomain, e.g., $\mathsf{T}_{D,i}$, $\mathsf{T}_{N,i}$, $\mathbb{T}_i$, etc., $i = 0, \dots, N$. Invariably, the induced orientation of $\partial\Omega_i$ is used.

Remark 2.8 (Radiation conditions). Dirichlet and Neumann traces occur in the formal statement of the radiation conditions mentioned in the statement of the transmission problems. Let us adopt the notation T_D^r and T_N^r for Dirichlet and Neumann traces onto the boundary of the ball $B_r := \{\boldsymbol{x} \in \mathbb{R}^d : |\boldsymbol{x}| < r\}$ (from inside). Then $\mathbf{u} \in \mathcal{H}_{\mathrm{loc}}(\mathsf{L}, \Omega)$ will satisfy the radiation condition, if

$$\lim_{r\to\infty} \int_{|\boldsymbol{x}|=r} \left| (\mathsf{T}_N^r\, \mathbf{u})(\boldsymbol{y}) - \imath\kappa (\mathsf{T}_D^r\, \mathbf{u})(\boldsymbol{y}) \right|^2 \mathrm{d}S(\boldsymbol{y}) = 0. \tag{2.20}$$

△

2.2 Potentials

Potential representations of solutions of (2.4) are the first step towards boundary integral equations. Here, by the term “potential” we refer to a linear operator that maps a function in a trace space on $\partial\Omega$ to a function defined everywhere on $\mathbb{R}^d \setminus \partial\Omega$. Further, potentials are supposed to provide radiating solutions of (2.4) in both Ω and $\mathbb{R}^d \setminus \overline{\Omega}$. Throughout, when acting on sufficiently smooth argument functions, potentials agree with integral operators with singular kernels spawned by fundamental solutions for L. The following key result can be found in [48, Section 3.11] and [39, Chapter 6] for Helmholtz’ equation (2.1) and in [12] for Maxwell’s equation (2.2).

Theorem 2.9 (Single domain representation formula). *There are continuous operators, depending on $\alpha > 0$ and $\kappa > 0$, the*

$$\begin{aligned} &\textit{single layer potential} \quad \mathsf{SL} : \mathcal{T}_N(\partial\Omega) \mapsto \mathcal{H}(\mathsf{L}, \Omega \cup \mathbb{R}^d \setminus \overline{\Omega}), \\ &\textit{double layer potential} \quad \mathsf{DL} : \mathcal{T}_D(\partial\Omega) \mapsto \mathcal{H}(\mathsf{L}, \Omega \cup \mathbb{R}^d \setminus \overline{\Omega}), \end{aligned}$$

such that

(i) $\mathsf{SL}(\varphi)$ *and* $\mathsf{DL}(u)$ *are radiating solutions of* (2.4) *in* $\Omega \cup \mathbb{R}^d \setminus \overline{\Omega}$ *for any* $\varphi \in \mathcal{T}_N(\partial\Omega)$, $u \in \mathcal{T}_D(\partial\Omega)$.

(ii) *every solution* $\mathbf{u} \in \mathcal{H}_{\mathrm{rad}}(\mathsf{L}, \Omega)$ *of* $\mathsf{L}\mathbf{u} = 0$ *can be written as*

$$\mathbf{u} = -\mathsf{DL}(\mathsf{T}_D\,\mathbf{u}) + \mathsf{SL}(\mathsf{T}_N\,\mathbf{u}), \tag{2.21}$$

A more compact way to write (2.21) is

$$\mathbf{u} = \mathbb{G}(\mathbb{T}\,\mathbf{u}), \text{ with } \quad \mathbb{G}(\mathrm{u}) := -\mathsf{DL}(u) + \mathsf{SL}(\varphi), \quad \mathrm{u} := \begin{pmatrix} u \\ \varphi \end{pmatrix} \in \mathcal{T}(\partial\Omega). \tag{2.22}$$

Concretization 2.10 (Helmholtz, $d = 3$). For the Helmholtz problem (2.1) with wave number κ and scalar coefficient $\alpha > 0$ the concrete integral formulas for the potentials are [48, Section 3.3]

$$\begin{aligned} \mathsf{SL}(\varphi)(\boldsymbol{x}) &= \frac{1}{\alpha}\int_{\partial\Omega} \Phi_{\kappa/\sqrt{\alpha}}(\boldsymbol{x}-\boldsymbol{y})\,\varphi(\boldsymbol{y})\,\mathrm{d}S, \\ \mathsf{DL}(u)(\boldsymbol{x}) &= \int_{\partial\Omega} \operatorname{grad}\Phi_{\kappa/\sqrt{\alpha}}(\boldsymbol{x}-\boldsymbol{y})\cdot\boldsymbol{n}(\boldsymbol{y})\,u(\boldsymbol{y})\,\mathrm{d}S, \end{aligned} \tag{2.23}$$

for $\boldsymbol{x} \notin \partial\Omega$, u a distribution on $\partial\Omega$, and with the fundamental solution

$$\Phi_\theta(\boldsymbol{z}) = \frac{\exp(\imath\theta\,|\boldsymbol{z}|)}{4\pi\,|\boldsymbol{z}|}, \qquad \theta \in \mathbb{R}. \tag{2.24}$$

△

Concretization 2.11 (Maxwell). In the case of Maxwell's equations (2.2) ($\kappa_i = \kappa$) we find from the fundamental Stratton–Zhu representation formula [12, Section 4], [20, Theorem 6.2] for $\boldsymbol{x} \notin \partial\Omega$

$$\begin{aligned} \mathsf{SL}(\varphi)(\boldsymbol{x}) &= \frac{1}{\alpha}\int_{\partial\Omega} \Phi_{\kappa/\sqrt{\alpha}}(\boldsymbol{x}-\boldsymbol{y})\varphi(\boldsymbol{y})\,\mathrm{d}S + \\ &\quad \frac{1}{\alpha\kappa^2}\,\mathbf{grad}_{\boldsymbol{x}} \int_{\Omega} \Phi_{\kappa/\sqrt{\alpha}}(\boldsymbol{x}-\boldsymbol{y})(\mathrm{div}_\Gamma\varphi)(\boldsymbol{y})\,\mathrm{d}S, \\ \mathsf{DL}(u)(\boldsymbol{x}) &= -\,\mathbf{curl}_{\boldsymbol{x}} \int_{\partial\Omega} \Phi_{\kappa/\sqrt{\alpha}}(\boldsymbol{x}-\boldsymbol{y})(\boldsymbol{n}\times u)(\boldsymbol{y})\,\mathrm{d}S, \end{aligned} \tag{2.25}$$

for tangential vector fields $\varphi, u : \partial\Omega \mapsto \mathbb{C}^3$. △

Inherent in the definition of potentials is that they generate a function defined both inside and outside of Ω. This enables us to take traces of potentials also from the exterior of $\partial\Omega$. The famous jump relations tell us that, thus, we will not get any new boundary integral operators.

Theorem 2.12 (Jump relations).

$$(\mathbb{T}^+ - \mathbb{X}\,\mathbb{T}) \circ \mathbb{G} = \begin{pmatrix} \mathsf{T}_D^+ - \mathsf{T}_D \\ \mathsf{T}_N^+ + \mathsf{T}_N \end{pmatrix} \circ \mathbb{G} = -\,\mathbb{X} \quad on \quad \mathcal{T}(\partial\Omega).$$

We refer to [48, Section 3.3.1], [39, Theorem 6.11] (Helmholtz) and [12, Theorem 7] (Maxwell) for proof and more details.

2.3 Calderón projectors

Obviously, applying the total trace operator $\mathbb{T}$ to the representation formula (2.21) exactly recovers the traces of a radiating solution of (2.4). The converse is also true. To formalize this, we introduce a pivotal operator, cf. [48, Section 3.6], [34, Section 5.6].

Definition 2.13 (Calderón projector). The Calderón projector $\mathbb{P}$ for L on Ω is defined by

$$\mathbb{P} := \mathbb{T} \circ \mathbb{G} : \mathcal{T}(\partial\Omega) \to \mathcal{T}(\partial\Omega).$$

Thanks to the mapping properties of the trace operators and the potentials, $\mathbb{P}$ is linear and continuous. It has a close link with the subspace of $\mathcal{T}(\partial\Omega)$ that contains all traces of functions in the kernel of L.

Definition 2.14 (Cauchy data). The space of Cauchy data $\mathcal{CD}(\partial\Omega) \subset \mathcal{T}(\partial\Omega)$ for the differential equation (2.4) is

$$\mathcal{CD}(\partial\Omega) := \{\mathfrak{v} \in \mathcal{T}(\partial\Omega) : \exists \mathbf{u} \in \mathcal{H}_{\mathrm{rad}}(\mathsf{L},\Omega),\ \mathsf{L}\mathbf{u} = 0,\ \mathfrak{v} = \mathbb{T}\,\mathbf{u}\}.$$

As corollary of Theorem 2.9 we infer the following fact.

Theorem 2.15. *The Calderón projector is a continuous projector onto $\mathcal{CD}(\partial\Omega)$, that is,*

$$\mathbb{P}^2 = \mathbb{P} \quad \textit{and} \quad \mathrm{Ker}(\mathsf{Id} - \mathbb{P}) = \mathcal{CD}(\partial\Omega).$$

Boundary integral operators are obtained by letting the trace operators T_D and T_N act on the potentials SL and DL. This is exactly what Definition 2.13 of the Calderón projector boils down to. Hence, we use it to define the continuous compound boundary integral operator

$$\mathbb{A} := \mathbb{P} - \frac{1}{2}\mathsf{Id} \overset{\text{(Theorem 2.12)}}{=} \frac{1}{2}\begin{pmatrix} \mathsf{T}_D + \mathsf{T}_D^+ \\ \mathsf{T}_N - \mathsf{T}_N^+ \end{pmatrix} \circ \mathbb{G} : \mathcal{T}(\partial\Omega) \to \mathcal{T}(\partial\Omega). \tag{2.26}$$

Recalling the definition of $\mathbb{G}$, (2.26) boils down to applying two traces to two potentials (single and double layers), which yields a total of four boundary integral operators. Yet, we never need to worry about the fine structure of $\mathbb{A}$, which can safely be regarded as a "black box operator" that is at our disposal (in discretized form, of course, see Section 3.3).

Concretization 2.16 (Helmholtz case). In the case of D = grad, we can disassemble $\mathbb{A}$ into

$$\mathbb{A} = \begin{pmatrix} -\mathsf{K} & \mathsf{V} \\ \mathsf{W} & \mathsf{K}' \end{pmatrix}, \tag{2.27}$$

where the individual boundary integral operators are known as single layer boundary integral operator V, double layer boundary integral operator K, adjoint double layer boundary integral operator K′, and hypersingular boundary integral operator W. Our notation follows [48, Chapter 3], but the reader should be aware that no universally accepted notational conventions have emerged. △

Concretization 2.17 (Maxwell case). For a splitting of $\mathbb{A}$ according to (2.27) in the case of Maxwell's equations we refer to [30, Section 3] for descriptions of the operator building blocks. Beware that [12] and the monograph [41] adopt conventions for the Neumann trace different from ours, which yields slightly different forms of the boundary integral operators. △

A deep result from the theory of boundary integral operators asserts a generalized Gårding inequality for $\mathbb{A}$, see [48, Proposition 3.5.5] for Helmholtz and [12, Lemma 10] for the Maxwell case.

Theorem 2.18 (Coercivity of boundary integral operators). *For any α and κ there exist an isomorphism $\mathbb{F} : \mathcal{T}(\partial\Omega) \to \mathcal{T}(\partial\Omega)$ and a compact operator $\mathbb{K} : \mathcal{T}(\partial\Omega) \to \mathcal{T}(\partial\Omega)$ such that for some constant $C > 0$*[5]

$$|[\![\mathbb{A}\,\mathfrak{v}, \mathbb{F}\,\overline{\mathfrak{v}}]\!]_{\mathcal{T}(\partial\Omega)} + [\![\mathbb{K}\,\mathfrak{v}, \overline{\mathfrak{v}}]\!]_{\mathcal{T}(\partial\Omega)}| \geq C\,\|\mathfrak{v}\|^2_{\mathcal{T}(\partial\Omega)} \qquad \forall \mathfrak{v} \in \mathcal{T}(\partial\Omega).$$

Remark 2.19 (Weak boundary integral equations). In order to pursue a Galerkin discretization of the boundary integral equations, these have to be cast in variational form. To this end the self-duality of $\mathcal{T}(\partial\Omega)$ with respect to the pairing $[\![\cdot,\cdot]\!]_{\mathcal{T}(\partial\Omega)}$ comes handy; for instance, to $\mathbb{A}$ we can associate the bilinear form

$$(\mathfrak{u}, \mathfrak{v}) \mapsto [\![\mathbb{A}\,\mathfrak{u}, \mathfrak{v}]\!]_{\mathcal{T}(\partial\Omega)}, \quad \mathfrak{u}, \mathfrak{v} \in \mathcal{T}(\partial\Omega). \tag{2.28}$$

In turns, the bilinear form gives rise to an operator $\mathcal{T}(\partial\Omega) \to \mathcal{T}(\partial\Omega)'$ mapping the Cauchy trace space to its dual. Occasionally we will tag such operators with a prime $'$, e.g., $\mathbb{A}'$. △

Notations. Above potentials and boundary integral operators have been introduced for a generic domain Ω and their dependence on the coefficients was suppressed in the notation. In the following, we will have to consider potentials defined on different boundaries and relying on different coefficients.

First, to distinguish their incarnations for a subdomain Ω_j we use a subscript j, e.g., $\mathbb{A}_j$, $\mathbb{G}_j$, etc. Doing so, we tacitly assume that all traces and potentials are based on the coefficients α_j and κ_j associated with Ω_j. Sometimes, though potentials and integral operators are defined on $\partial\Omega_j$ we will nevertheless base them on the coefficients α_i and κ_i from another subdomain Ω_i. If this is the case, we write $\mathbb{A}_j[i]$, $\mathbb{G}_j[i]$.

3 Classical single-trace integral equations

This section is dedicated to a review of a well established boundary integral equation formulation for the transmission problems (2.1) and (2.2). For Helmholtz' equation it was first analyzed in [54], but probably used earlier. For Maxwell's equations it agrees with the so-called Poggio–Miller–Chew–Harrington–Wu–Tsai (PMCHWT) integral equations [14, 27, 45, 55], whose numerical analysis was first accomplished in [12].

5 We use an overbar to designate complex conjugation.

3.1 Skeleton trace spaces

The skeleton multitrace space is simply defined as

$$\mathcal{MT}(\Sigma) := \mathcal{MT}_D(\Sigma) \times \mathcal{MT}_N(\Sigma), \tag{2.29}$$

$$\text{with} \quad \begin{aligned} \mathcal{MT}_D(\Sigma) &:= \mathcal{T}_D(\partial\Omega_0) \times \cdots \times \mathcal{T}_D(\partial\Omega_N)\,, \\ \mathcal{MT}_N(\Sigma) &:= \mathcal{T}_N(\partial\Omega_0) \times \cdots \times \mathcal{T}_N(\partial\Omega_N)\,. \end{aligned} \tag{2.30}$$

It owes its name to the fact that on each interface Γ_{ij} a function $\vec{\mathfrak{u}} \in \mathcal{MT}(\Sigma)$ comprises two pairs of Dirichlet and Neumann data, one contributed by the subdomain on either side. Obviously, $\mathcal{MT}(\Sigma)$ can be identified with the product of local Cauchy trace spaces

$$\mathcal{MT}(\Sigma) \cong \mathcal{T}(\partial\Omega_0) \times \cdots \times \mathcal{T}(\partial\Omega_N), \tag{2.31}$$

which illustrate the localized nature of $\mathcal{MT}(\Sigma)$. We chose the ordering of local trace components given by (2.29) for the sake of simpler notations in Section 3. To isolate the contribution of a single subdomain we rely on trivial localization operators[6]

$$\mathbb{L}_i : \mathcal{MT}(\Sigma) \to \mathcal{T}(\partial\Omega_i), \quad \mathbb{L}_i\,\vec{\mathfrak{u}} := \begin{pmatrix} u_i \\ v_i \end{pmatrix}, \quad \begin{aligned} &\vec{\mathfrak{u}} = (u_0, \ldots, u_N, v_0, \ldots, v_N), \\ &u_i \in \mathcal{T}_D(\partial\Omega_i),\; v_i \in \mathcal{T}_N(\partial\Omega_i). \end{aligned}$$

The multitrace spaces inherit all properties of their local components. For instance, a self-duality of $\mathcal{MT}(\Sigma)$ is obviously induced by the L^2-type bilinear pairing, cf. (2.14),

$$[\![\vec{\mathfrak{u}}, \vec{\mathfrak{v}}]\!]_{\mathcal{MT}(\Sigma)} := \sum_{i=0}^{N} [\![\mathbb{L}_i\,\vec{\mathfrak{u}}, \mathbb{L}_i\,\vec{\mathfrak{v}}]\!]_{\mathcal{T}(\partial\Omega_i)}, \quad \vec{\mathfrak{u}}, \vec{\mathfrak{v}} \in \mathcal{MT}(\Sigma). \tag{2.32}$$

Now, we introduce an important notion intimately linked to the transmission conditions that connect traces across interfaces. This connection is captured by the so-called *single-trace spaces*

$$\mathcal{ST}(\Sigma) := \mathcal{ST}_D(\Sigma) \times \mathcal{ST}_N(\Sigma) \subset \mathcal{MT}(\Sigma), \tag{2.33}$$

$$\begin{aligned} \mathcal{ST}_D(\Sigma) &:= \{(u_0, \ldots, u_N) \in \mathcal{MT}_D(\Sigma) : \exists \mathbf{u} \in \mathcal{H}(\mathsf{D}, \mathbb{R}^d),\; u_i = \mathsf{T}_{\mathsf{D},i}\,\mathbf{u}\}, \\ \mathcal{ST}_N(\Sigma) &:= \{(v_0, \ldots, v_N) \in \mathcal{MT}_N(\Sigma) : \exists \boldsymbol{\phi} \in \mathcal{H}(\mathsf{D}^*, \mathbb{R}^d),\; v_i = \mathsf{T}_{\mathsf{D}^*,i}\,\boldsymbol{\phi}\}. \end{aligned} \tag{2.34}$$

In words, functions in $\mathcal{ST}_D(\Sigma)$ and $\mathcal{ST}_N(\Sigma)$ are skeleton traces of functions defined on $\mathbb{R}^d$. Further explanations of the term "single trace" are given in Remark 2.25.

Concretization 2.20 (Helmholtz). For the case $\mathsf{D} = \text{grad}$, the space $\mathcal{ST}_D(\Sigma)$ contains restrictions of functions in $\mathcal{H}(\mathsf{D}, \mathbb{R}^d) = H^1(\mathbb{R}^d)$ to the skeleton Σ. We deem the

6 Functions in a multitrace space will be distinguished by an overset arrow, e.g., $\vec{\mathfrak{u}}$, $\vec{\mathfrak{v}}$.

mental image of continuous functions on Σ an appropriate view of $\mathcal{ST}_D(\Sigma)$ in this case.

The space $\mathcal{ST}_N(\Sigma)$ contains the normal component traces of vectorfields in $\boldsymbol{H}(\mathrm{div}, \mathbb{R}^3)$. However, the normal components rely on the local exterior unit vectors $\boldsymbol{n}_i$, which reflect the induced orientation of $\partial\Omega_i$. Note that the two induced orientations of an interface Γ_{ij} are opposite. Thus, an element of $\mathcal{ST}_N(\Sigma)$ will involve two Neumann traces on each interface that have opposite sign. △

Concretization 2.21 (Maxwell). Consider $d = 3$ and $\mathsf{D} = \mathbf{curl}$. Then $\mathcal{ST}_D(\Sigma)$ will contain tangential components on Σ of vectorfields in $\boldsymbol{H}(\mathbf{curl}, \mathbb{R}^3)$. Imagine these as vectorfields $\Sigma \mapsto \mathbb{R}^3$, tangential to the interfaces in Σ that, in addition, are tangentially continuous across the seams between genuine interfaces.

Again, elements of $\mathcal{ST}_N(\Sigma)$ will no longer support an interpretation as well-defined functions on Σ. They can be obtained as

$$\mathcal{ST}_N(\Sigma) = \{(v_1, \ldots, v_N) \in \mathcal{MT}_N(\Sigma) : \exists \vec{\mathrm{u}} \in \mathcal{ST}_D(\Sigma) : v_i = u_i \times \boldsymbol{n}_i\}, \tag{2.35}$$

that is, by taking a tangential vectorfield on Σ and rotating it counterclockwise (seen from inside Ω_i) on $\partial\Omega_i$. It is clear that the result will be two vectorfields of opposite sign on each interface. △

Remark 2.22. With single trace spaces at our disposal, we can eventually give a rigorous description of the *transmission conditions* already mentioned in (2.1) and (2.2). These are conditions to be satisfied by functions $\mathbf{u}_i$ that provide local solutions of (2.4) on Ω_i so that they qualify as solutions of (2.8). They read

$$(\mathsf{T}_{D,0}\,\mathbf{u}_0, \ldots, \mathsf{T}_{D,N}\,\mathbf{u}_N, \mathsf{T}_{N,0}\,\mathbf{u}_0, \ldots, \mathsf{T}_{D,N}\,\mathbf{u}_N) \in \mathcal{ST}(\Sigma), \tag{2.36}$$

which means, in the sense of distributions,

$$\mathsf{T}_{D,i}\,\mathbf{u}_i = \mathsf{T}_{D,j}\,\mathbf{u}_j \quad \text{and} \quad \mathsf{T}_{N,i}\,\mathbf{u}_i = -\,\mathsf{T}_{N,j}\,\mathbf{u}_j \quad \text{on}\, \Gamma_{ij}. \tag{2.37}$$

In light of (2.34), the equivalence between (2.36) and (2.37) is known in the form of compatibility conditions inherent in the definition of Sobolev spaces. For instance, in the Helmholtz case, if $\mathbf{u}$ fulfills $\mathbf{u}_{|\Omega_i} \in H^1(\Omega_i)$ it will belong to $H^1(\mathbb{R}^d)$ if and only if it features weak continuity (agreement of Dirichlet traces in the sense of distributions) across all genuine interfaces. Analogous facts are known for $\mathbf{u}_{|\Omega_i} \in \boldsymbol{H}(\mathrm{div}, \Omega_i)$ and $\mathbf{u}_{|\Omega_i} \in \boldsymbol{H}(\mathbf{curl}, \Omega_i)$. △

A fundamental result is the following "polar set" characterization of $\mathcal{ST}(\Sigma)$ as a subspace of $\mathcal{MT}(\Sigma)$ given in [19, Proposition 2.1].

Theorem 2.23.

$$\mathcal{ST}(\Sigma) = \{\mathfrak{u} \in \mathcal{MT}(\Sigma) : [\![\mathfrak{u}, \mathfrak{v}]\!]_{\mathcal{MT}(\Sigma)} = 0 \ \forall \mathfrak{v} \in \mathcal{ST}(\Sigma)\}.$$

Before we give a formal proof, let us explain the intuition behind this result; for sufficiently smooth functions we find

$$[\![\vec{\mathfrak{u}}, \vec{\mathfrak{v}}]\!]_{\mathcal{MT}(\Sigma)} = \sum_{i,j} \int_{\Gamma_{ij}} u_i \psi_i + v_i v_i + u_j \psi_j + v_j v_j \, \mathrm{d}S, \tag{2.38}$$

where $\vec{\mathfrak{u}} = (u_0, \ldots, u_N, v_0, \ldots, v_N)$, $\vec{\mathfrak{v}} = (v_0, \ldots, v_N, \psi_0, \ldots, \psi_N)$. This reflects the fact that each interface is visited twice when summing integrals over all subdomain boundaries. Guided by the insight that functions in $\mathcal{ST}_D(\Sigma)$ are single valued on each interface, whereas functions in $\mathcal{ST}_N(\Sigma)$ differ in sign on both sides of Γ_{ij}, cf. (2.37), we conclude complete cancelation of all terms in (2.38).

Proof of Theorem 2.23.
(i) First assume $\vec{\mathfrak{u}} = \binom{\vec{u}}{\vec{v}} \in \mathcal{ST}(\Sigma)$, that is, there are associated compactly supported functions $\mathbf{u} \in \mathcal{H}(\mathsf{D}, \mathbb{R}^d)$ and $\boldsymbol{\phi} \in \mathcal{H}(\mathsf{D}^*, \mathbb{R}^d)$ such that $u_i = \mathsf{T}_{\mathsf{D},i}\,\mathbf{u}$, $v_i = \mathsf{T}_{\mathsf{D}^*,i}\,\boldsymbol{\phi}$.

Pick $\vec{\mathfrak{v}} \in \mathcal{ST}(\Sigma)$ with associated compactly supported functions $\mathbf{v} \in \mathcal{H}(\mathsf{D}, \mathbb{R}^d)$, $\boldsymbol{\psi} \in \mathcal{H}(\mathsf{D}^*, \mathbb{R}^d)$, so that integration by parts according to (2.10) yields, $\vec{\mathfrak{v}} = \binom{\vec{v}}{\vec{\psi}}$,

$$\begin{aligned}
[\![\vec{\mathfrak{u}}, \vec{\mathfrak{v}}]\!]_{\mathcal{MT}(\Sigma)} &= \sum_{i=0}^{N} [u_i, \psi_i]_{\partial\Omega_i} + [v_i, v_i]_{\partial\Omega_i} \\
&= \sum_{i=0}^{N} [\mathsf{D}\mathbf{u}, \boldsymbol{\psi}]_{\Omega_i} - [\mathbf{u}, \mathsf{D}^*\boldsymbol{\psi}]_{\Omega_i} + [\mathsf{D}\mathbf{v}, \boldsymbol{\varphi}]_{\Omega_i} - [\mathbf{v}, \mathsf{D}^*\boldsymbol{\varphi}]_{\Omega_i} \\
&= [\mathsf{D}\mathbf{u}, \boldsymbol{\psi}]_{\mathbb{R}^d} - [\mathbf{u}, \mathsf{D}^*\boldsymbol{\psi}]_{\mathbb{R}^d} + [\mathsf{D}\mathbf{v}, \boldsymbol{\varphi}]_{\mathbb{R}^d} - [\mathbf{v}, \mathsf{D}^*\boldsymbol{\varphi}]_{\mathbb{R}^d} = 0.
\end{aligned}$$

(ii) Pick $\mathfrak{u} = (u_0, \ldots, u_N, v_0, \ldots, v_N) \in \mathcal{MT}(\Sigma)$ and local extensions $\mathbf{u}_i \in \mathcal{H}(\mathsf{D}, \Omega_i)$ and $\mathbf{f}_i \in \mathcal{H}(\mathsf{D}^*, \Omega_i)$ such that $u_i = \mathsf{T}_{\mathsf{D},i}\,\mathbf{u}_i$ and $v_i = \mathsf{T}_{\mathsf{D}^*,i}\,\mathbf{f}_i$. Then introduce the functions $\mathbf{u} \in L^2(\mathbb{R}^d)$ and $\mathbf{f} \in L^2(\mathbb{R}^d)$ defined piecewise according to

$$\mathbf{u}_{|\Omega_i} := \mathbf{u}_i, \quad \mathbf{f}_{|\Omega_i} = \mathbf{f}_i, \quad i = 0, \ldots, N. \tag{2.39}$$

We have to show that $\mathbf{u} \in \mathcal{H}(\mathsf{D}, \mathbb{R}^d)$ and $\mathbf{f} \in \mathcal{H}(\mathsf{D}^*, \mathbb{R}^d)$ provided that $[\![\mathfrak{u}, \mathfrak{v}]\!]_{\mathcal{T}(\Sigma)} = 0$ for all $\mathfrak{v} \in \mathcal{ST}(\Sigma)$. This follows, if we can establish

$$\mathsf{D}\mathbf{u}_{|\Omega_i} = \mathsf{D}\mathbf{u}_i, \quad \mathsf{D}^*\mathbf{f}_{|\Omega_i} = \mathsf{D}^*\mathbf{f}_i, \quad i = 0, \ldots, N, \tag{2.40}$$

in the sense of distributions, because a function that agrees with $\mathsf{D}\mathbf{u}_i \in L^2(\Omega_i)$ in each subdomain certainly belongs to $L^2(\mathbb{R}^d)$.

Let $\mathbf{w} \in C_0^\infty(\mathbb{R}^d) \cap \mathcal{H}(\mathsf{D}^*, \mathbb{R}^d)$ and $\mathbf{v} \in C_0^\infty(\mathbb{R}^d) \cap \mathcal{H}(\mathsf{D}, \mathbb{R}^d)$ be smooth compactly supported test functions. Then

$$\begin{aligned}
&-[\mathbf{u}, \mathsf{D}^*\mathbf{w}]_{\mathbb{R}^d} + [\mathbf{f}, \mathsf{D}\mathbf{v}]_{\mathbb{R}^d} = \sum_{i=0}^{N} -[\mathbf{u}_i, \mathsf{D}^*\mathbf{w}]_{\Omega_i} + [\mathbf{f}_i, \mathsf{D}\mathbf{v}]_{\Omega_i} \\
&= \sum_{i=0}^{N} -[\mathsf{D}\mathbf{u}_i, \mathbf{w}]_{\Omega_i} + [\mathsf{T}_{\mathsf{D},i}\,\mathbf{u}_i, \mathsf{T}_{\mathsf{D}^*,i}\,\mathbf{w}]_{\partial\Omega_i} + [\mathsf{D}^*\mathbf{f}_i, \mathbf{v}]_{\Omega_i} + [\mathsf{T}_{\mathsf{D},i}\,\mathbf{v}, \mathsf{T}_{\mathsf{D}^*,i}\,\mathbf{f}_i]_{\partial\Omega_i} \\
&= \sum_{i=0}^{N} -[\mathsf{D}\mathbf{u}_i, \mathbf{w}]_{\Omega_i} + [\mathsf{D}^*\mathbf{f}_i, \mathbf{v}]_{\Omega_i} + [u_i, \mathsf{T}_{\mathsf{D}^*,i}\,\mathbf{w}]_{\partial\Omega_i} + [\mathsf{T}_{\mathsf{D},i}\,\mathbf{v}, v_i]_{\partial\Omega_i} \\
&= \sum_{i=0}^{N} -[\mathsf{D}\mathbf{u}_i, \mathbf{w}]_{\Omega_i} + [\mathsf{D}^*\mathbf{f}_i, \mathbf{v}]_{\Omega_i} \\
&\quad + \underbrace{[\![\vec{\mathbf{u}}, (\mathsf{T}_{\mathsf{D},0}\,\mathbf{v}, \dots, \mathsf{T}_{\mathsf{D},N}\,\mathbf{v}, \mathsf{T}_{\mathsf{D}^*,0}\,\mathbf{w}, \dots, \mathsf{T}_{\mathsf{D}^*,N}\,\mathbf{w})]\!]_{\mathcal{T}(\Sigma)}}_{=0},
\end{aligned}$$

which amounts to (2.40). □

Let us mention an immediate consequence of Theorem 2.23:

Corollary 2.24. *$\mathcal{ST}(\Sigma)$ is a closed subspace of $\mathcal{MT}(\Sigma)$.*

Remark 2.25. The term "single-trace space" conjures up the image of functions $\Sigma \mapsto \mathbb{C}$. In fact, pairs of such functions can be identified with elements of $\mathcal{ST}(\Sigma)$. First, we are going to elaborate the connection for the Helmholtz problem (2.1), that is $\mathcal{T}_D(\partial\Omega_i) = H^{\frac{1}{2}}(\partial\Omega_i)$ and $\mathcal{T}_N(\partial\Omega_i) = H^{-\frac{1}{2}}(\partial\Omega_i)$. Remember that each interface Γ_{ij} has an intrinsic orientation given by a prescribed transversal direction. A potential mismatch with induced orientation is taken into account by "orientation adjustment functions" $\sigma_j \in L^\infty(\partial\Omega_j)$, $j = 0, \dots, N$, defined as

$$\sigma_j(\boldsymbol{x}) := \begin{cases} 1, & \text{if } \boldsymbol{x} \in \Gamma_{ij} \subset \partial\Omega_j \text{ and orientations of } \partial\Omega_j \text{ and } \Gamma_{ij} \text{ match,} \\ -1, & \text{if } \boldsymbol{x} \in \Gamma_{ij} \subset \partial\Omega_j \text{ and orientations of } \partial\Omega_j \text{ and } \Gamma_{ij} \text{ are opposite.} \end{cases}$$

Observe that $\sigma_i = -\sigma_j$ on Γ_{ij}. Given a *continuous* function $\hat{u}$ on Σ with $\hat{u}_{|\Gamma_{ij}} \in H^1(\Gamma_{ij})$ and $\hat{v} \in L^2(\Sigma)$, we can now define

$$u_j := \hat{u}_{|\partial\Omega_j} \in H^{\frac{1}{2}}(\partial\Omega_j), \quad v_j := \sigma_j \cdot \hat{v}_{|\partial\Omega_j} \in H^{-\frac{1}{2}}(\partial\Omega_j), \tag{2.41}$$

which gives $\vec{\mathrm{u}} = (u_0, \dots, u_N, v_0, \dots, v_N) \in \mathcal{ST}(\Sigma)$.

For electromagnetic scattering, see (2.2), a similar approach works. We start from two *tangential* vector fields $\Sigma \mapsto \mathbb{C}^3$, again denoted by $\hat{u}$, and $\hat{v}$, that are piecewise smooth on the interfaces and *tangentially continuous* across the seams of the skeleton, that is, they have a unique tangential component on the boundaries of the interfaces, recall Concretization 2.21. Then define

$$\begin{aligned} u_j &:= \hat{u}_{|\partial\Omega_j} \in \boldsymbol{H}^{-\frac{1}{2}}(\mathbf{curl}_\Gamma, \partial\Omega_j), \\ v_j &:= \hat{v}_{|\partial\Omega_j} \times \boldsymbol{n}_j \in \boldsymbol{H}^{-\frac{1}{2}}(\mathrm{div}_\Gamma, \partial\Omega_j). \end{aligned} \tag{2.42}$$

Here, a sign flip as in (2.41) is induced by the opposite directions of $\boldsymbol{n}_i$ and $\boldsymbol{n}_j$ on Γ_{ij}. One easily verifies that $(u_0, \dots, u_N, v_0, \dots, v_N) \in \boldsymbol{\mathcal{ST}}(\Sigma)$. △

3.2 A first-kind boundary integral equation

Assume that $\mathbf{u} \in \boldsymbol{\mathcal{H}}_{\mathrm{loc}}(\mathsf{L}, \mathbb{R}^d)$ solves the transmission problem (2.8). Obviously, $\mathbf{u}_{|\Omega_j}$ provides local solutions of the homogeneous PDE, see Definition 2.1, on the bounded subdomains $\Omega_1, \dots, \Omega_N$, and Theorem 2.15 permits us to conclude

$$(\mathsf{Id} - \mathbb{P}_j)\, \mathbb{T}_j \mathbf{u} = 0 \quad \text{in } \mathcal{T}(\partial\Omega_j), \quad j = 1, \dots, N. \tag{2.43}$$

As $\mathbf{u} - \mathbf{u}_{\mathrm{inc}}$ is a radiating solution on Ω_0, owing to (2.9) the same argument gives

$$(\mathsf{Id} - \mathbb{P}_0)\, \mathbb{T}_0 (\mathbf{u} - \mathbf{u}_{\mathrm{inc}}) = 0 \tag{2.44}$$

which implies that for all $\vec{\mathfrak{v}} \in \boldsymbol{\mathcal{ST}}(\Sigma)$

$$\sum_{j=0}^{N} \left[\!\left[(\mathsf{Id} - \mathbb{P}_j)\, \mathbb{T}_j \mathbf{u}, \mathbb{L}_j \vec{\mathfrak{v}} \right]\!\right]_{\mathcal{T}(\partial\Omega_j)} = \left[\!\left[(\mathsf{Id} - \mathbb{P}_0)\, \mathbb{T}_0 \mathbf{u}_{\mathrm{inc}}, \mathbb{L}_0 \vec{\mathfrak{v}} \right]\!\right]_{\mathcal{T}(\partial\Omega_0)}. \tag{2.45}$$

A key observation is that by definition $\mathbb{T}_\Sigma \mathbf{u} \in \boldsymbol{\mathcal{ST}}(\Sigma)$ so that, thanks to Theorem 2.23, the variational formulation (2.45) is *equivalent* to

$$\sum_{j=0}^{N} \left[\!\left[\mathbb{A}_j \mathbb{T}_j \mathbf{u}, \mathbb{L}_j \vec{\mathfrak{v}} \right]\!\right]_{\mathcal{T}(\partial\Omega_j)} = -\left[\!\left[(\mathsf{Id} - \mathbb{P}_0)\, \mathbb{T}_0 \mathbf{u}_{\mathrm{inc}}, \mathbb{L}_0 \vec{\mathfrak{v}} \right]\!\right]_{\mathcal{T}(\partial\Omega_0)}, \tag{2.46}$$

for all $\vec{\mathfrak{v}} \in \boldsymbol{\mathcal{ST}}(\Sigma)$. Another simplification results from the property (2.7), which implies $\mathbb{X}\, \mathbb{T}_0 \mathbf{u}_{\mathrm{inc}} \in \mathcal{CD}(\partial\Omega_*)$ (with respect to wave number κ_0), and by Theorem 2.15 we can conclude that $(\mathsf{Id} - \mathbb{P}_0)\, \mathbb{T}_0 \mathbf{u}_{\mathrm{inc}} = \mathbb{T}_0 \mathbf{u}_{\mathrm{inc}}$. Thus, the integral operator $\mathbb{P}_0$ on the right-hand side of (2.46) can be dropped.

We have derived a variational equation satisfied by $\mathbb{T}_\Sigma \mathbf{u} \in \boldsymbol{\mathcal{ST}}(\Sigma)$. Regarding the skeleton Cauchy traces as unknowns, this yields the classical single-trace boundary integral equation (STF): seek $\vec{\mathfrak{u}} \in \boldsymbol{\mathcal{ST}}(\Sigma)$ such that

$$\sum_{j=0}^{N} \left[\!\left[\mathbb{A}_j \mathbb{L}_j \vec{\mathfrak{u}}, \mathbb{L}_j \vec{\mathfrak{v}} \right]\!\right]_{\mathcal{T}(\partial\Omega_j)} = -\left[\!\left[\mathbb{T}_0 \mathbf{u}_{\mathrm{inc}}, \mathbb{L}_0 \vec{\mathfrak{v}} \right]\!\right]_{\mathcal{T}(\partial\Omega_0)} \quad \forall \vec{\mathfrak{v}} \in \boldsymbol{\mathcal{ST}}(\Sigma). \tag{2.47}$$

This STF variational problem can equivalently be expressed as the operator equation[7]

$$\mathbb{S}'_\Sigma \vec{\mathfrak{u}} := \sum_{j=0}^{N} \mathbb{L}_j^* \,\mathbb{A}_j \,\mathbb{L}_j \,\vec{\mathfrak{u}} = -\,\mathbb{L}_0^* \,\mathbb{T}_0 \,\mathbf{u}_{\text{inc}}, \tag{2.48}$$

with $\mathbb{S}'_\Sigma : \boldsymbol{\mathcal{ST}}(\Sigma) \to \boldsymbol{\mathcal{ST}}(\Sigma)'$. Here, $\mathbb{L}_j^* : \mathcal{T}(\partial\Omega_j) \to \boldsymbol{\mathcal{ST}}(\Sigma)'$ stands for an adjoint of the localization operator $\mathbb{L}_j : \boldsymbol{\mathcal{ST}}(\Sigma) \to \mathcal{T}(\partial\Omega_j)$ with respect to the respective L^2-type duality pairings

$$\left\langle \mathbb{L}_j^* \,\mathfrak{u}_j, \vec{\mathfrak{v}} \right\rangle_{\boldsymbol{\mathcal{ST}}(\Sigma)} := \left[\!\left[\mathfrak{u}_j, \mathbb{L}_j \vec{\mathfrak{v}} \right]\!\right]_{\mathcal{T}(\partial\Omega_j)} \quad \forall \mathfrak{u}_j \in \mathcal{T}(\partial\Omega_j),\ \vec{\mathfrak{v}} \in \boldsymbol{\mathcal{ST}}(\Sigma). \tag{2.49}$$

A consequence of Theorem 2.18 is that the bilinear form on the left-hand side of (2.47) fulfills a generalized Gårding inequality.

Theorem 2.26. *There is an isomorphism* $\mathbb{F} : \boldsymbol{\mathcal{ST}}(\Sigma) \mapsto \boldsymbol{\mathcal{ST}}(\Sigma)$ *and a compact operator* $\mathbb{K} : \boldsymbol{\mathcal{ST}}(\Sigma) \mapsto \boldsymbol{\mathcal{MT}}(\Sigma)$ *such that for some* $C > 0$

$$\left| \left\langle \mathbb{S}'_\Sigma \vec{\mathfrak{v}}, \mathbb{F}\,\overline{\vec{\mathfrak{v}}} \right\rangle + \left[\!\left[\mathbb{K}\,\vec{\mathfrak{v}}, \overline{\vec{\mathfrak{v}}} \right]\!\right]_{\mathcal{T}(\Sigma)} \right| \geq C \,\|\vec{\mathfrak{v}}\|^2_{\boldsymbol{\mathcal{MT}}(\Sigma)} \quad \forall \vec{\mathfrak{v}} \in \boldsymbol{\mathcal{ST}}(\Sigma). \tag{2.50}$$

Concretization 2.27. For the case of the Helmholtz equation we find $\mathbb{F} = \mathsf{Id}$ and the assertion of Theorem 2.26 is a corollary of Theorem 2.18. Note that in this case (2.50) is a standard Gårding inequality.

For electromagnetic wave propagation (2.2) a sophisticated $\mathbb{F}$ is needed based on a direct splitting of $\boldsymbol{\mathcal{ST}}(\Sigma)$, see [8]. △

The variational problem (2.47) will always possess unique solutions, cf. [54] and [19, Proposition A.1].

Lemma 2.28 (Uniqueness of solutions of STF). $\mathbb{S}'_\Sigma$ *from* (2.48) *is injective.*

Proof. Let $\vec{\mathfrak{w}} \in \boldsymbol{\mathcal{ST}}(\Sigma)$ satisfy

$$\langle \mathbb{S}'_\Sigma \vec{\mathfrak{w}}, \vec{\mathfrak{v}} \rangle = \sum_{j=0}^{N} \left[\!\left[\mathbb{A}_j \,\mathbb{L}_j \,\vec{\mathfrak{w}}, \mathbb{L}_j \,\vec{\mathfrak{v}} \right]\!\right]_{\mathcal{T}(\partial\Omega_j)} = 0 \quad \forall \vec{\mathfrak{v}} \in \boldsymbol{\mathcal{ST}}(\Sigma). \tag{2.51}$$

Define (radiating) local solutions

$$\mathbf{u}_j := \mathbb{G}_j(\mathbb{L}_j \,\vec{\mathfrak{w}})_{|\Omega_j} \in \boldsymbol{\mathcal{H}}_{\text{rad}}(\mathsf{L}, \Omega_j), \tag{2.52}$$

and recall that Definition 2.13 means

$$\mathbb{T}_j \,\mathbf{u}_j = \mathbb{P}_j(\mathbb{L}_j \,\vec{\mathfrak{w}}) = (\mathbb{A}_j + \tfrac{1}{2}\mathsf{Id})\,\mathbb{L}_j \,\vec{\mathfrak{w}}. \tag{2.53}$$

---

7 The prime in the notation $\mathbb{S}'_\Sigma$ is intended to highlight that this operator is defined via a bilinear form and, thus, must be regarded as a mapping of a space into a dual space.

Hence, (2.51) implies

$$\sum_{j=0}^{N} [\![\mathbb{T}_j \mathbf{u}_j, \mathbb{L}_j \vec{\mathfrak{v}}]\!]_{\mathcal{T}(\partial\Omega_j)} = \sum_{j=0}^{N} \left[\!\!\left[(\mathbb{A}_j + \frac{1}{2}\mathsf{Id})\, \mathbb{L}_j \vec{\mathfrak{w}}, \mathbb{L}_j \vec{\mathfrak{v}}\right]\!\!\right]_{\mathcal{T}(\partial\Omega_j)} = 0 \tag{2.54}$$

for all $\vec{\mathfrak{v}} \in \mathcal{ST}(\Sigma)$ by virtue of (2.51) and Theorem 2.23. This latter theorem, using its other inclusion, permits us to conclude

$$\left(\mathbb{T}_j \mathbf{u}_j\right)_{j=0}^{N} \in \mathcal{ST}(\Sigma). \tag{2.55}$$

In other words, the compound function

$$\mathbf{u} \in L^2_{\text{loc}}(\mathbb{R}^d), \quad \mathbf{u}_{|\Omega_j} := \mathbf{u}_j,$$

satisfies the transmission conditions, which renders $\mathbf{u} \in \mathcal{H}(\mathsf{L}, \mathbb{R}^d)$ an entire radiating solution. Thus, we conclude $\mathbf{u} = 0$ from the uniqueness result of Theorem 2.2.

Next, we introduce “exterior” radiating solutions

$$\mathbf{u}_j^+ := \mathbb{G}_j(\mathbb{L}_j \vec{\mathfrak{w}})_{|\mathbb{R}^d \setminus \overline{\Omega}_j} \in \mathcal{H}_{\text{rad}}(\mathsf{L}, \mathbb{R}^d \setminus \overline{\Omega}_j). \tag{2.56}$$

and examine their exterior traces

$$\mathbb{T}_j^+ \mathbf{u}_j^+ = (\mathsf{Id} - \mathbb{P}_j)(\mathbb{L}_j \vec{\mathfrak{w}}) = (\frac{1}{2}\mathsf{Id} - \mathbb{A}_j)\, \mathbb{L}_j \vec{\mathfrak{w}}. \tag{2.57}$$

Thus, the same arguments as in (2.54) imply

$$\left(\mathbb{T}_j^+ \mathbf{u}_j^+\right)_{j=0}^{N} \in \mathcal{ST}(\Sigma). \tag{2.58}$$

Relying on “Rellich’s lemma” [20, Lemma 2.11] we are going to show $\mathbf{u}_j^+ = 0$ for all $j = 0, \dots, N$. This is a standard technique, but important enough to justify a detailed presentation. Rellich’s lemma asserts that

$$\mathbf{u} \in \mathcal{H}_{\text{rad}}(\mathsf{L}, \Omega_0), \quad \mathsf{L}\mathbf{u} = 0, \quad \lim_{r\to\infty} \int_{|\boldsymbol{x}|=r} |\,\mathsf{T}_D^r \mathbf{u}|^2 \, dS = 0 \tag{2.59}$$

implies $\mathbf{u} = 0$ in Ω_0. Recall the notation T_D^r and T_N^r for Dirichlet- and Neumann traces onto the boundary of the ball $B_r := \{\boldsymbol{x} \in \mathbb{R}^d : |\boldsymbol{x}| < r\}$. Applying integration by parts (2.19) we get for r large enough and $j = 1, \dots, N$

$$\begin{aligned} &\int_{|\boldsymbol{x}|=r} \mathsf{T}_D^r \mathbf{u}_j^+ \cdot \mathsf{T}_N^r \overline{u}_j^+ \, dS \\ &= \int_{B_r \setminus \overline{\Omega}_j} \alpha \left|\mathsf{D}\mathbf{u}_j^+\right|^2 - \kappa^2 \left|\mathbf{u}_j^+\right|^2 d\boldsymbol{x} - \int_{\partial\Omega_j} \mathsf{T}_{D,j}^+ \mathbf{u}_j^+ \cdot \mathsf{T}_{N,j}^+ \overline{u}_j^+ \, dS, \end{aligned} \tag{2.60}$$

and for $j = 0$

$$0 = \int_{\mathbb{R}^d \setminus \overline{\Omega}_0} \alpha \, |\mathrm{D}\mathbf{u}_0^+|^2 - \kappa^2 \, |\mathbf{u}_0^+|^2 \, \mathrm{d}\boldsymbol{x} - \int_{\partial\Omega_0} \mathsf{T}_{D,0}^+ \mathbf{u}_0^+ \cdot \mathsf{T}_{N,0}^+ \overline{\mathcal{u}_0^+} \, \mathrm{d}S. \tag{2.61}$$

Summing these equations for $j = 0, \dots, N$ and taking the imaginary part yields

$$\begin{aligned}\sum_{j=1}^{N} - \operatorname{Im} \int_{|\boldsymbol{x}|=r} \mathsf{T}_D^r \mathbf{u}_j^+ \cdot \mathsf{T}_N^r \overline{\mathcal{u}_j^+} \, \mathrm{d}S &= \sum_{j=0}^{N} \operatorname{Im} \int_{\partial\Omega_j} \mathsf{T}_{D,j}^+ \mathbf{u}_j^+ \cdot \mathsf{T}_{N,j}^+ \overline{\mathcal{u}_j^+} \, \mathrm{d}S \\ &= - \operatorname{Im} \left[\!\left[\left(\mathbb{T}_j^+ \mathbf{u}_j^+ \right)_{j=0}^{N}, \left(\mathbb{T}_j^+ \overline{\mathcal{u}_j^+} \right)_{j=0}^{N} \right]\!\right]_{\mathcal{T}(\Sigma)} = 0,\end{aligned} \tag{2.62}$$

because of (2.58), which allows us to apply Theorem 2.23.

Simple algebra shows that the radiation conditions (2.20) for $\mathbf{u}_j^+$ combined with (2.62) gives $\lim_{r\to\infty} \sum_{j=1}^{N} \int_{|\boldsymbol{x}|=r} |\, \mathsf{T}_D^r \mathbf{u}_j^+ |^2 \, \mathrm{d}S = 0$ as required by Rellich's lemma, and, consequently, $\mathbf{u}_j^+ = 0$, $j = 1, \dots, N$. Next, we appeal to the jump relations of Theorem 2.12 to see

$$\mathbb{X}_j \, \mathbb{L}_j \vec{\mathfrak{w}} = \mathbb{T}_j \mathbf{u}_j - \mathbb{T}_j^+ \mathbf{u}_j^+ = 0. \tag{2.63}$$

As $\vec{\mathfrak{w}} \in \boldsymbol{\mathcal{ST}}(\Sigma)$ this immediately implies $\mathbb{L}_j \vec{\mathfrak{w}} = 0$, that is, $\vec{\mathfrak{w}} = 0$. □

Combining Theorem 2.26 and Lemma 2.28 provides the essential prerequisite for applying the Fredholm alternative to (2.47), which immediately confirms existence and uniqueness of solutions of (2.47): using the STF operator $\mathbb{S}'_\Sigma$, we can give a concise statement of this main result.

Corollary 2.29. $\mathbb{S}'_\Sigma : \boldsymbol{\mathcal{ST}}(\Sigma) \mapsto \boldsymbol{\mathcal{ST}}(\Sigma)'$ *from* (2.48) *is an isomorphism.*

3.3 Boundary element Galerkin discretization

Abstract Galerkin discretization of (2.47) restricts the variational equation to a finite-dimensional subspace $\boldsymbol{\mathcal{ST}}_h(\Sigma) \subset \boldsymbol{\mathcal{ST}}(\Sigma)$ [48, Section 4.2.2]. Here we take a closer look only at piecewise polynomial constructions with respect to a triangulation Σ_h of the skeleton Σ. For $d = 2$ this amounts to partitioning all interfaces Γ_{ij} into (curved) segments, separated by so-called nodes. We demand that multiple junctions of interfaces will always coincide with nodes. In three dimensions, each interface is equipped with a (curved) triangular mesh. These meshes are supposed to be compatible across interface boundaries ("no hanging nodes"). The requirements imply that the restriction of Σ_h to $\partial\Omega_j$ is a valid triangulation $\Gamma_{j,h} := \Sigma_{h\,|\partial\Omega_j}$ of this boundary.

Suitable piecewise polynomial (with respect to $\Gamma_{j,h}$) subspaces $\mathcal{T}_{D,h}(\partial\Omega_j)$ and $\mathcal{T}_{N,h}(\partial\Omega_j)$ of $\mathcal{T}_D(\partial\Omega_j)$ and $\mathcal{T}_N(\partial\Omega_j)$, respectively, can be obtained by applying the trace operators $\mathsf{T}_{\mathsf{D},j}$ and $\mathsf{T}_{\mathsf{D}^*,j}$ to piecewise polynomial finite element subspaces of $\mathcal{H}(\mathsf{D},\Omega_j)$ and $\mathcal{H}(\mathsf{D}^*,\Omega_j)$, respectively. The finite element spaces are built upon some triangular ($d = 2$) or tetrahedral ($d = 3$) mesh of Ω_j, whose restriction to $\partial\Omega_j$ coincides with $\Gamma_{j,h}$.

Concretization 2.30 (Helmholtz). For the Helmholtz transmission problem (2.1) we have $\mathcal{T}_D(\partial\Omega_j) = H^{\frac{1}{2}}(\partial\Omega_j)$ and the simplest choice is continuous piecewise linear functions on $\Gamma_{j,h}$. Obviously, this space can be obtained by taking the point trace of continuous piecewise linear functions on a suitable finite element mesh. The natural tent function basis of this $\mathcal{T}_{D,h}(\partial\Omega_j)$ is dual to the point evaluations in the nodes of $\Gamma_{j,h}$.

The Neumann trace space for the Helmholtz transmission problem (2.1) is $\mathcal{T}_N(\partial\Omega_j) = H^{-\frac{1}{2}}(\partial\Omega_j)$, which can be approximated by means of piecewise constant discontinuous functions on $\Gamma_{j,h}$. The characteristics functions of mesh cells provide a natural basis.

This and more general constructions of piecewise polynomial boundary element subspaces of $H^{\frac{1}{2}}(\partial\Omega_j)$ and $H^{-\frac{1}{2}}(\partial\Omega_j)$ are discussed in [48, Section 4.1]. △

Concretization 2.31 (Maxwell). In the Maxwell case, that is, $d = 3$ and $\mathcal{T}_D(\partial\Omega_j) = \boldsymbol{H}^{-\frac{1}{2}}(\mathbf{curl}_\Gamma,\partial\Omega_j)$, surface edge elements supply the simplest space $\mathcal{T}_{D,h}(\partial\Omega_j)$. They are piecewise linear tangentially continuous tangential vector fields. Locally supported basis functions are associated with edges of the surface mesh $\Gamma_{j,h}$. In computational electromagnetics these are known as Rao–Wilton–Glisson (RWG) basis functions [46].

As for the approximation of Neumann traces, we have the simple formula

$$\mathcal{T}_{N,h}(\partial\Omega_j) = \mathcal{T}_{D,h}(\partial\Omega_j) \times \boldsymbol{n}_j,$$

that is, Neumann traces can be approximated by the same finite dimensional spaces of tangential vector fields apart from a rotation by $\frac{\pi}{2}$. This matches the identity $\boldsymbol{H}^{-\frac{1}{2}}(\mathbf{curl}_\Gamma,\partial\Omega_j) = \boldsymbol{H}^{-\frac{1}{2}}(\mathrm{div}_\Gamma,\partial\Omega_j) \times \boldsymbol{n}_j$. The space thus constructed is closely related to the so-called lowest order Raviart–Thomas finite element space [5, 6, 47]. △

With $\mathcal{T}_{D,h}(\partial\Omega_j)$ and $\mathcal{T}_{N,h}(\partial\Omega_j)$, $j = 0,\dots,N$, at our disposal, finite dimensional subspaces of $\boldsymbol{\mathcal{MT}}(\Sigma)$ can be obtained by taking products. The construction of $\boldsymbol{\mathcal{ST}}_h(\Sigma)$ is not so straightforward and is guided by Remark 2.25. We describe it separately for the Helmholtz and Maxwell case.

Concretization 2.32 (Helmholtz). In the case of the Helmholtz transmission problem (2.1) $\boldsymbol{\mathcal{ST}}_h(\Sigma)$ can be obtained by combining two spaces,
(i) a space $\mathcal{F}_{D,h}$ of *continuous* Σ_h-piecewise polynomial functions,
(ii) and a space $\mathcal{F}_{N,h}$ of *possibly discontinuous* Σ_h-piecewise polynomial functions,

from which we get

$$\boldsymbol{\mathcal{ST}}_h(\Sigma) = \bigotimes_{j=0}^{N} \mathcal{F}_{D,h\,|\partial\Omega_j} \times \bigotimes_{j=0}^{N} \sigma_j\,\mathcal{F}_{N,h\,|\partial\Omega_j}\,,$$

where the "orientation adjustment functions" σ_j are defined in Remark 2.25. We point out that even if $\mathcal{F}_{N,h}$ contained only continuous functions, we will usually wind up with a discontinuous approximation space for Neumann data in $\boldsymbol{\mathcal{ST}}_h(\Sigma)$, because the orientation adjustment function σ_j may jump.

Clearly, for this simplest choice of piecewise linear and piecewise constant boundary element spaces as explained in Concretization 2.30, the dimension of $\boldsymbol{\mathcal{ST}}_h(\Sigma)$ will agree with the sum of the number of vertices and cells of Σ_h. △

Concretization 2.33 (Maxwell). The construction of $\boldsymbol{\mathcal{ST}}_h(\Sigma)$ for the Maxwell case $\mathsf{D} = \mathbf{curl}$ and $d = 3$ starts from a space $\boldsymbol{\mathcal{F}}_h$ of Σ_h-piecewise polynomial tangential and tangentially continuous vector fields, cf. Concretization 2.21. This means that for an $\vec{u}_h \in \boldsymbol{\mathcal{F}}_h$ its orthogonal projection onto any edge of Σ_h must be unique. The simplest specimen is the edge element skeleton space, whose functions are piecewise linear and uniquely characterized by their path integrals along edges, cf. Concretization 2.31.

Then, taking the cue from (2.42), we find

$$\boldsymbol{\mathcal{ST}}_h(\Sigma) = \bigotimes_{j=0}^{N} \boldsymbol{\mathcal{F}}_{h\,|\partial\Omega_j} \times \bigotimes_{j=0}^{N} (\boldsymbol{\mathcal{F}}_{h\,|\partial\Omega_j} \times \boldsymbol{n}_j).$$

Using surface edge elements to define $\boldsymbol{\mathcal{F}}_h$, we thus end up with a space $\boldsymbol{\mathcal{ST}}_h(\Sigma)$, whose dimension is twice the number of edges in Σ_h. △

Remark 2.34. Galerkin discretization of (2.47) using the standard choices for $\boldsymbol{\mathcal{ST}}_h(\Sigma)$ explained above will lead to a linear system of equations, whose matrix can be obtained by adding the matrices arising from a boundary element Galerkin discretization of the boundary integral operators $\mathbb{A}_j$, $j = 0, \dots, N$, after suitably padding them with zero rows and columns, in order to take into account the localization to $\partial\Omega_j$. △

It turns out that all piecewise polynomial boundary element spaces give rise to asymptotically quasi-optimal Galerkin approximations. This result is stated as a discrete inf-sup condition in the next theorem. In the case of Helmholtz' equation (2.1) its proof boils down to a classical duality argument, see [49] and [48, Section 4.2.3]. For Maxwell's equation (2.2) it requires more subtle considerations centering around a uniform gap property of a discrete Hodge-type decomposition. We are not going to give details on this technique, but refer to [12, Section 9.1] for principal ideas and [8] for the application to STF.

Theorem 2.35. *Let $\boldsymbol{\mathcal{ST}}_h(\Sigma) \subset \boldsymbol{\mathcal{ST}}(\Sigma)$ be a space of piecewise polynomial boundary element functions of fixed polynomial degree based on a skeleton mesh Σ_h with mesh-width $h > 0$.*

There is $h_0 > 0$ and a constant $c > 0$ depending only on Σ, the type and polynomial degree of the boundary element space, and the shape-regularity of Σ_h, such that the following discrete inf-sup condition holds

$$\sup_{\vec{\mathfrak{v}}_h \in \boldsymbol{\mathcal{ST}}_h(\Sigma)} \frac{|\langle \mathbb{S}'_\Sigma \vec{\mathfrak{u}}_h, \vec{\mathfrak{v}}_h \rangle|}{\|\vec{\mathfrak{v}}_h\|_{\mathcal{MT}(\Sigma)}} \geq c \, \|\vec{\mathfrak{u}}_h\|_{\mathcal{MT}(\Sigma)} \quad \forall \vec{\mathfrak{u}}_h \in \boldsymbol{\mathcal{ST}}_h(\Sigma), \quad \forall h < h_0. \tag{2.64}$$

This implies that the trace norm of the Galerkin discretization error is proportional to the best approximation error of $\boldsymbol{\mathcal{ST}}_h(\Sigma)$ in $\boldsymbol{\mathcal{ST}}(\Sigma)$ provided that Σ_h is *sufficiently fine*; hence the attribute "asymptotic" attached to this estimate. More details can be found in [48, Section 4.2.2].

4 Preconditioning

According to the results reported in the previous section, the STF operator $\mathbb{S}'_\Sigma$ defined in (2.48) induces an isomorphism between the trace space $\boldsymbol{\mathcal{ST}}(\Sigma)$ and its dual. This is typical of *first-kind* boundary integral equations and the STF belongs to this class. Unfortunately, the standard local bases of the piecewise polynomial boundary element spaces discussed in Section 3.3 are not stable with respect to the norm of $\boldsymbol{\mathcal{ST}}(\Sigma)$. Indeed, writing $\{\vec{\mathfrak{b}}_i\}_{i=1}^M$, $M := \dim \boldsymbol{\mathcal{ST}}_h(\Sigma)$, for this basis of the boundary element space $\boldsymbol{\mathcal{ST}}_h(\Sigma) \subset \boldsymbol{\mathcal{ST}}(\Sigma)$, we find $r, s \in \mathbb{Z}$ satisfying $s - r = 2$, such that

$$\underline{C} h^s \left\| \sum_{i=1}^M \gamma_i \vec{\mathfrak{b}}_i \right\|^2_{\mathcal{MT}(\Sigma)} \leq \sum_{i=1}^M \gamma_i^2 \left\| \vec{\mathfrak{b}}_i \right\|^2_{\mathcal{MT}(\Sigma)} \leq \overline{C} h^r \left\| \sum_{i=1}^M \gamma_i \vec{\mathfrak{b}}_i \right\|^2_{\mathcal{MT}(\Sigma)}, \tag{2.65}$$

for all $\gamma_i \in \mathbb{C}$, where $h > 0$ stands for the meshwidth of the skeleton mesh Σ_h and the constants $\underline{C}, \overline{C} > 0$ depend only on the geometry of Σ, the ratio of the sizes of the largest and smallest panel, and, for $d = 3$, on the shape regularity measure of the triangles of Σ_h. For both Helmholtz and Maxwell the estimate (2.65) can be proved by means of local inverse inequalities and interpolation in Sobolev scales [48, Section 6.4.1].

As a consequence, on a sequence of meshes created by uniform regular refinement the Galerkin matrix $\mathbf{A} \in \mathbb{C}^{M,M}$ arising from a standard boundary element Galerkin discretization of (2.47) will display a growth of its condition number (defined through the Euclidean norm on $\mathbb{C}^M$) like

$$\operatorname{cond}_2(\mathbf{A}) = O(h^{-2}), \tag{2.66}$$

cf. [48, Cor. 6.4.14 & Lemma 4.5.1]. This can be blamed for the commonly observed severely degraded performance of iterative Krylov-subspace solvers such as GMRES or BiCGStab on fine meshes, though a rigorous theoretical link between (2.66) and speed of convergence has not yet been established, cf. [4, 13], unless more precise information about the spectrum of $\mathbf{A}$ is available.

Slow convergence of iterative solvers has become a major issue, since local low rank matrix compression algorithms have paved the way for using boundary element discretizations with many degrees of freedoms (10^4–10^6). These algorithms are known as fast multipole methods [21, 25], $\mathcal{H}$-matrix compression [26], or adaptive cross approximation [3, Chapter 3].

Compressed matrices merely allow fast matrix×vector multiplication, which is sufficient for applying iterative solvers, but precludes direct elimination techniques. The bottom line is that it becomes essential to accelerate the convergence of the Krylov-subspace iterative solvers, which is usually achieved by preconditioning. The next two sections will outline a powerful heuristics that, nowadays, plays an important role in boundary element implementations.

4.1 Operator products

Temporarily we focus on the single-trace formulation (2.47) for the case $N = 1$, that is, the case of a single homogeneous scatterer. We elaborate how in this very special situation the discrete variational STF can be preconditioned efficiently. A first particular feature of the $N = 1$ case is that $\Sigma = \partial\Omega_0 = \partial\Omega_1$; from now we are going to write Γ for this single interface. We endow it with a positive orientation with respect to Ω_1. Then the single-trace variational formulation (2.47) becomes

$$[\![\mathbb{A}_0\,\mathbb{L}_0\,\vec{\mathfrak{u}},\mathbb{L}_0\,\vec{\mathfrak{v}}]\!]_{\mathcal{T}(\Gamma)}+[\![\mathbb{A}_1\,\mathbb{L}_1\,\vec{\mathfrak{u}},\mathbb{L}_1\,\vec{\mathfrak{v}}]\!]_{\mathcal{T}(\Gamma)}=-[\![\mathbb{T}_\Gamma\,\mathbf{u}_{\mathrm{inc}},\mathbb{L}_0\,\vec{\mathfrak{v}}]\!]_{\mathcal{T}(\Gamma)}\quad\forall\vec{\mathfrak{v}}\in\boldsymbol{\mathcal{ST}}(\Sigma),\tag{2.67}$$

with equivalent operator form, cf. (2.48),

$$\mathbb{S}'_\Gamma\vec{\mathfrak{u}}:=(\mathbb{L}_0^*\,\mathbb{A}_0\,\mathbb{L}_0+\mathbb{L}_1^*\,\mathbb{A}_1\,\mathbb{L}_1)\,\vec{\mathfrak{u}}=-\,\mathbb{L}_0^*\,\mathbb{T}_\Gamma\,\mathbf{u}_{\mathrm{inc}}.\tag{2.68}$$

The particular situation makes possible the straightforward identification of $\boldsymbol{\mathcal{ST}}(\Sigma)$ with the Cauchy trace space $\mathcal{T}(\Gamma)$ via

$$\boldsymbol{\mathcal{ST}}(\Sigma)=\{(u,u,-\varphi,\varphi):\ u\in\mathcal{T}_D(\Gamma),\ \varphi\in\mathcal{T}_N(\Gamma)\},\tag{2.69}$$

which introduces the isometric isomorphism

$$\mathbb{C}:\mathcal{T}(\Gamma)\mapsto\boldsymbol{\mathcal{ST}}(\Sigma),\quad\mathfrak{u}=\begin{pmatrix}u\\\varphi\end{pmatrix}\mapsto\frac{1}{2}(u,u,-\varphi,\varphi).\tag{2.70}$$

Thus, the L^2-type pairing providing self-duality of $\mathcal{T}(\Gamma)$ indirectly induces a self-duality of $\boldsymbol{\mathcal{ST}}(\Sigma)$. The related isometric isomorphism is denoted by $\mathbb{D}':\boldsymbol{\mathcal{ST}}(\Sigma)\to\boldsymbol{\mathcal{ST}}(\Sigma)'$ and defined through

$$\langle\mathbb{D}'\vec{\mathfrak{u}},\vec{\mathfrak{v}}\rangle_{\boldsymbol{\mathcal{MT}}(\Sigma)}:=[\![\mathbb{C}^{-1}\vec{\mathfrak{u}},\mathbb{C}^{-1}\vec{\mathfrak{v}}]\!]_{\mathcal{T}(\Gamma)},\quad\forall\vec{\mathfrak{u}},\vec{\mathfrak{v}}\in\boldsymbol{\mathcal{ST}}(\Sigma).\tag{2.71}$$

As is evident from (2.70) and (2.14), $\mathbb{D}'$ is spawned by an L^2-type local pairing on Γ. Since $\mathbb{S}'_\Sigma : \boldsymbol{\mathcal{ST}}(\Sigma) \mapsto \boldsymbol{\mathcal{ST}}(\Sigma)'$ is continuous, we find that the operator product

$$(\mathbb{D}'^{-1})^* \mathbb{S}'_\Gamma \mathbb{D}'^{-1} \mathbb{S}'_\Gamma : \boldsymbol{\mathcal{ST}}(\Sigma) \mapsto \boldsymbol{\mathcal{ST}}(\Sigma) \tag{2.72}$$

gives rise to a *continuous* mapping within the *same* space. The relevance of this observation for accelerating iterative solvers will be elaborated in Section 4.3.

4.2 Calderón identities

Now we explore an interesting property of the operator product from (2.72) in the special case $\alpha_1 = \alpha_0$; equal coefficients $\alpha_0 = \alpha_1$ will be taken for granted throughout this subsection. The starting point is the projector property of the Calderón projectors stated in Theorem 2.15. It can be used to extract from (2.26) that

$$\mathbb{A}^2 = \frac{1}{4}\mathsf{Id}. \tag{2.73}$$

This equation comprises several so-called *Calderón identities* for boundary integral operators, see [48, Proposition 3.6.4]. Similar relationships hold for the operator product from (2.72) and we will also call them "Calderón identities". To derive these we rely on the identification of $\boldsymbol{\mathcal{ST}}(\Sigma)$ and $\mathcal{T}(\Gamma)$ according to (2.70). It enables us to convert (2.67) into: seek $\mathfrak{u} \in \mathcal{T}(\Gamma)$ such that

$$\langle \mathbb{S}'_\Gamma \mathbb{C}\,\mathfrak{u}, \mathbb{C}\,\mathfrak{v}\rangle_{\boldsymbol{\mathcal{ST}}(\Sigma)} = [\![\mathbb{A}_0\,\mathbb{X}\,\mathfrak{u}, \mathbb{X}\,\mathfrak{v}]\!]_{\mathcal{T}(\Gamma)} + [\![\mathbb{A}_1\,\mathfrak{u}, \mathfrak{v}]\!]_{\mathcal{T}(\Gamma)} = -[\![\mathbb{T}_\Gamma\,\mathbf{u}_{\mathrm{inc}}, \mathbb{X}\,\mathfrak{v}]\!]_{\mathcal{T}(\Gamma)}, \tag{2.74}$$

for all $\mathfrak{v} \in \mathcal{T}(\Gamma)$.

Now, we aim to relate $\mathbb{A}_0$ and $\mathbb{A}_1$. To do so, it is important to note that both potentials $\mathbb{G}_0$ and $\mathbb{G}_1$ map between the same spaces and satisfy[8]

$$\mathbb{G}_0 = -\mathbb{G}_1[0] \circ \mathbb{X} \quad \text{on } \mathcal{T}(\Gamma), \tag{2.75}$$

due to the fact that the orientation of the boundary matters for the kernel of the double layer potential, cf. the formulas (2.23) and (2.25). In addition, we resort to the *jump relations* of Theorem 2.12 that tell us

$$\mathbb{T}_0\,\mathbb{G}_1[0] = \begin{pmatrix} \mathsf{T}^+_{D,1} \\ \mathsf{T}^+_{N,1} \end{pmatrix} \mathbb{G}_1[0] = \begin{pmatrix} \mathsf{T}_{D,1} \\ -\mathsf{T}_{N,1} \end{pmatrix} \mathbb{G}_1[0] - \mathbb{X} = \mathbb{X}\,\mathbb{T}_1\,\mathbb{G}_1[0] - \mathbb{X}\,. \tag{2.76}$$

8 We remind that square brackets following a potential or an integral operator indicate that those are based on the coefficients from a subdomain different from the subdomain on whose boundary they live, see explanations on Page 64.

Since $\mathbb{X}$ is an involution and $[\![\mathbb{X}\,\mathfrak{u}, \mathbb{X}\,\mathfrak{v}]\!]_{\mathcal{T}(\partial\Omega)} = -[\![\mathfrak{u}, \mathfrak{v}]\!]_{\mathcal{T}(\partial\Omega)}$ (from (2.17) and (2.14)) we have the equalities

$$\begin{aligned}
[\![\mathbb{A}_0\,\mathbb{X}\,\mathfrak{u}, \mathbb{X}\,\mathfrak{v}]\!]_{\mathcal{T}(\Gamma)} &\overset{(2.26)}{=} \left[\!\!\left[(\mathbb{T}_0\,\mathbb{G}_0 - \tfrac{1}{2}\mathsf{Id})\,\mathbb{X}\,\mathfrak{u}, \mathbb{X}\,\mathfrak{v}\right]\!\!\right]_{\mathcal{T}(\Gamma)} \\
&\overset{(2.75)}{=} [\![-\mathbb{T}_0\,\mathbb{G}_1[0]\mathfrak{u}, \mathbb{X}\,\mathfrak{v}]\!]_{\mathcal{T}(\Gamma)} - \frac{1}{2}[\![\mathbb{X}\,\mathfrak{u}, \mathbb{X}\,\mathfrak{v}]\!]_{\mathcal{T}(\Gamma)} \\
&\overset{(2.76)}{=} [\![-(\mathbb{X}\,\mathbb{T}_1\,\mathbb{G}_1[0] - \mathbb{X})\mathfrak{u}, \mathbb{X}\,\mathfrak{v}]\!]_{\mathcal{T}(\Gamma)} + \frac{1}{2}[\![\mathfrak{u}, \mathfrak{v}]\!]_{\mathcal{T}(\Gamma)} \\
&= [\![(\mathbb{T}_1\,\mathbb{G}_1[0] - \mathsf{Id})\mathfrak{u}, \mathfrak{v}]\!]_{\mathcal{T}(\Gamma)} + \frac{1}{2}[\![\mathfrak{u}, \mathfrak{v}]\!]_{\mathcal{T}(\Gamma)} \\
&\overset{(2.26)}{=} [\![\mathbb{A}_1[0]\mathfrak{u}, \mathfrak{v}]\!]_{\mathcal{T}(\Gamma)}.
\end{aligned} \tag{2.77}$$

This means that the single trace boundary integral equations for a homogeneous scatterer can be rephrased as: seek $\mathfrak{u} \in \mathcal{T}(\Gamma)$ such that

$$[\![\mathbb{S}'_\Gamma \mathbb{C}\,\mathfrak{u}, \mathbb{C}\,\mathfrak{v}]\!]_{\mathcal{T}(\Gamma)} = [\![(\mathbb{A}_1 + \mathbb{A}_1[0])\,\mathfrak{u}, \mathfrak{v}]\!]_{\mathcal{T}(\Gamma)} = -[\![\mathbb{T}_\Gamma\,\mathbf{u}_{\mathrm{inc}}, \mathbb{X}\,\mathfrak{v}]\!]_{\mathcal{T}(\Gamma)}, \tag{2.78}$$

for all $\mathfrak{v} \in \mathcal{T}(\Gamma)$. Here we relied on the identification (2.70) through the isomorphism $\mathbb{C}$ again.

A closer inspection of $\mathbb{A}_1$ and $\mathbb{A}_1[0]$ for the case of *equal coefficients* $\alpha_0 = \alpha_1$ reveals that these boundary integral operators feature kernels with the same singular behavior for $\boldsymbol{x} = \boldsymbol{y}$, cf. (2.23). Forming their difference, these singularities will cancel, resulting in an operator that is smoothing compared to $\mathbb{A}_1$ and $\mathbb{A}_1[0]$. Hence, the Rellich embedding theorem [39, Theorem 3.27] allows the following conclusion, see [48, Lemma 3.9.8].

Lemma 2.36 (Compact modification). *Provided that $\alpha_j = \alpha_k$, the operator $\mathbb{A}_i[j] - \mathbb{A}_i[k] : \mathcal{T}(\partial\Omega_i) \mapsto \mathcal{T}(\partial\Omega_i)$ is compact.*

As a consequence of this result and (2.73) applied to $\mathbb{A}_1$ we find the desired Calderón identity for the operator product from (2.72):

$$\mathbb{A}_1 \circ (\mathbb{A}_1 + \mathbb{A}_1[0]) = \frac{1}{2}\mathsf{Id} + \{\text{compact}\} \quad \text{in} \quad \mathcal{T}(\Gamma) \tag{2.79}$$

$$\Downarrow \quad (2.78)$$

$$(\mathbb{D}'^{-1})^* \mathbb{S}'_\Gamma \mathbb{D}'^{-1} \mathbb{S}'_\Gamma = \mathsf{Id} + \{\text{compact}\} \quad \text{in} \quad \boldsymbol{S}\mathcal{T}(\Sigma). \tag{2.80}$$

We learn that $\mathbb{B} := (\mathbb{D}'^{-1})^* \mathbb{S}'_\Gamma \mathbb{D}'^{-1}$ is an inverse of $\mathbb{S}'_\Gamma$ up to a compact perturbation, which will make the eigenvalues of the operator product in (2.80) cluster around 1. In a sense, $\mathbb{B}$ qualifies as a *preconditioning operator* for $\mathbb{S}'_\Gamma$! Why this parlance is justified and how to transfer this insight to Ritz–Galerkin discretization will be explained in the next section.

4.3 Operator preconditioning

The policy of operator preconditioning is best understood in an abstract framework that we present following [31, Section 2]. We consider a Hilbert space V with dual V' and two bounded operators $\mathsf{A} : V \mapsto V'$ and $\mathsf{B} : V \mapsto V'$ with associated sesquilinear forms $\mathsf{a} : V \times V \mapsto \mathbb{C}$ and $\mathsf{b} : V \times V \mapsto \mathbb{C}$. The spaces V and V' are also dual with respect to the pairing $[\cdot,\cdot] : V \times V \mapsto \mathbb{C}$, which is supposed to give rise to an *isometric* isomorphism $V \mapsto V'$. In other words, the norm on V' is defined via $[\cdot,\cdot]$.

Galerkin discretization of A and B is based on finite-dimensional subspaces $V_h \subset V$ and $\widetilde{V}_h \subset V$. Stability of the discretization is assumed, which amounts to discrete inf-sup-conditions of the form

$$\sup_{v_h \in V_h} \frac{|\mathsf{a}(u_h, v_h)|}{\|v_h\|_V} \geq c_A \|u_h\|_V \quad \forall u_h \in V_h. \tag{2.81}$$

$$\sup_{\widetilde{v}_h \in \widetilde{V}_h} \frac{|\mathsf{b}(\widetilde{u}_h, \widetilde{v}_h)|}{\|\widetilde{v}_h\|_V} \geq c_B \|\widetilde{u}_h\|_V \quad \forall \widetilde{u}_h \in \widetilde{V}_h. \tag{2.82}$$

The crucial assumption is the stability of the duality pairing for the Galerkin trial spaces

$$\sup_{\widetilde{v}_h \in \widetilde{V}_h} \frac{[\widetilde{v}_h, u_h]}{\|\widetilde{v}_h\|_V} \geq c_D \|u_h\|_V \quad \forall u_h \in V_h. \tag{2.83}$$

A necessary condition for (2.83) is $M := \dim V_h = \dim \widetilde{V}_h$. Choosing *arbitrary* bases of V_h and $\widetilde{V}_h$ we can represent the sesquilinear forms a, b, and $[\cdot,\cdot]$ by means of their $M \times M$ Galerkin matrices $\mathbf{A}, \mathbf{B}, \mathbf{D} \in \mathbb{C}^{M,M}$. Then [31, Theorem 2.1] asserts that

$$\kappa(\mathbf{D}^{-H}\mathbf{B}\mathbf{D}^{-1}\mathbf{A}) \leq \delta := \frac{\|\mathsf{A}\| \, \|\mathsf{B}\|}{c_A c_B c_D^2}, \tag{2.84}$$

where κ stands for the spectral condition number of a matrix, that is, the ratio of the moduli of the largest (in modulus) and smallest eigenvalue.

Estimate (2.84) suggests that $\mathbf{D}^{-H}\mathbf{B}\mathbf{D}^{-1}$ can serve as a preconditioner of $\mathbf{A}$. Indeed, for symmetric sesquilinear forms a and b the estimate (2.84) immediately implies that the rate of linear convergence of a preconditioned iterative Krylov-subspace solver (CG or MINRES) can be bounded in terms of δ alone. Provided that the stability constants c_A, c_B, and c_D can be uniformly bounded away from zero for a family of trial spaces, "uniformly fast"/"h-independent" convergence of the iterative solver is ensured.

Without symmetry of a and b, (2.84) does not permit us to predict the speed of convergence of an iterative solver like GMRES. In this case using $\mathbf{D}^{-H}\mathbf{B}\mathbf{D}^{-1}$ as a preconditioner is heuristics, which has proved to be highly successful in practice, though.

4.4 Stable duality pairing for boundary elements

The single trace formulation for $N = 1$, discretized by means of the lowest order piecewise polynomial boundary elements as discussed in Section 4.1, can be fit into the abstract framework of operator preconditioning as introduced in the previous section in a straightforward way; the incarnations of the operators and pairings are immediate from (2.72) and (2.80). Based on the Hilbert space $V := \boldsymbol{ST}(\Sigma)$, the role of A and B is played by the continuous STF operator $\mathbb{S}'_\Gamma : \boldsymbol{ST}(\Sigma) \to \boldsymbol{ST}(\Sigma)'$. The isometric duality pairing is provided by the operator $\mathbb{D}'$ according to (2.71).

The discrete inf-sup conditions (2.81) and (2.82) are guaranteed by Theorem 2.35 on sufficiently fine interface meshes with constants independent of the meshwidth h. However, no h-uniform discrete inf-sup condition (2.83) is available for the pairing on $\boldsymbol{ST}_h(\Sigma) \times \boldsymbol{ST}_h(\Sigma)$ effected by the operator $\mathbb{D}'$!

In the case of D = grad (Helmholtz) and lowest degree boundary elements, see Concretization 2.30, this is obvious, because in general $\dim \mathcal{T}_{N,h}(\Gamma) \neq \dim \mathcal{T}_{D,h}(\Gamma)$, whereas the duality pairing relies on products of functions of these two spaces. Even in the case of D = **curl** (Maxwell) and surface edge elements (2.83) fails to hold uniformly in h, though in this case $\dim \mathcal{T}_{N,h}(\Gamma)$ and $\dim \mathcal{T}_{D,h}(\Gamma)$ both agree with the number of edges of Γ_h. This was demonstrated in [17, Proposition 3.1].

A remedy for this conundrum is to use *different meshes* for the construction of $\mathcal{T}_{N,h}(\Gamma)$ and $\mathcal{T}_{D,h}(\Gamma)$, thereby ensuring stability and, of course, matching dimensions $\dim \mathcal{T}_{N,h}(\Gamma) = \dim \mathcal{T}_{D,h}(\Gamma)$. For second-order diffusion problems the pioneering work [51] investigated a suitable construction for the lowest order polynomial BEM. It relies on *dual meshes*. An application to acoustic scattering (Helmholtz case) is covered in [16]. The idea was adapted to surface edge elements in [9] and we refer to this chapter for a comprehensive analysis and the proof of uniform stability of standard duality pairings on the primal-dual boundary element spaces. Thus, all pieces are in place for operator preconditioning of the discretized STF for $N = 1$!

Remark 2.37 (Riesz pairing). The alert reader will have realized that on the Hilbert space $\boldsymbol{ST}(\Sigma)$ a suitable alternative candidate for $\mathbb{D}'$ in (2.71) would be the Riesz isomorphism $\boldsymbol{ST}(\Sigma) \to \boldsymbol{ST}(\Sigma)'$ induced by the inner product of $\boldsymbol{ST}(\Sigma)$. This choice would preserve the continuity properties of $(\mathbb{D}'^{-1})^* \mathbb{S}'_\Gamma \mathbb{D}'^{-1} \mathbb{S}'_\Gamma$.

At second glance, the Riesz isomorphism turns out to be utterly useless for the purpose of operator preconditioning; the matrix $\mathbf{D}$ would be the Galerkin matrix of another boundary integral operator of the first kind, structurally similar to $\mathbb{S}'_\Sigma$. Therefore, applying the preconditioner would entail solving two ill-conditioned linear systems of equations as hard as the one tackled by the preconditioner. △

Remark 2.38 (Sparse pairing matrix). Duality inducing pairings based on the pivot space L^2 are inherently local; inserting two functions with nonoverlapping supports will yield zero. Thus, using boundary element spaces on dual meshes for the stable discretization of the pairing along with standard local basis functions will lead to *sparse* Galerkin matrices $\mathbf{D}$, cf. (2.84). Thus, direct Gaussian elimination becomes a competitive option for computing the action of $\mathbf{D}^{-1}$ and $\mathbf{D}^{-H}$ in the preconditioner. The efficiency of iterative solvers also hinges on sparsity. Thus, strengthening Remark 2.37 we can make the statement that

> operator preconditioning of Galerkin matrices arising from boundary integral equations will be feasible, only if the underlying pairing is of L^2-type, hence local. △

Remark 2.39 (Varying coefficients α_i). Provided that the uniform stability assumptions are met, operator preconditioning for the discretized STF will be robust with respect to the choice of skeleton meshes. No assumptions on α_i other than $\alpha_i > 0$ are necessary. Of course, the heights of the jumps of $\alpha(\boldsymbol{x})$ will affect the stability constants.

In the case $\alpha_i = \alpha_0$, $i = 1, \dots, N$, we conclude from the results of Section 4.2 that the eigenvalues of the operator product from (2.80) have $1 \in \mathbb{C}$ as sole accumulation point. Therefore, one may hope that a similar property holds for the spectrum of the matrix product from (2.84). Yet, so far there is no rigorous proof that this is the case. △

4.5 The challenge

Unfortunately, the case of a homogeneous scatterer $N = 1$ is very special. In this case $\mathcal{T}(\Gamma)$ offers an isomorphic model of $\boldsymbol{\mathcal{ST}}(\Sigma)$, see (2.70), and, thus, a local L^2-type duality pairing has become available. Whenever $N > 1$ and there are material junction points, it seems impossible to find a way to express the duality of $\boldsymbol{\mathcal{ST}}(\Sigma)$ through an L^2-type pairing with another trace space. Also note that Theorem 2.23 rules out that $[\![\cdot, \cdot]\!]_{\mathcal{MT}(\Sigma)}$ can serve as a duality pairing for $\boldsymbol{\mathcal{ST}}(\Sigma)$.

Hence, efficient operator preconditioning of the single-trace boundary integral equations becomes elusive, remember Remark 2.38. Hitherto, this problem has not been overcome. This prompted us to

> seek alternatives to the single trace formulation that are amenable to operator preconditioning.

Our search has been trained on formulations with straightforward L^2-type duality pairings. It led to the development of new boundary integral equations belonging to the class of multitrace formulations (MTF). They will be presented in the remainder of the paper.

5 Global multitrace formulation

5.1 Separated subdomains

In Section 4.1 we have seen that the classical single trace formulation allows operator preconditioning for $N = 1$. Asking the question, what was exceptional about this setting, we arrived at the conclusion that it is the availability of a simple L^2-pairing bringing about the duality of $\mathcal{ST}(\Sigma)$ and $\mathcal{ST}(\Sigma)'$. This was the gift of the isomorphism $\mathbb{C}$ from (2.70).

Obviously, an analogous isometric isomorphism

$$\mathcal{ST}(\Sigma) \cong \mathcal{T}(\partial\Omega_1) \times \cdots \times \mathcal{T}(\partial\Omega_N), \tag{2.85}$$

still exists for $N > 1$, if all subdomains Ω_j are *separated* as illustrated in Figure 2.2 for $N = 2$. In this setting the isomorphism $\mathbb{C} : \mathcal{T}(\partial\Omega_1) \times \cdots \times \mathcal{T}(\partial\Omega_N) \to \mathcal{ST}(\Sigma)$ underlying (2.85) reads

$$\mathbb{C} : \begin{cases} \mathcal{T}(\partial\Omega_1) \times \mathcal{T}(\partial\Omega_2) & \to \quad \mathcal{ST}(\Sigma) \\ \begin{pmatrix} \binom{u_1}{\varphi_1} \\ \binom{u_2}{\varphi_2} \end{pmatrix} & \mapsto \quad (u_1 \vee u_2, u_1, u_2, -(\varphi_1 \vee \varphi_2), \varphi_1, \varphi_2), \end{cases} \tag{2.86}$$

where $\vee$ indicates that two functions on $\partial\Omega_1$ and $\partial\Omega_2$ are combined into a function on $\partial\Omega_0 = \partial\Omega_1 \cup \partial\Omega_2$:

$$(\mathbf{u}_1 \vee \mathbf{u}_2)(\boldsymbol{x}) := \mathbf{u}_i(\boldsymbol{x}), \text{ if } \quad \boldsymbol{x} \in \Omega_i, \quad i = 1,2.$$

We remark that this isomorphism relies on the induced orientation of $\partial\Omega_j$ from Ω_j, see Figure 2.2.

A duality pairing for the right-hand side of (2.85) is furnished by the local L^2-type pairings on the subdomain boundaries $\partial\Omega_j$. As pointed out in Section 4.4, related h-uniformly stable discrete duality pairings for standard BEM can be achieved by employing dual meshes on each $\partial\Omega_j$. Summing up, standard operator preconditioning remains feasible for separated subdomains.

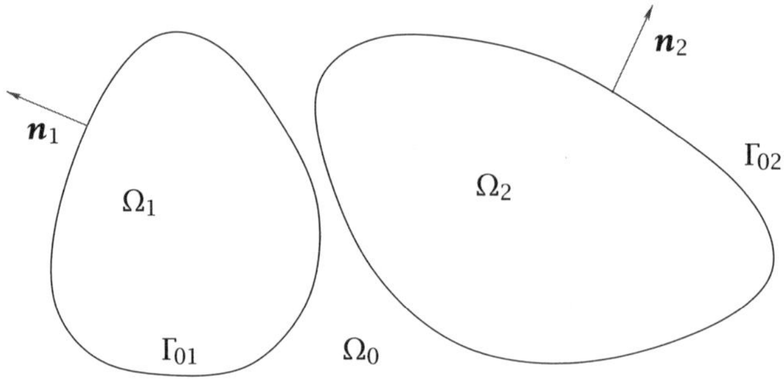

Figure 2.2: $N = 2$, isolated subdomains

Instead of using $\mathbb{C}$ as a tool to graft a duality pairing onto $\boldsymbol{\mathcal{ST}}(\Sigma)$ we may as well use it to define a "new" operator

$$\mathbb{M}_G' := \mathbb{C}^* \, \mathbb{S}_\Sigma' \, \mathbb{C} : \widehat{\boldsymbol{\mathcal{MT}}}(\Sigma) \to \widehat{\boldsymbol{\mathcal{MT}}}(\Sigma)', \tag{2.87}$$

where we wrote

$$\widehat{\boldsymbol{\mathcal{MT}}}(\Sigma) := \mathcal{T}(\partial\Omega_1) \times \cdots \times \mathcal{T}(\partial\Omega_N), \tag{2.88}$$

for a "clipped multitrace space," that differs from $\boldsymbol{\mathcal{MT}}(\Sigma)$ by a missing contribution from $\mathbb{T}_{\partial\Omega_0}$. The operator $\mathbb{C}^*$ is the formal adjoint of $\mathbb{C}$. By means of $\mathbb{M}_G'$ we can state a variational problem completely equivalent to (2.47): seek $\hat{\mathfrak{u}} \in \widehat{\boldsymbol{\mathcal{MT}}}(\Sigma)$ such that

$$\begin{aligned} \left\langle \mathbb{M}_G' \hat{\mathfrak{u}}, \hat{\mathfrak{v}} \right\rangle_{\widehat{\boldsymbol{\mathcal{MT}}}(\Sigma)} &= -[\![\mathbb{T}_0 \, \mathbf{u}_{\mathrm{inc}}, \mathbb{X}_1 \mathfrak{v}_1 \vee \mathbb{X}_2 \mathfrak{v}_2 \vee \cdots \vee \mathbb{X}_N \mathfrak{v}_N]\!]_{\mathcal{T}(\partial\Omega_0)} \\ &\overset{(2.17)}{=} \sum_{i=1}^{N} [\![\mathbb{T}_i[0] \mathbf{u}_{\mathrm{inc}}, \mathfrak{v}_i]\!]_{\mathcal{T}(\partial\Omega_i)}. \end{aligned} \tag{2.89}$$

for all $\hat{\mathfrak{v}} = (\mathfrak{v}_1, \ldots, \mathfrak{v}_N) \in \widehat{\boldsymbol{\mathcal{MT}}}(\Sigma)$.

Restricting the pairing $[\![\cdot,\cdot]\!]_{\boldsymbol{\mathcal{MT}}(\Sigma)}$ to $\widehat{\boldsymbol{\mathcal{MT}}}(\Sigma)$ gives another L^2*-type pairing*

$$[\![\vec{\mathfrak{u}}, \vec{\mathfrak{v}}]\!]_{\widehat{\boldsymbol{\mathcal{MT}}}(\Sigma)} := \sum_{i=1}^{N} [\![\mathbb{L}_i \vec{\mathfrak{u}}, \mathbb{L}_i \vec{\mathfrak{v}}]\!]_{\mathcal{T}(\partial\Omega_i)}, \quad \mathfrak{u}, \mathfrak{v} \in \widehat{\boldsymbol{\mathcal{MT}}}(\Sigma), \tag{2.90}$$

that will induce a self-duality of $\widehat{\boldsymbol{\mathcal{MT}}}(\Sigma)$. Harking back to (2.71), let us designate the isomorphism induced by the pairing $[\![\cdot,\cdot]\!]_{\widehat{\boldsymbol{\mathcal{MT}}}(\Sigma)}$ by $\widehat{\mathbb{D}}' : \widehat{\boldsymbol{\mathcal{MT}}}(\Sigma) \to \widehat{\boldsymbol{\mathcal{MT}}}(\Sigma)'$. It can be used to convert $\mathbb{M}_G'$ into a genuine boundary integral operator $\mathbb{M}_G : \widehat{\boldsymbol{\mathcal{MT}}}(\Sigma) \to \widehat{\boldsymbol{\mathcal{MT}}}(\Sigma)$ according to $\mathbb{M}_G := (\widehat{\mathbb{D}}')^{-1} \circ \mathbb{M}_G'$.

Working with $\mathbb{M}_G$ has the benefit that its structure can be made explicit and we do so for the case $N = 2$. To begin with we recast $\mathbb{A}_0$, which acts on functions in $\mathcal{T}(\partial\Omega_0) = \mathcal{T}(\partial\Omega_1) \times \mathcal{T}(\partial\Omega_2) = \widehat{\boldsymbol{\mathcal{MT}}}(\Sigma)$. For $\hat{\mathfrak{u}} = (\mathfrak{u}_1, \mathfrak{u}_2)$, $\hat{\mathfrak{v}} = (\mathfrak{v}_1, \mathfrak{v}_2)$ we find

$$\begin{aligned} &\left[\!\!\left[\mathbb{A}_0 \, \mathbb{L}_0 \, \mathbb{C} \begin{pmatrix} \mathfrak{u}_1 \\ \mathfrak{u}_2 \end{pmatrix}, \mathbb{L}_0 \, \mathbb{C} \begin{pmatrix} \mathfrak{v}_1 \\ \mathfrak{v}_2 \end{pmatrix} \right]\!\!\right]_{\mathcal{T}(\partial\Omega_0)} \\ &= [\![\mathbb{A}_1[0] \mathfrak{u}_1, \mathfrak{v}_1]\!]_{\mathcal{T}(\partial\Omega_1)} + [\![\mathbb{T}_1[0] \mathbb{G}_2[0] \mathfrak{u}_2, \mathfrak{v}_1]\!]_{\mathcal{T}(\partial\Omega_1)} \\ &\quad + [\![\mathbb{T}_2[0] \mathbb{G}_1[0] \mathfrak{u}_1, \mathfrak{v}_2]\!]_{\mathcal{T}(\partial\Omega_2)} + [\![\mathbb{A}_2[0] \mathfrak{u}_2, \mathfrak{v}_2]\!]_{\mathcal{T}(\partial\Omega_2)}. \end{aligned} \tag{2.91}$$

To see this, we advise the reader to study the derivation of (2.77) by means of the jump relations again, because similar manipulations yield the identity

$$
\begin{aligned}
&\mathbb{A}_0\,\mathbb{L}_0\,\mathbb{C}\begin{pmatrix}\mathfrak{u}_1\\ \mathfrak{u}_2\end{pmatrix}\\
&= (\mathbb{T}_0\,\mathbb{G}_0 - \frac{1}{2}\mathsf{Id})(\mathbb{X}_1\,\mathfrak{u}_1 \vee \mathbb{X}_2\,\mathfrak{u}_2)\\
&\overset{(2.75)}{=} \mathbb{T}_1^+(-\mathbb{G}_1[0]\mathfrak{u}_1 - \mathbb{G}_2[0]\mathfrak{u}_2) \vee \mathbb{T}_2^+(-\mathbb{G}_1[0]\mathfrak{u}_1 - \mathbb{G}_2[0]\mathfrak{u}_2)\\
&\qquad - \frac{1}{2}(\mathbb{X}_1\,\mathfrak{u}_1 \vee \mathbb{X}_2\,\mathfrak{u}_2)\\
&\overset{(2.76)}{=} (-\mathbb{X}_1\,\mathbb{T}_1[0]\mathbb{G}_1[0] + \frac{1}{2}\mathbb{X}_1)\mathfrak{u}_1 \vee (-\mathbb{X}_2\,\mathbb{T}_2[0]\mathbb{G}_2[0] + \frac{1}{2}\mathbb{X}_2)\mathfrak{u}_2\\
&\qquad - \mathbb{X}_1\,\mathbb{T}_1[0]\mathbb{G}_2[0]\mathfrak{u}_2 \vee \mathbb{X}_2\,\mathbb{T}_2[0]\mathbb{G}_1[0]\mathfrak{u}_1\\
&= -\mathbb{X}_1(\mathbb{T}_1[0]\mathbb{G}_1[0] - \frac{1}{2}\mathsf{Id})\mathfrak{u}_1 \vee \mathbb{X}_2(\mathbb{T}_2[0]\mathbb{G}_2[0] - \frac{1}{2}\mathsf{Id})\mathfrak{u}_2\\
&\qquad \mathbb{X}_1\,\mathbb{T}_1[0]\mathbb{G}_2[0]\mathfrak{u}_2 \vee \mathbb{X}_2\,\mathbb{T}_2[0]\mathbb{G}_1[0]\mathfrak{u}_1.\\
&\overset{(2.26)}{=} -\mathbb{X}_1(\mathbb{A}_1[0]\mathfrak{u}_1 + \mathbb{T}_1[0]\mathbb{G}_2[0]\mathfrak{u}_2) - \mathbb{X}_2(\mathbb{T}_2[0]\mathbb{G}_1[0]\mathfrak{u}_1 + \mathbb{A}_2[0]\mathfrak{u}_2).
\end{aligned}
\tag{2.92}
$$

We conclude (2.91) because of $[\![\mathbb{X}_i\,\mathfrak{u}, \mathbb{X}_i\,\mathfrak{v}]\!]_{\mathcal{T}(\partial\Omega_i)} = -[\![\mathfrak{u}, \mathfrak{v}]\!]_{\mathcal{T}(\partial\Omega_i)}$. For the sake of a compact operator notation we may abbreviate the "remote coupling integral operators"

$$\mathbb{R}_i^j := \mathbb{T}_j[0]\mathbb{G}_i[0], \quad i \neq j, \tag{2.93}$$

which evaluate a potential defined on one subdomain boundary on the other. Then, for $N = 2$, $\mathbb{M}_G$ can be rephrased as

$$\mathbb{M}_G\begin{pmatrix}\mathfrak{u}_1\\ \mathfrak{u}_2\end{pmatrix} = \begin{pmatrix}\mathbb{A}_1[0] + \mathbb{A}_1 & \mathbb{R}_2^1\\ \mathbb{R}_1^2 & \mathbb{A}_2[0] + \mathbb{A}_2\end{pmatrix}\begin{pmatrix}\mathfrak{u}_1\\ \mathfrak{u}_2\end{pmatrix}, \tag{2.94}$$

for $\mathfrak{u}_1 \in \mathcal{T}(\partial\Omega_1)$, $\mathfrak{u}_2 \in \mathcal{T}(\partial\Omega_2)$.

In Section 4.2 we established the Calderón identity (2.80) for the STF operator in the case $N = 1$ and $\alpha_0 = \alpha_1$. Hardly surprising, a similar result remains valid for $N > 1$ and separated subdomains. For the sake of simplicity we largely restrict the discussion to $N = 2$, see Figure 2.2.

First, we recall the null field property of the total potential $\mathbb{G}$ belonging to a generic domain Ω with Lipschitz boundary $\partial\Omega$.

Lemma 2.40 (Null field property). *If $\mathfrak{u} \in \mathcal{T}(\partial\Omega)$ belongs to the space of Cauchy data $\mathcal{CD}(\partial\Omega)$ for the differential equation* (2.4), *see Definition* 2.14, *then $\mathbb{G}\mathfrak{u} = 0$ in $\mathbb{R}^d \setminus \overline{\Omega}$.*

An immediate consequence are important identities for the remote coupling operators.

Lemma 2.41 (Annihilation of remote coupling operators). *The remote coupling operators $\mathbb{R}_i^j$ defined in (2.93) satisfy for $i \neq j$*

$$\begin{aligned} \mathbb{R}_i^j\,\mathbb{R}_j^i &= 0, \\ \mathbb{R}_i^j\,\mathbb{A}_i[0] &= -\frac{1}{2}\,\mathbb{R}_i^j, \\ \mathbb{A}_j[0]\,\mathbb{R}_i^j &= \frac{1}{2}\,\mathbb{R}_i^j . \end{aligned} \tag{2.95}$$

Proof. The product of two different remote coupling operators vanishes:

$$\mathbb{R}_i^j\,\mathbb{R}_j^i = \mathbb{T}_j[0]\mathbb{G}_i[0]\,\mathbb{T}_i[0]\mathbb{G}_j[0] = \mathbb{T}_j[0](0) = 0,$$

because Theorem 2.9 asserts $\mathbb{T}_i[0]\mathbb{G}_j[0](\mathcal{T}(\partial\Omega_j)) \subset \mathcal{CD}(\partial\Omega_i)$ and, thus, by Lemma 2.40

$$\mathbb{G}_i[0]\,\mathbb{T}_i[0]\mathbb{G}_j[0]_{\,|\overline{\Omega}_j} = 0.$$

The same arguments can be applied to the other operator products, for instance,

$$\mathbb{R}_i^j\,\mathbb{A}_i[0] = \mathbb{T}_j[0]\mathbb{G}_i[0]\left(\mathbb{T}_i[0]\mathbb{G}_i[0] - \frac{1}{2}\mathsf{Id}\right) = -\frac{1}{2}\,\mathbb{R}_i^j . \qquad \square$$

Returning to $N = 2$, owing to the assumption $\alpha_0 = \alpha_1 = \alpha_2$, we can appeal to Lemma 2.36 and obtain from (2.94)

$$\mathbb{M}_G = \mathbb{M}_G[0] + \{\text{compact}\}, \quad \mathbb{M}_G[0] := \begin{pmatrix} 2\,\mathbb{A}_1[0] & \mathbb{R}_2^1 \\ \mathbb{R}_1^2 & 2\,\mathbb{A}_2[0] \end{pmatrix}. \tag{2.96}$$

An immediate consequence of Lemma 2.41 and (2.73) is that $(\mathbb{M}_G[0])^2 = \mathsf{Id}$, which implies the Calderón identity

$$\mathbb{M}_G \circ \mathbb{M}_G = \mathsf{Id} + \{\text{compact}\}, \tag{2.97}$$

provided that $\alpha_i = \alpha_0$ for all $i = 1, \dots, N$.

5.2 The gap idea

The situation of separated subdomains is of little interest in itself, but it will serve as a crucial stepping stone for the generalization of the STF to a first type of multitrace boundary integral formulation.

Consider the situation sketched in Figure 2.3 (left) with the subdomains separated by a narrow gap of width $\delta > 0$. We investigate what will happen to the operator $\mathbb{R}_2^1[0] := \mathbb{T}_2[0]\mathbb{G}_1[0]$ as $\delta \to 0$. Keep in mind that

$$(\mathbb{G}_1[0]\mathfrak{u}_1)_{\,|\mathbb{R}^d\setminus\Omega_1} \in \mathcal{H}_{\mathrm{loc}}(\mathsf{L}, \mathbb{R}^d \setminus \Omega_1), \quad \mathfrak{u}_1 \in \mathcal{T}(\partial\Omega_1). \tag{2.98}$$

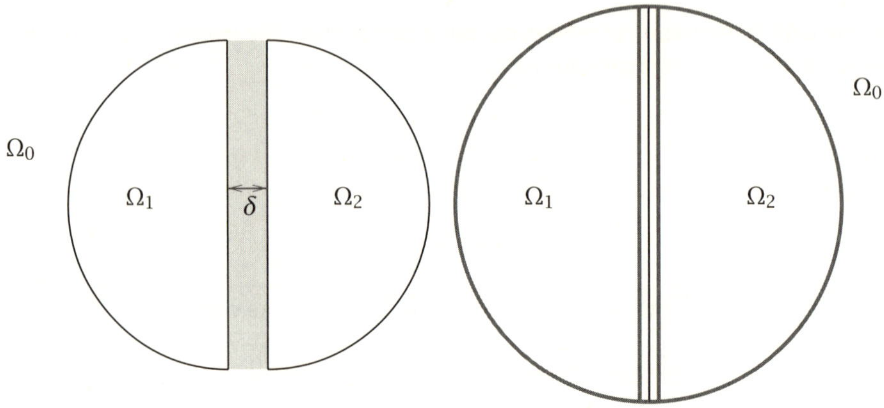

Figure 2.3: Illustration of the "gap idea" for $N = 2$. Left: subdomains separated by gap of width $\delta > 0$. Right: closed gap

Hence, $\mathbb{T}_2[0]\mathbb{G}_1[0]\mathfrak{u}_1$ is well defined as a function in $\mathcal{T}(\partial\Omega_2)$ whenever $\partial\Omega_2 \subset \mathbb{R}^d \setminus \Omega_1$. These conditions are still met even when $\delta = 0$. Thus the operator $\mathbb{M}'_G : \widehat{\mathcal{MT}}(\Sigma) \to \widehat{\mathcal{MT}}(\Sigma)'$ as defined in (2.94) for $N = 2$§ and in (2.87) for general separated subdomains remains meaningful and will give rise to the *global multitrace integral equation formulation* (global MTF) for the transmission problem (2.8). Manipulations completely parallel to the derivation of (2.92) give the final variational form of the global MTF [19, Section 8.2]: seek $\hat{\mathfrak{u}} = (\mathfrak{u}_1, \ldots, \mathfrak{u}_N) \in \widehat{\mathcal{MT}}(\Sigma)$

$$[\![\mathbb{M}_G\,\hat{\mathfrak{u}}, \hat{\mathfrak{v}}]\!]_{\widehat{\mathcal{MT}}(\Sigma)} = \sum_{i=1}^{N} [\![\mathbb{T}_i[0]\mathbf{u}_{\mathrm{inc}}, \mathbb{L}_i\,\hat{\mathfrak{v}}]\!]_{\mathcal{T}(\partial\Omega_i)}\,, \tag{2.99}$$

for all $\hat{\mathfrak{v}} = (\mathfrak{v}_1, \ldots, \mathfrak{v}_N) \in \widehat{\mathcal{MT}}(\Sigma)$ with $\mathbb{M}_G : \widehat{\mathcal{MT}}(\Sigma) \to \widehat{\mathcal{MT}}(\Sigma)$ given by

$$\mathbb{M}_G := \begin{pmatrix} \mathbb{A}_1[0] + \mathbb{A}_1 & \mathbb{R}_2^1 & \cdots & \cdots & \mathbb{R}_N^1 \\ \mathbb{R}_1^2 & \mathbb{A}_2[0] + \mathbb{A}_2 & & & \vdots \\ \vdots & & \ddots & & \vdots \\ \vdots & & & \mathbb{A}_{N-1}[0] + \mathbb{A}_{N-1} & \mathbb{R}_N^{N-1} \\ \mathbb{R}_1^N & \cdots & \cdots & \mathbb{R}_{N-1}^N & \mathbb{A}_N[0] + \mathbb{A}_N \end{pmatrix}. \tag{2.100}$$

5.3 Properties of global MTF

Owing to the derivation of the global MTF from the STF via a "zero gap reasoning," as explained in Section 5.2, we expect that

> all properties of the classical STF for separated subdomains carry over to the global MTF.

The global MTF lives up to this expectation. Concrete results are given in the following theorems. Firstly, we adapt Theorem 2.26 to the global MTF:

Theorem 2.42 (Coercivity of global MTF, [19, Section 10], [18, Section 9.5]). *The bilinear form of the global MTF from* (2.99) *satisfies a (generalized) Gårding inequality on* $\widehat{\mathcal{MT}}(\Sigma)$.

The assertion made in Lemma 2.28 for the STF remains valid for the global MTF.

Theorem 2.43 (Uniqueness for global MTF, [19, Section 9], [18, Theorem 8.1]). *The global MTF operator from* (2.100) *is injective.*

By virtue of a Fredholm alternative argument, the last two theorems imply existence of solutions of (2.99) and the existence of a bounded inverse of $\mathbb{M}_G$. The special Calderón identity (2.97) holds for the global MTF without restrictions on the geometry.

Theorem 2.44 (Calderón identity, [19, Section 11], [18, Section 11]). *Provided that* $\alpha_i = \alpha_0$, $i = 1, \dots, N$, *the global MTF operator from* (2.100) *satisfies*

$$(\mathbb{M}_G)^2 = \mathsf{Id} + \{compact\}.$$

5.4 Galerkin discretization

In Section 3.3 we introduced finite dimensional subspaces $\mathcal{T}_{D,h}(\partial\Omega_i) \subset \mathcal{T}_D(\partial\Omega_i)$ and $\mathcal{T}_{N,h}(\partial\Omega_i) \subset \mathcal{T}_N(\partial\Omega_i)$ generated by piecewise polynomial functions on meshes of $\partial\Omega_i$. Writing $\mathcal{T}_h(\partial\Omega_i) := \mathcal{T}_{D,h}(\partial\Omega_i) \times \mathcal{T}_{N,h}(\partial\Omega_i)$ the definition (2.88) immediately suggests that we use

$$\widehat{\mathcal{MT}}_h(\Sigma) := \mathcal{T}_h(\partial\Omega_1) \times \cdots \times \mathcal{T}_h(\partial\Omega_N) \tag{2.101}$$

as trial and test space for a boundary element Galerkin discretization of (2.99). Again, the assertion of Theorem 2.35 for the STF has a counterpart for the global MTF.

Theorem 2.45 (Stability of Galerkin discretization, [19, Proposition 10.4], [18, Proposition 10.1]). *The Galerkin discretization of the global MTF* (2.99) *based on standard piecewise polynomial boundary element spaces satisfies an* h*-uniform asymptotic discrete inf-sup condition.*

Remark 2.46. We emphasize that, in principle, the spaces $\mathcal{T}_h(\partial\Omega_i)$ need not match at all, which allows the combinations of different types of boundary elements on different subdomains. △

Remark 2.47. It goes without saying that the system matrix $\mathbf{M}$ of the resulting linear systems of equations inherits the structure of (2.100); it is the ith diagonal block that will agree with the usual Galerkin matrix corresponding to $\mathbb{A}_i[0] + \mathbb{A}_i$ on $\mathcal{ST}_h(\partial\Omega_i)$. For the off-diagonal block related to the operators $\mathbb{R}_i^j[0]$ recall that, in case Ω_i and Ω_j

have a common interface, according to (2.93) the traces to be used on Γ_{ij} are exterior traces with respect to Ω_i. Hence, the jump relations of Theorem 2.12 have to be taken into account and the integral representation formulas have to be supplemented by $\frac{1}{2}$-weighted L^2-pairings. Details can be found in [35, Section 5]. △

The global MTF (2.99) readily lends itself to efficient operator preconditioning as discussed in Section 4.3:

(i) Obviously, $\mathbb{M}_G : \widehat{\mathcal{MT}}(\Sigma) \to \widehat{\mathcal{MT}}(\Sigma)$ is continuous, and so will be $(\mathbb{M}_G)^2 : \widehat{\mathcal{MT}}(\Sigma) \to \widehat{\mathcal{MT}}(\Sigma)$. In terms of $\mathbb{M}_G' := \widehat{\mathbb{D}'} \circ \mathbb{M}_G$ this states the continuity of $(\widehat{\mathbb{D}'}^{-1})^* \mathbb{M}_G' \widehat{\mathbb{D}'}^{-1} \mathbb{M}_G' : \widehat{\mathcal{MT}}(\Sigma) \to \widehat{\mathcal{MT}}(\Sigma)$. Further, by Theorem 2.44 the latter operator will be a compact perturbation of the identity in the case of constant α, compare (2.97).
(ii) The duality pairing $[\![\cdot,\cdot]\!]_{\widehat{\mathcal{MT}}(\Sigma)}$ from (2.90) is a local L^2-type pairing, which will yield a sparse Galerkin matrix **D**, cf. Remark 2.38 for an explanation why this is essential.
(iii) Uniformly stable discretizations of the pairing on $\widehat{\mathcal{MT}}_h(\partial\Omega_i)$ can be constructed based on dual meshes, see Section 4.4.

Remark 2.48. In the context of operator preconditioning we may replace $\mathbb{M}_G'$ with the fully decoupled local integral operators, that is, we may use the "diagonal preconditioner"

$$\widehat{\mathbb{D}'}^{-1} \mathbb{A}_G' \widehat{\mathbb{D}'}^{-1} : \widehat{\mathcal{MT}}(\Sigma)' \to \widehat{\mathcal{MT}}(\Sigma),$$

with

$$\langle \mathbb{A}_G' \hat{\mathfrak{u}}, \hat{\mathfrak{v}} \rangle_{\widehat{\mathcal{MT}}(\Sigma)} := \sum_{i=1}^{N} [\![\mathbb{A}_i \mathfrak{u}_i, \mathfrak{v}_i]\!]_{\mathcal{T}(\partial\Omega_i)}, \quad \hat{\mathfrak{u}}, \hat{\mathfrak{v}} \in \widehat{\mathcal{MT}}(\Sigma). \tag{2.102}$$

Given a uniformly stable discrete duality pairing this will also provide a preconditioner that is robust with respect to the resolution of the surface meshes. △

6 Local multitrace formulation

The idea to decouple the skeleton Cauchy traces into local contributions that guided the development of the global multitrace formulation of Section 5 can also be pursued in a fairly different spirit. This led to the so-called local multitrace formulation, which will be derived and explored in this section.

The local MTF was first presented in [32] for the Helmholtz transmission problems (2.1). Its extension to electromagnetic transmission problems (2.2) is work in progress and has not been published yet (as of summer 2012). Formally, the adaptation of the method to Maxwell's transmission problems is straightforward and in the next section we will also include this case.

6.1 Partial transmission conditions

First, let us examine the situation $N = 2$, that is, a scatterer composed of only two subdomains, as sketched in Figure 2.4 for $d = 2$. For the purpose of presenting the local multitrace formulation this case is generic, completely captures the essence of the methods, but still helps to avoid excessive notational complexity. With $\mathbf{u} \in \mathcal{H}_{\mathrm{loc}}(\mathsf{L}, \mathbb{R}^d)$ denoting the weak solution of the transmission problem (2.8) we write $\vec{\mathfrak{u}} := \mathbb{T}_\Sigma \mathbf{u} \in \mathcal{ST}(\Sigma) \subset \mathcal{MT}(\Sigma)$ for its skeleton Cauchy trace as defined in Section 3.1. As in (2.43), from Theorem 2.15 we immediately conclude

$$(\frac{1}{2}\mathsf{Id} - \mathbb{A}_i)\,\mathbb{L}_i\,\vec{\mathfrak{u}} = 0 \qquad \text{in } \mathcal{T}(\partial\Omega_i), \quad i = 1,2, \tag{2.103}$$

$$(\frac{1}{2}\mathsf{Id} - \mathbb{A}_0)(\mathbb{L}_0\,\vec{\mathfrak{u}} - \mathbb{T}_0\,\mathbf{u}_{\mathrm{inc}}) = 0 \qquad \text{in } \mathcal{T}(\partial\Omega_0). \tag{2.104}$$

The same simplification that permitted us to switch from (2.46) to (2.47) can be applied to (2.104) and yields

$$(\frac{1}{2}\mathsf{Id} - \mathbb{A}_0)\,\mathbb{L}_0\,\vec{\mathfrak{u}} = -\,\mathbb{T}_0\,\mathbf{u}_{\mathrm{inc}}. \tag{2.105}$$

The next key manipulation of (2.103) and (2.104) starts from the local transmission conditions

$$\mathsf{T}_{D,i}\,\mathbf{u} = \mathsf{T}_{D,j}\,\mathbf{u} \quad \text{and} \quad \mathsf{T}_{N,i}\,\mathbf{u} = -\,\mathsf{T}_{N,j}\,\mathbf{u} \quad \text{on } \Gamma_{ij} \quad \text{in } \mathcal{T}(\partial\Omega_0). \tag{2.37}$$

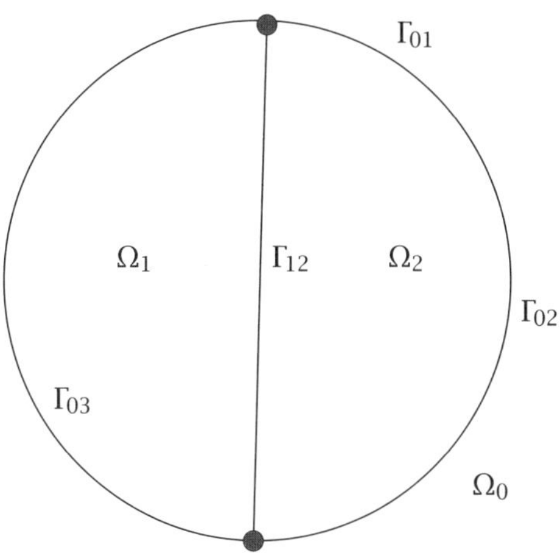

Figure 2.4: Generic geometric setting for $N = 2$

They are used to substitute the terms $\frac{1}{2}\,\mathbb{L}_i\,\vec{\mathfrak{u}}$ in (2.103) and (2.105) with Cauchy traces from adjacent subdomains. This is done on each interface Γ_{ij} and yields

$$\begin{aligned} \mathbb{A}_0\,\mathbb{L}_0\,\vec{\mathfrak{u}} \quad & -\frac{1}{2}\,\mathbb{X}_{1\to 0}\,\mathbb{L}_1\,\vec{\mathfrak{u}} - \frac{1}{2}\,\mathbb{X}_{2\to 0}\,\mathbb{L}_2\,\vec{\mathfrak{u}} = -\,\mathbb{T}_0\,\mathbf{u}_{\mathrm{inc}}, \\ -\frac{1}{2}\,\mathbb{X}_{0\to 1}\,\mathbb{L}_0\,\vec{\mathfrak{u}} + \quad & \mathbb{A}_1\,\mathbb{L}_1\,\vec{\mathfrak{u}} \quad - \frac{1}{2}\,\mathbb{X}_{2\to 1}\,\mathbb{L}_2\,\vec{\mathfrak{u}} = 0, \\ -\frac{1}{2}\,\mathbb{X}_{0\to 2}\,\mathbb{L}_0\,\vec{\mathfrak{u}} - & \frac{1}{2}\,\mathbb{X}_{1\to 2}\,\mathbb{L}_1\,\vec{\mathfrak{u}} + \quad \mathbb{A}_2\,\mathbb{L}_2\,\vec{\mathfrak{u}} \quad = 0. \end{aligned} \tag{2.106}$$

Here, we have introduced the *local transmission operators* $\mathbb{X}_{i\to j}$, which, if Γ_{ij} is a genuine interface,

(i) take a pair of functions on $\partial\Omega_i$ as argument, corresponding to Dirichlet and Neumann traces,
(ii) restrict both functions to Γ_{ij},
(iii) flip the sign of the second function in order to take into account the transmission conditions (2.37),
(iv) and, finally, extend both functions by zero to functions on $\partial\Omega_j$.

Formally, we can define

$$\mathbb{X}_{i\to j} : \begin{cases} L^2(\partial\Omega_i)\times L^2(\partial\Omega_i) \to L^2(\partial\Omega_j)\times L^2(\partial\Omega_j) \\ \mathfrak{u} := \begin{pmatrix} u \\ \varphi \end{pmatrix} \quad \mapsto (\mathbb{X}_{i\to j}\,\mathfrak{u})(\boldsymbol{x}) := \begin{cases} \begin{pmatrix} u(\boldsymbol{x}) \\ -\varphi(\boldsymbol{x}) \end{pmatrix} & \text{for } \boldsymbol{x}\in\Gamma_{ij}, \\ 0 & \text{elsewhere on } \partial\Omega_j. \end{cases} \end{cases} \tag{2.107}$$

The reader might be wondering, what made us rely on L^2-spaces here; we are going to answer this question in the next section.

Temporarily glossing over issues of proper function spaces, we have found that Cauchy traces of solutions of the transmission problems (2.8) satisfy the boundary integral equation (2.106). Turning the tables, now we boldly convert (2.106) into novel boundary integral equations for unknown *local* Cauchy traces: find $\mathfrak{u}_i \in \mathcal{T}(\partial\Omega_i)$ satisfying

$$\begin{aligned} \mathbb{A}_0\,\mathfrak{u}_0 \quad & -\frac{1}{2}\,\mathbb{X}_{1\to 0}\,\mathfrak{u}_1 - \frac{1}{2}\,\mathbb{X}_{2\to 0}\,\mathfrak{u}_2 = -\,\mathbb{T}_0\,\mathbf{u}_{\mathrm{inc}}, \\ -\frac{1}{2}\,\mathbb{X}_{0\to 1}\,\mathfrak{u}_0 + \quad & \mathbb{A}_1\,\mathfrak{u}_1 \quad - \frac{1}{2}\,\mathbb{X}_{2\to 1}\,\mathfrak{u}_2 = 0, \\ -\frac{1}{2}\,\mathbb{X}_{0\to 2}\,\mathfrak{u}_0 - & \frac{1}{2}\,\mathbb{X}_{1\to 2}\,\mathfrak{u}_1 + \quad \mathbb{A}_2\,\mathfrak{u}_2 \quad = 0. \end{aligned} \tag{2.108}$$

We have arrived at a boundary integral equation, which is set on the skeleton multitrace space $\boldsymbol{\mathcal{MT}}(\Sigma)$! We call (2.108) the *local multitrace formulation*, due to the fact, that, in contrast to the global multitrace formulation (2.99) introduced in Section 5, coupling of subdomains relies on local transmission operators alone.

It is not difficult to state the local MTF for an arbitrary number of subdomains: find $\vec{\mathfrak{u}} \in \mathcal{MT}(\Sigma)$ that solves

$$\mathbb{A}_i \mathbb{L}_i \vec{\mathfrak{u}} - \frac{1}{2} \sum_{\substack{j=0 \\ j \neq i}}^{N} \mathbb{X}_{j \to i} \mathbb{L}_j \vec{\mathfrak{u}} = \begin{cases} -\mathbb{T}_0 \mathbf{u}_{\text{inc}} & \text{for } i = 0, \\ 0 & \text{else}, \end{cases} \qquad i = 0, \ldots, N. \tag{2.109}$$

Of course, we have set $\mathbb{X}_{i \to j} := 0$ in case there is no genuine interface Γ_{ij}.

Having departed from a solution of the transmission problem tells us that its Cauchy traces will satisfy (2.109). There is no guarantee yet that a solution of (2.109) has anything to do with a solution of the transmission problem. This reverse conclusion holds thanks to uniqueness of solutions of (2.109).

Theorem 2.49 (Uniqueness of solutions of local MTF boundary integral equations, [32, Section 3.2.6]).
For $\mathbf{u}_{\text{inc}} = 0$ *the local MTF integral equations* (2.109) *have the only solution* $\vec{\mathfrak{u}} = 0$.

6.2 Local MTF: variational formulation

A rigorous statement of the local MTF (2.108) in trace spaces encounters massive obstacles, because the local transmission operators $\mathbb{X}_{i \to j}$ fail to be continuous mappings $\mathcal{T}(\partial\Omega_i) \to \mathcal{T}(\partial\Omega_j)$. We elaborate this for the case of the Helmholtz transmission problem and the corresponding trace spaces given in Concretization 2.5. First, we remind that the mapping

$$L^2(\partial\Omega_i) \to L^2(\partial\Omega_i), \quad w \mapsto \chi_{ij} w, \quad \chi_{ij}(\mathbf{x}) = \begin{cases} 1, & \text{if } \boldsymbol{x} \in \Gamma_{ij}, \\ 0, & \text{elsewhere}, \end{cases} \tag{2.110}$$

does not have a continuous restriction or extension as a mapping $H^{\frac{1}{2}}(\partial\Omega_i) \to H^{\frac{1}{2}}(\partial\Omega_i)$ or $H^{-\frac{1}{2}}(\partial\Omega_i) \to H^{-\frac{1}{2}}(\partial\Omega_i)$, respectively, if $\Gamma_{ij} \neq \partial\Omega_i$. This peculiar property of the trace spaces forces us to resort to a slightly modified function space framework in order to obtain a meaningful variational formulation of (2.109).

To begin with, let the space $\mathcal{T}_*(\Gamma_{ij})$, $* = D, N$, contain the restrictions of Dirichlet/Neumann traces in $\mathcal{T}_*(\partial\Omega_i)$ to Γ_{ij}. It is endowed with the natural norm

$$\|u\|_{\mathcal{T}_*(\Gamma_{ij})} := \inf\{\|v\|_{\mathcal{T}_*(\partial\Omega_i)} : v \in \mathcal{T}_*(\partial\Omega_i),\ v_{|\Gamma_{ij}} = u\}, \quad * = D, N.$$

These spaces can be assembled into "broken trace spaces" on $\partial\Omega_i$

$$\mathcal{T}_{*,\text{pw}}(\partial\Omega_i) := \bigotimes\nolimits_j \mathcal{T}_*(\Gamma_{ij}), \quad * = D, N,$$

where, of course, the product of spaces covers only genuine interfaces. We also write $\mathcal{T}_{\text{pw}}(\partial\Omega_i) := \mathcal{T}_{D,\text{pw}}(\partial\Omega_i) \times \mathcal{T}_{N,\text{pw}}(\partial\Omega_i)$ for a space of "broken Cauchy traces" on $\partial\Omega_i$

and it is convenient to introduce the "broken multitrace space"

$$\mathcal{MT}_{\mathrm{pw}}(\Sigma) := \mathcal{T}_{\mathrm{pw}}(\partial\Omega_i) \times \cdots \times \mathcal{T}_{\mathrm{pw}}(\partial\Omega_N). \tag{2.111}$$

By restriction to interfaces, $\mathcal{T}_*(\partial\Omega_i)$ can be embedded into $\mathcal{T}_{*,\mathrm{pw}}(\partial\Omega_i)$. The latter may be a strictly larger space, and the L^2 duality pairing between $\mathcal{T}_D(\partial\Omega_i)$ and $\mathcal{T}_N(\partial\Omega_i)$ usually cannot be extended to a continuous bilinear form on $\mathcal{T}_{D,\mathrm{pw}}(\partial\Omega_i) \times \mathcal{T}_N(\partial\Omega_i)$ or $\mathcal{T}_D(\partial\Omega_i) \times \mathcal{T}_{N,\mathrm{pw}}(\partial\Omega_i)$, respectively. When feeding broken traces into one slot of the L^2 duality pairing, the other slot has to be supplied with traces satisfying certain constraints. These constraints are met by localized Dirichlet and Neumann traces belonging to the spaces

$$\widetilde{\mathcal{T}}_*(\Gamma_{ij}) = \{\phi \in \mathcal{T}_*(\Gamma_{ij}) : \tilde{\phi} \in \mathcal{T}_*(\partial\Omega_j)\}, \tag{2.112}$$

with norm

$$\|\phi\|_{\widetilde{\mathcal{T}}_*(\Gamma_{ij})} = \left\|\tilde{\phi}\right\|_{\mathcal{T}_*(\partial\Omega_i)}, \quad * = D, N, \quad * = D, N, \tag{2.113}$$

where $\tilde{\phi}$ is the extension of ϕ to $\partial\Omega_j$ by zero. In general, the norm of $\widetilde{\mathcal{T}}_*(\Gamma_{ij})$ is strictly stronger than that of $\mathcal{T}_*(\Gamma_{ij})$. Most importantly, the pairs of spaces $\mathcal{T}_D(\Gamma_{ij})$–$\widetilde{\mathcal{T}}_N(\Gamma_{ij})$ and $\widetilde{\mathcal{T}}_D(\Gamma_{ij})$–$\mathcal{T}_N(\Gamma_{ij})$ are in duality with respect to the L^2-pairing on Γ_{ij}, cf. (2.11). Gluing the spaces of localized Dirichlet and Neumann traces yields spaces with "zero boundary conditions at the boundaries of the interfaces"

$$\widetilde{\mathcal{T}}_*(\partial\Omega_i) = \bigotimes_j \widetilde{\mathcal{T}}_*(\Gamma_{ij}) \subset \mathcal{T}_*(\partial\Omega_i), \quad * = D, N, \tag{2.114}$$

which can be assembled into the constrained Cauchy trace space $\widetilde{\mathcal{T}}(\partial\Omega_i) := \widetilde{\mathcal{T}}_D(\partial\Omega_i) \times \widetilde{\mathcal{T}}_N(\partial\Omega_i)$. In analogy to (2.29) we set

$$\widetilde{\mathcal{MT}}(\Sigma) := \widetilde{\mathcal{T}}(\partial\Omega_0) \times \cdots \times \widetilde{\mathcal{T}}(\partial\Omega_N). \tag{2.115}$$

The relationships of the above spaces can be summarized as

$$\widetilde{\mathcal{T}}_D(\partial\Omega_i) \subset \mathcal{T}_D(\partial\Omega_i) \subset \mathcal{T}_{D,\mathrm{pw}}(\partial\Omega_i) \qquad \widetilde{\mathcal{T}}_N(\partial\Omega_i) \subset \mathcal{T}_N(\partial\Omega_i) \subset \mathcal{T}_{N,\mathrm{pw}}(\partial\Omega_i), \tag{2.116}$$

where the arrows connect spaces that are in duality with respect to the L^2-pairing on $\partial\Omega_i$. Combine this with the continuity of the transmission operators

$$\mathbb{X}_{i\to j} : \mathcal{T}(\partial\Omega_i) \to \mathcal{T}_{\mathrm{pw}}(\partial\Omega_j), \tag{2.117}$$

to see that

$$(\vec{\mathfrak{u}}, \vec{\mathfrak{v}}) \to [\![\mathbb{X}_{i\to j}\,\mathbb{L}_i\,\vec{\mathfrak{u}}, \mathbb{L}_j\,\vec{\mathfrak{v}}]\!]_{\mathcal{T}(\partial\Omega_j)}$$

becomes a continuous bilinear form on $\mathcal{MT}(\Sigma) \times \widetilde{\mathcal{MT}}(\Sigma)$.

Thus, $\widetilde{\mathcal{MT}}(\Sigma)$ is a suitable *test space* for the weak variational formulation of (2.108), because the resulting bilinear form will be continuous. For the special situation $N = 2$ we arrive at the variational equation: seek $(\mathfrak{u}_1, \mathfrak{u}_2, \mathfrak{u}_3) \in \mathcal{MT}(\Sigma)$ such that for all $(\mathfrak{v}_1, \mathfrak{v}_2, \mathfrak{v}_3) \in \widetilde{\mathcal{MT}}(\Sigma)$

$$\begin{aligned}
&[\![\mathbb{A}_0\,\mathfrak{u}_0, \mathfrak{v}_0]\!]_{\mathcal{T}(\partial\Omega_0)} - \left[\!\!\left[\frac{1}{2}\,\mathbb{X}_{1\to 0}\,\mathfrak{u}_1, \mathfrak{v}_0\right]\!\!\right]_{\mathcal{T}(\partial\Omega_0)} - \left[\!\!\left[\frac{1}{2}\,\mathbb{X}_{2\to 0}\,\mathfrak{u}_2, \mathfrak{v}_0\right]\!\!\right]_{\mathcal{T}(\partial\Omega_0)} = -[\![\mathbb{T}_0\,\mathbf{u}_{\mathrm{inc}}, \mathfrak{v}_0]\!]_{\mathcal{T}(\partial\Omega_0)},\\
&-\left[\!\!\left[\frac{1}{2}\,\mathbb{X}_{0\to 1}\,\mathfrak{u}_0, \mathfrak{v}_1\right]\!\!\right]_{\mathcal{T}(\partial\Omega_1)} + [\![\mathbb{A}_1\,\mathfrak{u}_1, \mathfrak{v}_1]\!]_{\mathcal{T}(\partial\Omega_1)} - \left[\!\!\left[\frac{1}{2}\,\mathbb{X}_{2\to 1}\,\mathfrak{u}_2, \mathfrak{v}_1\right]\!\!\right]_{\mathcal{T}(\partial\Omega_1)} = 0,\\
&-\left[\!\!\left[\frac{1}{2}\,\mathbb{X}_{0\to 2}\,\mathfrak{u}_0, \mathfrak{v}_2\right]\!\!\right]_{\mathcal{T}(\partial\Omega_2)} - \left[\!\!\left[\frac{1}{2}\,\mathbb{X}_{1\to 2}\,\mathfrak{u}_1, \mathfrak{v}_2\right]\!\!\right]_{\mathcal{T}(\partial\Omega_2)} + [\![\mathbb{A}_2\,\mathfrak{u}_2, \mathfrak{v}_2]\!]_{\mathcal{T}(\partial\Omega_2)} = 0.
\end{aligned} \tag{2.118}$$

In a general situation the variational problem for the local MTF reads [32, Problem 6] as follows.

Seek $\vec{\mathfrak{u}} \in \mathcal{MT}(\Sigma)$ that satisfies

$$\langle \mathbb{M}_L'\,\vec{\mathfrak{u}}, \vec{\mathfrak{v}}\rangle = -[\![\mathbb{T}_0\,\mathbf{u}_{\mathrm{inc}}, \mathbb{L}_0\,\vec{\mathfrak{v}}]\!]_{\mathcal{T}(\partial\Omega_0)} \quad \forall \vec{\mathfrak{v}} \in \widetilde{\mathcal{MT}}(\Sigma), \tag{2.119}$$

with a *continuous* operator

$$\mathbb{M}_L' := \sum_{i=0}^{N} \left[\mathbb{L}_i^*\,\mathbb{A}_i\,\mathbb{L}_i - \frac{1}{2}\sum_{\substack{j=0\\ j\neq i}}^{N} \mathbb{L}_i^*\,\mathbb{X}_{j\to i}\,\mathbb{L}_j \right] : \mathcal{MT}(\Sigma) \to \widetilde{\mathcal{MT}}(\Sigma)'.$$

Continuity of $\mathbb{M}_L'$ is an immediate consequence of the L^2-type dualities expressed in (2.116).

Corollary 2.50 (Continuity of local MTF bilinear form, [32, Sect 3.2.4]). *There is a constant $C > 0$ such that*

$$|\langle \mathbb{M}_L'\,\vec{\mathfrak{u}}, \vec{\mathfrak{v}}\rangle| \le C\,\|\vec{\mathfrak{u}}\|_{\mathcal{MT}(\Sigma)}\,\|\vec{\mathfrak{v}}\|_{\widetilde{\mathcal{MT}}(\Sigma)} \quad \forall \vec{\mathfrak{u}} \in \mathcal{MT}(\Sigma),\ \vec{\mathfrak{v}} \in \widetilde{\mathcal{MT}}(\Sigma).$$

6.3 Local MTF: Stability

A result in the theory of Sobolev spaces states that the function space $\mathcal{H}(\mathsf{D}, \Omega)$ as introduced in (2.5) does not support a continuous trace onto the “wire basket,” that is, onto the union of all boundaries of the interfaces. In other words, smooth functions whose supports do not intersect $\bigcup_{i,j} \partial\Gamma_{ij}$ are dense in $\mathcal{H}(\mathsf{D},\Omega)$. This translates into following density result for trace spaces.

Lemma 2.51. *The constrained trace space* $\widetilde{\mathcal{MT}}(\Sigma)$ *is a dense subspace of* $\mathcal{MT}(\Sigma)$.

Thus, sloppily speaking, the space $\widetilde{\mathcal{MT}}(\Sigma)$ provides enough test functions to let the variational equation (2.119) imply the integral equations (2.109). Solutions of the latter are unique and so this remains true for the variational equation.

Theorem 2.52 (Uniqueness of solutions of variational local MTF). *Solutions of the variational problem* (2.119) *are unique.*

The local MTF bilinear form also enjoys surprising coercivity properties.

Theorem 2.53 (Coercivity of local MTF bilinear form, [32, Sect 3.2.7]). *There exist an isomorphism* $\mathbb{F} : \mathcal{MT}(\Sigma) \to \mathcal{MT}(\Sigma)$ *and a compact operator* $\mathbb{K} : \mathcal{MT}(\Sigma) \to \mathcal{MT}(\Sigma)$ *such that for some constant* $C > 0$ *the following (generalized) Gårding inequality holds*

$$\operatorname{Re}\left\{\left\langle \mathbb{M}'_L\,\vec{\mathfrak{u}}, \mathbb{F}\overline{\vec{\mathfrak{u}}}\right\rangle + + [\![\mathbb{K}\,\mathfrak{v},\overline{\mathfrak{v}}]\!]_{\mathcal{T}(\partial\Omega)}\right\} \geq C\,\|\vec{\mathfrak{v}}\|^2_{\mathcal{MT}(\Sigma)} \quad \forall \vec{\mathfrak{u}} \in \widetilde{\mathcal{MT}}(\Sigma). \tag{2.120}$$

Proof. We restrict ourselves to the Helmholtz case, for which $\mathbb{F} = \mathsf{Id}$. Then note that for $\mathfrak{u}_i = \binom{u_i}{\varphi_i} \in \widetilde{\mathcal{T}}(\partial\Omega_i)$ and $\mathfrak{u}_j = \binom{u_j}{\varphi_j} \in \widetilde{\mathcal{T}}(\partial\Omega_j)$

$$[\![\mathbb{X}_{i\to j}\,\mathfrak{u}_i, \overline{\mathfrak{u}}_j]\!]_{\mathcal{T}(\partial\Omega_j)} = \left[u_i, \overline{\varphi}_j\right]_{\Gamma_{ij}} - \left[u_j, \overline{\varphi}_i\right]_{\Gamma_{ij}} = -\overline{[\![\mathbb{X}_{j\to i}\,\mathfrak{u}_j, \overline{\mathfrak{u}}_i]\!]_{\mathcal{T}(\partial\Omega_i)}}.$$

We observe a cancellation of the real parts of "off-diagonal terms" in the local MTF bilinear form when written as in (2.118), which permits us to conclude

$$\operatorname{Re}\left\langle \mathbb{M}'_L\,\vec{\mathfrak{u}}, \overline{\vec{\mathfrak{u}}}\right\rangle = \sum_{j=1}^{N} \operatorname{Re}\,[\![\mathbb{A}_i\,\mathfrak{u}_i, \overline{\mathfrak{u}}_i]\!]_{\mathcal{T}(\partial\Omega_i)},$$

for $\vec{\mathfrak{u}} = (\mathfrak{u}_1, \dots, \mathfrak{u}_N) \in \widetilde{\mathcal{MT}}(\Sigma)$, and the assertion of the theorem is implied by Theorem 2.18. □

At first glance, Theorems 2.52 and 2.53 seem to pave the way for an application of a Fredholm alternative argument as in Section 3.2. A closer scrutiny, however, reveals a fundamental *mismatch* between the norms that support continuity of $\mathbb{M}'_L$, see Corollary 2.50, and the $\mathcal{MT}(\Sigma)$-norm, for which we have coercivity. More sophisticated tools are needed to cope with this mismatch and they are summarized in the following lemma, which is an extension of a results first obtained by J.L. Lions [38, Chapter III, Theorem 1.1], see also [23].

Lemma 2.54 (Lemma 10 in [32]). *Let* H *be a Hilbert space and* Φ *a subspace of* H *(not necessarily closed). Moreover, let* $\mathsf{m} : H \times \Phi \to \mathbb{C}$ *and* $\mathsf{k} : H \times H \to \mathbb{C}$ *be bilinear forms satisfying the following properties:*

(i) *For every* $\varphi \in \Phi$, *the linear form* $u \mapsto \mathsf{m}(u, \varphi)$ *is continuous in* H.

(ii) *The linear operator* $\mathsf{K} : H \to H'$ *associated to the bilinear form* $\mathsf{k}(\cdot,\cdot)$ *is compact and continuous.*

(iii) *There exists* $\alpha > 0$ *such that*

$$\operatorname{Re}\{\mathsf{m}(\varphi,\overline{\varphi}) + \mathsf{k}(\varphi,\overline{\varphi})\} \geq \alpha \, \|\varphi\|_H^2 , \quad \forall\, \varphi \in \Phi. \tag{2.121}$$

(iv) *The bilinear form* $u \mapsto \mathsf{m}(u,\varphi)$ *is injective, i.e.* $\mathsf{m}(u,\varphi) = 0$ *for all* $\varphi \in \Phi$, *implies* $u = 0$.

Then, for $l \in H'$ *there exists* $u_0 \in H$ *solution of*

$$\mathsf{m}(u,\varphi) = \langle l, \varphi \rangle , \quad \forall\, \varphi \in \Phi, \tag{2.122}$$

satisfying the stability estimate

$$\|u_0\|_H \leq \frac{C_{\mathsf{m}}}{\alpha} \, \|l\|_{H'} . \tag{2.123}$$

where $C_{\mathsf{m}} > 0$ *is independent of* l.

We apply this abstract lemma with

- the spaces $H := \mathcal{MT}(\Sigma)$, $\Phi := \widetilde{\mathcal{MT}}(\Sigma) \subset \mathcal{MT}(\Sigma)$,
- the bilinear form m given by $(\vec{\mathfrak{u}}, \vec{\mathfrak{v}}) \to \langle \mathbb{M}_L' \vec{\mathfrak{u}}, \vec{\mathfrak{v}} \rangle$,
- the operator $\mathbb{K}$ from Theorem 2.53 providing K,
- (2.120) in place of the estimate (2.121),
- Theorem 2.52 guaranteeing injectivity of m.

This gives existence of solutions of the local MTF variational problem (2.119). In addition, Theorem 2.52 ensures uniqueness, which permits us to conclude the following main result, which amounts to the well-posedness of the local MTF variational problem.

Theorem 2.55. *The local MTF operator* $\mathbb{M}_L' : \mathcal{MT}(\Sigma) \to \widetilde{\mathcal{MT}}(\Sigma)'$ *possesses a continuous inverse* $(\mathbb{M}_L')^{-1} : \mathcal{MT}(\Sigma)' \to \mathcal{MT}(\Sigma)$.

6.4 Boundary element Galerkin discretization

On an algorithmic level the boundary element Galerkin discretization of the local MTF variational formulation (2.119) largely runs parallel to that of the global MTF explained in Section 3.3. Both trial and test space in (2.119) are replaced with the piecewise polynomial boundary element spaces introduced in Section 3.3, that is, we rely on

$$\mathcal{MT}_h(\Sigma) := \mathcal{T}_h(\partial\Omega_0) \times \cdots \times \mathcal{T}_h(\partial\Omega_N). \tag{2.124}$$

Note the difference with the multitrace boundary element space $\widehat{\mathcal{MT}}_h(\Sigma)$ from (2.101) used for the global MTF: now $\partial\Omega_0$ also bears two discrete Cauchy traces.

Also be aware that $\mathcal{MT}_h(\Sigma) \not\subset \widetilde{\mathcal{MT}}(\Sigma)$, because $\mathcal{T}_{D,h}(\partial\Omega_i) \subset \widetilde{\mathcal{T}}_D(\partial\Omega_i)$ would entail setting all basis functions associated with vertices or edges on boundaries of interfaces Γ_{ij} to zero. Nevertheless, we can just go ahead and plug functions in $\mathcal{MT}_h(\Sigma)$ into the variational equations (2.119), since they are all square integrable and the terms involving the transmission operators remain well defined. In fact, all we have to integrate are products of piecewise polynomials.

Eventually, we arrive at a linear system of equations, whose system matrix is of size $\sum_{i=0}^{N} m_i$, $m_i := \dim \mathcal{T}_{D,h}(\partial\Omega_i) + \dim \mathcal{T}_{N,h}(\partial\Omega_i)$, and has the form

$$\begin{pmatrix} \mathbf{A}_1 & \mathbf{X}_{21} & \cdots & & \cdots & \mathbf{X}_{N1} \\ \mathbf{X}_{12} & \mathbf{A}_2 & \mathbf{X}_{32} & & & \vdots \\ \vdots & \ddots & \ddots & \ddots & & \\ & & \ddots & \ddots & \ddots & \vdots \\ \vdots & & & \mathbf{X}_{N-2,N-1} & \mathbf{A}_{N-1} & \mathbf{X}_{N,N-1} \\ \mathbf{X}_{1N} & \cdots & & \cdots & \mathbf{X}_{N-1,1} & \mathbf{A}_N \end{pmatrix}, \tag{2.125}$$

where the matrices $\mathbf{A}_i \in \mathbb{C}^{m_i,m_i}$, $i = 0,\ldots,N$, arise from standard Galerkin BEM discretization of the boundary integral operators $\mathbb{A}_i$ as defined in (2.26). The matrices $\mathbf{X}_{ij} \in \mathbb{R}^{m_j,m_i}$ are *sparse* and $\mathbf{X}_{ij} = 0$, if Ω_i and Ω_j have no common genuine interface.

Remark 2.56. Contrasting (2.125) with (2.100) reveals a substantial advantage of the local MTF compared with the global version: since the global MTF relies on nonlocal boundary integral operators to take into account the coupling between subdomains, its final system matrix will feature fully populated off-diagonal blocks throughout. Even though they will allow efficient compression, handling them requires much more effort than dealing with the sparse $\mathbf{X}_{ij}$ blocks in (2.125). △

The downside of the local MTF is its failure to fit natural trace spaces as discussed at length in Section 6.2. This also compounds difficulties for the a priori convergence analysis, which is still incomplete. A preliminary result is given in [32, Section 4] for the case of the Helmholtz transmission problem in 2D only.

Theorem 2.57 (Convergence of Galerkin discretization of local MTF, [32, Theorem 13]). *Consider the Galerkin discretization of* (2.119) *in the Helmholtz case with the lowest order piecewise polynomial (piecewise linear continuous/piecewise constant, cf. Concretization* 2.30*) boundary elements.*

If $\mathbb{M}_L'$ preserves the regularity of Cauchy traces (see [32, Conjecture 1] for a precise statement), then the bilinear form of the local MTF variational formulation satisfies a *discrete inf-sup condition uniform in the meshwidth, provided that the resolution of the boundary element spaces is large enough.*

Operator preconditioning of the discretized local MTF is a straightforward application of the recipes outlined in Section 4 and, in particular, Section 4.4. It naturally relies on the self-duality of $\boldsymbol{\mathcal{MT}}(\Sigma)$ with respect to the L^2-type pairing $[\![\cdot,\cdot]\!]_{\boldsymbol{\mathcal{MT}}(\Sigma)}$ from (2.32). As explained in Section 4.4, uniformly stable (with respect to the standard trace norm of $\boldsymbol{\mathcal{MT}}(\Sigma)$) discrete duality pairings are available for $\boldsymbol{\mathcal{MT}}_h(\Sigma)$ according to (2.124). The role of the operator B can be played by

$$\mathsf{B} \sim \sum_{i=0}^{N} \mathbb{L}_i^* \, \mathbb{A}_i \, \mathbb{L}_i : \boldsymbol{\mathcal{MT}}(\Sigma) \to \boldsymbol{\mathcal{MT}}(\Sigma)', \tag{2.126}$$

which amounts to the operator of the STF, see (2.48), but now considered on the fully decoupled multitrace space $\boldsymbol{\mathcal{MT}}(\Sigma)$, recall Remark 2.48.

Yet, operator preconditioning is also haunted by the nonstandard function space framework needed for local MTF; since A in (2.84) is the local MTF operator $\mathbb{M}_L'$ it is actually unbounded on $\boldsymbol{\mathcal{MT}}(\Sigma)$. Thus A has to be replaced with $\mathsf{A}_h : V_h \to V_h'$ in (2.84) and we will inevitably incur a blow-up of $\|\mathsf{A}_h\|$ as the meshwidth $h \to 0$. Inverse inequalities linking the $\mathcal{T}(\partial\Omega_i)$-norm and $\widetilde{\mathcal{T}}(\partial\Omega_i)$-norm of boundary element functions allow to quantify this blow-up. In the case of $H^{\frac{1}{2}}$-conforming boundary elements such estimates are known, see [40] and the so-called edge and face lemmas in the theory of domain decomposition methods, see [56, Section 4.2.4] and [53, Section 4.6]. They suggest a behavior like

$$\|\mathsf{A}_h\| \approx O(|\log h|^\alpha), \quad \alpha \text{ a small integer.} \tag{2.127}$$

Hence, a slight dependence of the performance of the preconditioner on the meshwidth can be expected, but will be very moderate and hardly noticeable in numerical computations. This expectation is bolstered by 2D numerical results reported in [32, Section 5].

References

[1] John M. Ball, Yves Capdeboscq, and Basang Tsering Xiao, On uniqueness for time harmonic anisotropic Maxwell's equations with piecewise regular coefficients, *Math. Models Meth. Appl. Sci.* (2012), arXiv:1201.2006v1 [math.AP].

[2] M. Bebendorf, Approximation of boundary element matrices, *Numer. Math.* 86 (2000), 565–589.

[3] M. Bebendorf, *Hierarchical Matrices: A Means to Efficiently Solve Elliptic Boundary Value Problems*, Lecture Notes in Computational Science and Engineering (LNCSE) 63, Springer-Verlag, 2008.

[4] B. Beckermann, S. A. Goreinov, and E. E. Tyrtyshnikov, Some remarks on the Elman estimate for GMRES, *SIAM J. Matrix Anal. Appl.* 27 (2005), 772–778.

[5] A. Bendali, Numerical analysis of the exterior boundary value problem for time harmonic Maxwell equations by a boundary finite element method. Part 1: The continuous problem, *Math. Comp.* 43 (1984), 29–46.

[6] A. Bendali, Numerical analysis of the exterior boundary value problem for time harmonic Maxwell equations by a boundary finite element method. Part 2: The discrete problem, *Math. Comp.* 43 (1984), 47–68.

[7] A. Buffa, *Traces theorems on non-smooth boundaries for functional spaces related to Maxwell equations: An overwiew*, Computational Electromagnetics (C. Carstensen, S. Funken, W. Hackbusch, R. H. W. Hoppe, and P. Monk, eds.), Lecture Notes in Computational Science and Engineering 28, Springer, Berlin, 2003, pp. 23–34.

[8] A. Buffa, Remarks on the discretization of some non-positive operators with application to heterogeneous Maxwell problems, *SIAM J. Numer. Anal.* 43 (2005), 1–18.

[9] A. Buffa and S. H. Christiansen, A dual finite element complex on the barycentric refinement, *Math. Comp.* 76 (2007), 1743–1769.

[10] A. Buffa and P. Ciarlet, On traces for functional spaces related to Maxwell's Equations. Part I: An integration by parts formula in Lipschitz polyhedra., *Math. Meth. Appl. Sci.* 24 (2001), 9–30.

[11] A. Buffa, M. Costabel, and D. Sheen, On traces for $\mathbf{H}(\mathbf{curl}, \Omega)$ in Lipschitz domains, *J. Math. Anal. Appl.* 276 (2002), 845–867.

[12] A. Buffa and R. Hiptmair, *Galerkin Boundary Element Methods for Electromagnetic Scattering*, Topics in Computational Wave Propagation. Direct and inverse Problems (M. Ainsworth, P. Davis, D. Duncan, P. Martin, and B. Rynne, eds.), Lecture Notes in Computational Science and Engineering 31, Springer, Berlin, 2003, pp. 83–124.

[13] S. L. Campbell, I. C. F. Ipsen, C. T. Kelley, and C. D. Meyer, GMRES and the minimal polynomial, *BIT Numerical Mathematics* 36 (1996), 664–675.

[14] Y. Chang and R. F. Harrington, A surface formulation for characteristic modes of material bodies, *IEEE Trans. Antennas and Propagation* 25 (1977), 789–795.

[15] Hongwei Cheng, William Y. Crutchfield, Zydrunas Gimbutas, Leslie F. Greengard, J. Frank Ethridge, Jingfang Huang, Vladimir Rokhlin, Norman Yarvin, and Junsheng Zhao, A wideband fast multipole method for the Helmholtz equation in three dimensions, *Journal of Computational Physics* 216 (2006), 300 – 325.

[16] S. H. Christiansen and J.-C. Nédélec, Des préconditionneurs pour la résolution numérique des équations intégrales de frontiére de l'acoustique, *C.R. Acad. Sci. Paris, Ser. I Math* 330 (2000), 617–622.

[17] S. H. Christiansen and J.-C. Nédélec, A preconditioner for the electric field integral equation based on Calderón formulas, *SIAM J. Numer. Anal.* 40 (2002), 1100–1135.

[18] X. Claeys and R. Hiptmair, Electromagnetic scattering at composite objects: A novel multi-trace boundary integral formulation, *Math. Modelling Numer. Anal.* 46 (2012), 1421–1445.

[19] X. Claeys and R. Hiptmair, Boundary integral formulation of the first kind for acoustic scattering by composite structures, *Comm. Pure Applied Math.* (2013), To appear.

[20] D. Colton and R. Kress, *Inverse Acoustic and Electromagnetic Scattering Theory*, 2nd ed, Applied Mathematical Sciences 93, Springer, Heidelberg, 1998.

[21] E. Darve and P. Have, Efficient fast multipole method for low-frequency scattering, *J. Comp. Phys.* 197 (2004), 341–363.

[22] C. Epstein and L. Greengard, Debye sources and the numerical solution of the time-harmonic Maxwell equations, *Comm. Pure Appl. Math.* 63 (2010), 413–463.

[23] F. Frühauf, B. Gebauer, and O. Scherzer, Detecting interfaces in a parabolic-elliptic problem from surface measurements, *SIAM J. Numer. Anal.* 45 (2007), 810–836.

[24] M. Ganesh, S. C. Hawkins, and D. Volkov, *An All-Frequency Weakly-Singular Surface Integral Equation For Electromagnetism In Dielectric Media: Reformulation And Well-Posedness Analysis*, Department of Applied Mathematics and Statistics, Colorado School of Mines, Preprint, Golden, CO, 2012.

[25] L. Greengard, J.-F. Huang, V. Rokhlin, and S. Wandzura, Accelerating fast multipole methods for the Helmholtz equation at low frequencies, *IEEE Computational Science and Engineering* 5 (1998), 32–38.

[26] W. Hackbusch, *Hierarchische Matrizen. Algorithmen und Analysis*, Springer, Heidelberg, 2009.

[27] R. F. Harrington, Boundary integral formulations for homogeneous material bodies, *J. Electromagnetic Waves and Applications* 3 (1989), 1–15.

[28] C. Hazard and M. Lenoir, On the solution of time-harmonic scattering problems for Maxwell's equations, *SIAM J. Math. Anal.* 27 (1996), 1597–1630.

[29] R. Hiptmair, Finite elements in computational electromagnetism, *Acta Numerica* 11 (2002), 237–339.

[30] R. Hiptmair, Coupling of finite elements and boundary elements in electromagnetic scattering, *SIAM J. Numer. Anal.* 41 (2003), 919–944.

[31] R. Hiptmair, Operator preconditioning, *Computers and Mathematics with Applications* 52 (2006), 699–706.

[32] R. Hiptmair and C. Jerez-Hanckes, Multiple traces boundary integral formulation for Helmholtz transmission problems, *Adv. Appl. Math.* 37 (2012), 39–91.

[33] R. Hiptmair, C. Jerez-Hanckes, J.-F. Lee, and Z. Peng, *Domain Decomposition for Boundary Integral Equations via Local Multi-Trace Formulations*, SAM, ETH Zürich, Report no. 2013–08, 2013, Submitted to Proceedings of XXI International Conference on Domain Decomposition Methods, Rennes, France, June 25–29, 2012.

[34] George C. Hsiao and Wolfgang L. Wendland, *Boundary integral equations*, Applied Mathematical Sciences 164, Springer-Verlag, Berlin, 2008.

[35] L. Kielhorn, *BETL - A generic boundary element template library*, SAM, ETH Zurich, Report, Zürich, Switzerland, 2012.

[36] Stefan Kurz and Bernhard Auchmann, *Differential Forms and Boundary Integral Equations for Maxwell-Type Problems*, Fast Boundary Element Methods in Engineering and Industrial Applications (Ulrich Langer, Martin Schanz, Olaf Steinbach, and Wolfgang L. Wendland, eds.), Lecture Notes in Applied and Computational Mechanics 63, Springer Berlin / Heidelberg, 2012, pp. 1–62.

[37] U. Langer and O. Steinbach, Boundary Element Tearing and Interconnecting Methods, *Computing* 71 (2003), 205–228.

[38] J.-L. Lions, *Équations différentielles opérationnelles et problèmes aux limites*, Die Grundlehren der mathematischen Wissenschaften, Bd. 111, Springer-Verlag, Berlin, 1961.

[39] W. McLean, *Strongly Elliptic Systems and Boundary Integral Equations*, Cambridge University Press, Cambridge, UK, 2000.

[40] W. McLean and O. Steinbach, Boundary element preconditioners for a hypersingular integral equations on an interval, *Adv. Comp. Math.* 11 (1999), 271–286.

[41] J.-C. Nédélec, *Acoustic and Electromagnetic Equations: Integral Representations for Harmonic Problems*, Applied Mathematical Sciences 44, Springer-Verlag, Berlin, 2001.

[42] Z. Peng, K.-H. Lim, and J.-F. Lee, Nonconformal Domain Decomposition Methods for Solving Large Multiscale Electromagnetic Scattering Problems, *Proceedings of the IEEE* electronic (2012), 1 –22.

[43] Z. Peng, X.-C. Wang, and J.-F. Lee, Integral equation domain decomposition method for solving electromagnetic wave scattering from non-penetrable objects, in: *Electromagnetics in Advanced Applications (ICEAA), 2010 International Conference on*, pp. 821 –824, sept. 2010.

[44] Z. Peng, X.-C. Wang, and J.-F. Lee, Integral Equation Based Domain Decomposition Method for Solving Electromagnetic Wave Scattering From Non-Penetrable Objects, *IEEE Trans. Antennas and Propagation* 59 (2011), 3328–3338.

[45] A. J. Poggio and E. K. Miller, *Integral equation solution of three-dimensional scattering problems*, Computer Techniques for Electromagnetics (R. Mittra, ed.), Pergamon, New York, 1973, pp. 159–263.

[46] S. M. Rao, D. R. Wilton, and A. W. Glisson, Electromagnetic scattering by surfaces of arbitrary shape, *IEEE Trans. Antennas and Propagation* 30 (1986), 409–418.

[47] P. A. Raviart and J. M. Thomas, *A Mixed Finite Element Method for Second Order Elliptic Problems*, Springer Lecture Notes in Mathematics 606, pp. 292–315, Springer, Ney York, 1977.

[48] S. Sauter and C. Schwab, *Boundary Element Methods*, Springer Series in Computational Mathematics 39, Springer, Heidelberg, 2010.

[49] A. H. Schatz, An observation concerning Ritz-Galerkin methods with indefinite bilinear forms, *Math. Comp.* 28 (1974), 959–962.

[50] Y. Shao, Z. Peng, K.-H. Lim, and J.-F. Lee, Non-conformal domain decomposition methods for time-harmonic Maxwell equations, *Proc. R. Soc. A* 468 (2012), 2433–2460, doi: 10.1098/rspa.2012.0028.

[51] O. Steinbach and W. Wendland, The construction of some efficient preconditioners in the boundary element method, *Adv. Comput. Math.* 9 (1998), 191–216.

[52] O. Steinbach and M. Windisch, Stable boundary element domain decomposition methods for the Helmholtz equation, *Numer. Math.* 118 (2011), 171–195.

[53] Andrea Toselli and Olof Widlund, *Domain decomposition methods – algorithms and theory*, Springer Series in Computational Mathematics 34, Springer-Verlag, Berlin, 2005.

[54] T. von Petersdorff, Boundary integral equations for mixed Dirichlet, Neumann and transmission problems, *Math. Meth. Appl. Sci.* 11 (1989), 185–213.

[55] T.-K. Wu and L.-L. Tsai, Scattering from arbitrarily-shaped lossy dielectric bodies of revolution, *Radio Science* 12 (1977), 709–718.

[56] J. Xu and J. Zou, Some non-overlapping domain decomposition methods, *SIAM Review* 40 (1998), 857–914.

Johannes Elschner and Guanghui Hu

Direct and Inverse Elastic Scattering Problems for Diffraction Gratings

Abstract: This chapter is concerned with the direct and inverse scattering of time-harmonic plane elastic waves by unbounded periodic structures (diffraction gratings). We present a variational approach to forward scattering problems with Lipschitz grating profiles and give a survey of recent uniqueness and existence results. We also report on recent global uniqueness results within the class of piecewise linear grating profiles for the corresponding inverse elastic scattering problems. Moreover, a discrete Galerkin method is presented to efficiently approximate solutions of direct scattering problems via an integral equation approach. Finally, an optimization method for solving the inverse problem of recovering a 2D periodic structure from scattered elastic waves measured above the structure is discussed.

Keywords: diffraction gratings, elastic waves, variational formulation, inverse elastic scattering, uniqueness, reconstruction method

2010 Mathematics Subject Classification: 74J20, 74J25, 74B05, 35B27, 35R30, 65J22, 65R20

Johannes Elschner: Weierstrass Institute for Applied Analysis and Stochastics, Mohrenstr. 39, 10117 Berlin, Germany, johannes.elschner@wias-berlin.de
Guanghui Hu: Weierstrass Institute for Applied Analysis and Stochastics, Mohrenstr. 39, 10117 Berlin, Germany, guanghui.hu@wias-berlin.de

1 Introduction

Diffraction gratings are widely used in many areas of science and technology and have been investigated since more than 100 years [62]. In recent years, the interest in these areas has grown immensely because of many industrial applications, for example, in radar imaging, nondestructive testing, micro-optics, or solar energy absorption. In particular, the scattering of acoustic and electromagnetic waves in periodic structures has been studied extensively concerning theoretical analysis and numerical approximation, using integral equation methods [32, 64, 67, 68] or variational methods [19–21, 30, 33, 42, 54, 70]. We refer to the monographs [72] for a comprehensive mathematical study and to [15] for the details of the applications of electromagnetic diffraction gratings. However, there have been only a few papers studying the

---

This work is supported by the German Research Foundation (DFG) under Grant No. EL 584/1–2.

scattering of elastic waves by unbounded surfaces. The aim of this survey chapter is to report on recent progress in the analysis and numerics of direct and inverse elastic scattering problems for diffraction gratings.

The relevant phenomena for elastic waves have a wide field of application, particularly in geophysics, seismology and nondestructive testing. For instance, the problem of elastic pulse transmission and reflection through the earth is fundamental to the investigation of earthquakes and the utility of seismic waves in search for oil and ore bodies [2, 48, 49, 71]. Moreover, identifying parallel vertical fractures (which can be modeled as periodic structures) in sedimentary rocks can have significant impact on the production of underground gas and liquids by employing controlled explosions [65]. Inverse elastic scattering problems also arise from detecting cracks and flaws in concrete structures, such as bridges, buildings, dams, highways and so on; see [69, 73] for applications in nondestructive testing.

Compared to acoustic and electromagnetic scattering, the elasticity problem is more complicated due to the coexistence of compressional and shear waves that propagate at different speeds. These two waves are coupled at interfaces where boundary conditions depending on the elastic medium are imposed. The first attempt to rigorously prove solvability of direct elastic scattering problems for unbounded surfaces is due to Arens; see [7, 8] for two-dimensional diffraction gratings and [9, 10] for the more general case of rough surfaces. In [7], existence and uniqueness of quasiperiodic solutions to the Dirichlet problem was established in the case that the grating profile is given by the graph of a smooth periodic function. The existence proof is based on the boundary integral equation method where the solution is sought as the superposition of single and double layer potentials. We also refer to [29] for earlier studies on the scattering of elastic waves in a locally perturbed isotropic half-space with a free boundary using the limiting absorption principle.

In this chapter, we focus on the variational approach, which reduces the forward scattering problem to an equivalent variational formulation in a truncated periodic cell involving a nonlocal boundary (Dirichlet-to-Neumann) operator; see Section 3. We refer to [1, 21, 54] for the variational approach applied to electromagnetic diffraction gratings. This approach appears to be well adapted to the analytical and numerical treatment of rather general two-dimensional and three-dimensional periodic structures involving nonsmooth interfaces and transmission conditions. The variational method can also be applied to elastic scattering by rough surfaces; see [40] for recent existence and uniqueness results in the case of the Dirichlet problem.

The inverse problem of recovering a grating profile from scattered elastic fields is of great practical importance. In contrast to the widespread belief of the uniqueness in bounded obstacle scattering problems with only one incident plane wave, we cannot expect the same global uniqueness in wave diffraction by periodic structures (see, e.g. [41] and [45]). This is mainly due to the Rayleigh expansion radiation condition (RERC) imposed on the scattered field, which involves several propagating modes without decay as $x_2 \to \infty$. In general it is difficult to characterize the exceptional grating profiles that cannot be identified by one incident plane wave. However, if the grating profiles are piecewise linear, by the reflection principles for the Helmholtz and Maxwell equations, global uniqueness by finitely many incident waves can be proved for the inverse scattering of time-harmonic electromagnetic waves, including TE and TM polarization [16, 17, 35]. In particular, one can classify and characterize all unidentifiable sets of grating profiles corresponding to one incident plane wave. In Section 4 we review corresponding global uniqueness results in elasticity within the class of polygonal grating profiles under the boundary conditions of the third or fourth kind, which are based on the reflection principle for the Navier system developed in [47]. It is a challenging open problem to extend these results to the case of the boundary conditions of the first or second kind.

The numerical solution of the direct elastic scattering problem under the Dirichlet boundary condition is addressed in Section 5.1 where we present a discrete Galerkin method for solving an equivalent first-kind integral equation. From the numerical view point, the implementation of this method is easier than the integral equation method with a second-kind integral equation that involves the computation of the stress operator on the profile.

The numerical treatment of the corresponding inverse problem is challenging since it is nonlinear and severely ill-posed. There is already a vast literature on the reconstruction of a perfectly conducting profile for the two-dimensional Helmholtz equation. Here we mention a conjugate gradient algorithm based on analytic continuation [53], an iterative regularization method [51], a factorization method [11, 12], and an optimization method [22–24] that follows an approach first developed by Kirsch and Kress for acoustic obstacle scattering (see [27, Chapter 5] and the references therein). In Section 5.2, the two-step algorithm of [22] for the reconstruction of one-dimensional electromagnetic grating profiles is extended to the more complicated case of elastic scattering. The first step of this method is to reconstruct the scattered field from near-field measurements by solving a first-kind integral equation. This step is the linear severely ill-posed part and requires Tikhonov regularization involving the singular value decomposition of the integral operator. The second step is to approximate the grating profile by solving a finite-dimensional least squares problem, which is nonlinear but well-posed.

2 Mathematical formulation of direct and inverse scattering problems

We assume that a periodic surface divides the three-dimensional space into two non-locally perturbed half-spaces filled with homogeneous and isotropic elastic media. Unless otherwise stated, this surface is always assumed to be invariant in the x_3-

direction, and its cross-section in the (x_1, x_2)-plane is to be represented by a curve Λ which is 2π-periodic in x_1. We suppose further that all elastic waves propagate perpendicular to the x_3-axis, so that the problem can be treated as a problem of plane elasticity.

Denote by Ω_Λ the region above the grating in $\mathbb{R}^2$. Suppose that a time-harmonic plane elastic wave with the incident angle $\theta \in (-\pi/2, \pi/2)$ is incident on Λ from Ω_Λ. The incident wave is allowed to be either an incident pressure wave taking the form

$$u^{in} = u_p^{in}(x) = \hat{\theta} \exp\left(ik_p x \cdot \hat{\theta}\right) \quad \text{with} \quad \hat{\theta} := (\sin\theta, -\cos\theta)^\top, \tag{3.1}$$

or an incident shear wave taking the form

$$u^{in} = u_s^{in}(x) = \hat{\theta}^\perp \exp\left(ik_s x \cdot \hat{\theta}\right) \quad \text{with} \quad \hat{\theta}^\perp := (\cos\theta, \sin\theta)^\top, \tag{3.2}$$

where

$$k_p := \omega/\sqrt{2\mu + \lambda}, \quad k_s := \omega/\sqrt{\mu}$$

are the compressional and shear wave numbers, respectively, λ and μ are the Lamé constants satisfying $\mu > 0$ and $\lambda + \mu > 0$, $\omega > 0$ denotes the angular frequency of the harmonic motion and the symbol $(\cdot)^\top$ indicates the transpose of a vector in $\mathbb{R}^2$. For simplicity we assume the mass density of the elastic medium is equal to one, so that the total displacement $u(x_1, x_2)$, which can be decomposed as the sum of the incident field u^{in} and the scattered field u^{sc}, satisfies the Navier equation (or system):

$$(\Delta^* + \omega^2)u = 0 \quad \text{in} \quad \Omega_\Lambda, \quad \Delta^* := \mu\Delta + (\lambda + \mu)\,\operatorname{grad}\operatorname{div}. \tag{3.3}$$

We first assume that the grating is impenetrable, and that one of the following boundary conditions is imposed on Λ.

First-kind boundary condition:	$u = 0,$	(3.4)
Second-kind boundary condition:	$Tu = 0,$	(3.5)
Third-kind boundary conditions:	$\mathbf{n} \cdot u = 0,\ \tau \cdot Tu = 0,$	(3.6)
Fourth-kind boundary conditions:	$\tau \cdot u = 0,\ \mathbf{n} \cdot Tu = 0,$	(3.7)

where Tu stands for the *stress vector* or *traction* having the form

$$Tu = 2\mu\,\partial_{\mathbf{n}} u + \lambda\,\mathbf{n}\,\operatorname{div} u + \mu\tau(\partial_2 u_1 - \partial_1 u_2), \tag{3.8}$$

with the exterior unit normal $\mathbf{n} = (n_1, n_2)^\top$ and the unit tangential vector $\tau = (-n_2, n_1)^\top$ on Λ. Here and in the following the notation $\partial_j v = \frac{\partial v}{\partial x_j}$ is used. Note that the boundary condition of the first (second) kind is also referred to as the Dirichlet (Neumann) boundary condition, and that vanishing normal displacement and tangential stress (normal stress and tangential displacement) correspond to the third (fourth)-kind boundary conditions. We refer to the monograph [57] for a comprehensive treatment of the boundary value problems of elasticity (including the boundary conditions of the third and fourth kind).

The periodicity of the structure together with the form of the incident waves implies that the solution u must be α-quasiperiodic, i.e.

$$u(x_1 + 2\pi, x_2) = \exp(2i\alpha\pi)\, u(x_1, x_2), \quad (x_1, x_2) \in \Omega_\Lambda, \tag{3.9}$$

where $\alpha := k_p \sin\theta$ is the incident pressure wave (3.1), and $\alpha := k_s \sin\theta$ the incident shear wave (3.2). To ensure well-posedness of the boundary value problem (3.3)–(3.9), a radiation condition must be imposed as $x_2 \to +\infty$. Observing that the scattered field u^{sc} also satisfies the Navier equation (3.3), we can decompose it into the compressional and shear parts

$$u^{sc} = \frac{1}{i}\,(\operatorname{grad}\varphi + \overrightarrow{\operatorname{curl}}\psi) \quad \text{with} \quad \varphi := -\frac{i}{k_p^2}\operatorname{div} u^{sc}, \quad \psi := \frac{i}{k_s^2}\operatorname{curl} u^{sc},$$

where the two curl operators in $\mathbb{R}^2$ are defined by

$$\operatorname{curl} u := \partial_1 u_2 - \partial_2 u_1, \quad u = (u_1, u_2)^\top \quad \text{and} \quad \overrightarrow{\operatorname{curl}} v := (\partial_2 v, -\partial_1 v)^\top,$$

and the scalar functions φ, ψ satisfy the homogeneous Helmholtz equations

$$(\Delta + k_p^2)\,\varphi = 0 \quad \text{and} \quad (\Delta + k_s^2)\,\psi = 0 \quad \text{in} \quad \Omega_\Lambda.$$

Applying the Rayleigh expansion for the scalar Helmholtz equation (see e.g. [54]) to φ and ψ, respectively, we finally obtain a corresponding expansion of u^{sc} into outgoing plane elastic waves,

$$\begin{aligned} u^{sc}(x) = &\sum_{n\in\mathbb{Z}} A_{p,n}(\alpha_n, \beta_n)^\top \exp(i\alpha_n x_1 + i\beta_n x_2) \\ &+ \sum_{n\in\mathbb{Z}} A_{s,n}(\gamma_n, -\alpha_n)^\top \exp(i\alpha_n x_1 + i\gamma_n x_2), \end{aligned} \tag{3.10}$$

for $x_2 > \Lambda^+ := \max_{(x_1,x_2)\in\Lambda} x_2$, where the constants $A_{p,n}, A_{s,n} \in \mathbb{C}$ are called the Rayleigh coefficients and

$$\alpha_n := \alpha + n, \quad \beta_n = \beta_n(\theta) := \begin{cases} \sqrt{k_p^2 - \alpha_n^2} & \text{if } |\alpha_n| \le k_p, \\ i\sqrt{\alpha_n^2 - k_p^2} & \text{if } |\alpha_n| > k_p. \end{cases} \tag{3.11}$$

The parameter $\gamma_n := \gamma_n(\theta)$ is defined analogously as β_n with k_p replaced by k_s. The expansion (3.10) is also called the RERC (see, e.g. [6–8]). We refer to [54] for the RERC of the scalar Helmholtz equation and to [1, 15, 25] in the case of Maxwell's equations in periodic structures.

Since β_n and γ_n in (3.10) are real for at most a finite number of indices n, only a finite number of plane waves in (3.10) propagate into the far field, with the remaining evanescent waves (or surface waves) decaying exponentially as $x_2 \to +\infty$. The above expansion converges uniformly with all derivatives in the half-plane $\{x \in \mathbb{R}^2 : x_2 \geq b\}$ for any $b > \Lambda^+$. Now, our direct diffraction problem can be formulated as the following boundary value problem.

(DP): Given a grating profile curve $\Lambda \subset \mathbb{R}^2$ (which is 2π-periodic in x_1) and an incident field u^{in} of the form (3.1) or (3.2), find a vector function $u = u(x;\theta) = u^{in} + u^{sc} \in H^1_{loc}(\Omega_\Lambda)^2$ that satisfies the Navier equation (3.3), one of the boundary conditions in (3.4)–(3.7), the quasiperiodicity (3.9) and the RERC (3.10).

Let $u(x;\theta_j)$ denote solutions of (DP) corresponding to N incident pressure or shear waves u^{in} of the form (3.1) or (3.2) with distinct incident angles $\theta_j \in (-\frac{\pi}{2}, \frac{\pi}{2})$ $(j = 1,2,\ldots,N)$. The inverse problem which involves near-field measurements $u(x_1, b)$ for some fixed $b > \Lambda^+$ can be formulated as follows.

(IP): Given N incident angles θ_j, determine the grating profile Λ from the knowledge of the near-field data $u(x_1, b; \theta_j)$ for all $x_1 \in (0, 2\pi)$, $j = 1,2,\ldots,N$.

As mentioned earlier, only a finite number of propagating modes of the compressional and shear parts can be measured far away from the grating surface. From the practical point of view, it is quite natural to reconstruct the unknown grating profile from the far-field data $u_b^\infty(x_1)$ of $u^{sc}(x)$ defined by

$$u_b^\infty(x_1) = \sum_{n \in \mathcal{U}_p} A_{p,n}(\alpha_n, \beta_n)^\top \exp(i\alpha_n x_1 + i\beta_n b) + \sum_{n \in \mathcal{U}_s} A_{s,n}(\gamma_n, -\alpha_n)^\top \exp(i\alpha_n x_1 + i\gamma_n b),$$

where the sets $\mathcal{U}_p$ and $\mathcal{U}_s$ are defined by

$$\mathcal{U}_p = \{n \in \mathbb{Z} : |\alpha_n| \leq k_p\}, \quad \mathcal{U}_s = \{n \in \mathbb{Z} : |\alpha_n| \leq k_s\}.$$

Therefore we also consider the following inverse problem:

(IP):* Determine the grating profile from the far-field data $u_b^\infty(x_1;\theta_j)$, $x_1 \in (0, 2\pi)$, for N incident pressure or shear waves u^{in} with distinct incident angles θ_j, $j = 1,2,\ldots,N$.

Of course, from these data one can only hope to reconstruct a finite number of parameters of the profile (see Section 5.2).

3 Solvability results for direct scattering problems: variational method

3.1 An equivalent variational formulation and its Fredholm property

Following the approach of [54] in the case of the scalar Helmholtz equation, we propose an equivalent variational formulation of the boundary value problem (DP), which is posed in a bounded periodic cell in $\mathbb{R}^2$ and is enforcing the radiation condition. Introduce an artificial boundary

$$\Gamma_b := \{(x_1, b) : 0 \le x_1 \le 2\pi\}, \quad b > \Lambda^+ = \max_{x \in \Lambda} x_2,$$

and the bounded domain

$$\Omega_b := \{(x_1, x_2) \in \Omega_\Lambda : 0 < x_1 < 2\pi,\ x_2 < b\},$$

lying between the segment Γ_b and one period of the grating profile curve which is denoted by Λ again. We assume that Λ is a Lipschitz curve, so that Ω_b is a bounded Lipschitz domain.

Let $H^1_\alpha(\Omega_b)$ denote the subspace of the Sobolev space $H^1(\Omega_b)$ of scalar functions on Ω_b which are α-quasiperiodic with respect to x_1. Introduce the variational space

$$V_\alpha = \{u \in H^1_\alpha(\Omega_b)^2 : u \text{ satisfies the corresponding essential boundary condition of (3.4)–(3.7) on } \Lambda\}$$

equipped with the norm in the usual Sobolev space $H^1(\Omega_b)^2$ of vector functions. Note that for the Neumann condition $V_\alpha = H^1_\alpha(\Omega_b)^2$, whereas for the third- and fourth-kind boundary conditions only the first condition of (3.6) and (3.7), respectively, is involved. By the first Betti formula, it follows that for $u, \varphi \in V_\alpha$,

$$-\int_{\Omega_b} \left(\Delta^* + \omega^2\right) u \cdot \overline{\varphi}\, dx = \int_{\Omega_b} \left(a(u, \overline{\varphi}) - \omega^2 u \cdot \overline{\varphi}\right) dx - \int_{\Gamma_b} \overline{\varphi} \cdot Tu\, ds, \tag{3.12}$$

where the bar indicates the complex conjugate, and $a(\cdot,\cdot)$ is the symmetric bilinear form defined by

$$\begin{aligned} a(u, \varphi) := {} & (2\mu + \lambda)\,(\partial_1 u_1\, \partial_1 \varphi_1 + \partial_2 u_2\, \partial_2 \varphi_2) + \mu\,(\partial_2 u_1\, \partial_2 \varphi_1 + \partial_1 u_2\, \partial_1 \varphi_2) \\ & + \lambda\,(\partial_1 u_1\, \partial_2 \varphi_2 + \partial_2 u_2\, \partial_1 \varphi_1) + \mu\,(\partial_2 u_1\, \partial_1 \varphi_2 + \partial_1 u_2\, \partial_2 \varphi_1). \end{aligned} \tag{3.13}$$

Let $H^s_\alpha(\Gamma_b)$ denote the Sobolev space of order $s \in \mathbb{R}$ of functions on Γ_b that are α-quasiperiodic. Now we introduce the DtN map $\mathcal{T}$ on the artificial boundary Γ_b.

Definition 3.1. For any $v \in H^{1/2}_\alpha(\Gamma_b)^2$, we define $\mathcal{T}v$ as the traction Tu^{sc} on Γ_b where u^{sc} is the unique α-quasiperiodic solution of the homogeneous Navier equation in $\{x_2 > b\}$ which satisfies (3.10) and $u^{sc} = v$ on Γ_b.

The next lemma is devoted to an explicit representation of $\mathcal{T}$ and its properties.

Lemma 3.2 ([34]).
(i) *For* $v = \sum_{n\in\mathbb{Z}} \hat{v}_n \exp(i\alpha_n x_1) \in H_\alpha^{1/2}(\Gamma_b)^2$, *we have*

$$\mathcal{T} v = \mathcal{T}(\omega,\alpha)v = -\sum_{n\in\mathbb{Z}} W_n \hat{v}_n \exp(i\alpha_n x_1), \tag{3.14}$$

where W_n *is the* 2×2 *matrix defined by*

$$W_n = W_n(\omega,\alpha) := \frac{1}{i}\begin{pmatrix} \omega^2\beta_n/d_n & 2\mu\alpha_n - \omega^2\alpha_n/d_n \\ -2\mu\alpha_n + \omega^2\alpha_n/d_n & \omega^2\gamma_n/d_n \end{pmatrix} \tag{3.15}$$

with $d_n := \alpha_n^2 + \beta_n\gamma_n$.
(ii) $\mathcal{T}$ *is a bounded linear map from* $H_\alpha^{1/2}(\Gamma_b)^2$ *to* $H_\alpha^{-1/2}(\Gamma_b)^2$.
(iii) *For* $M_0 > 0$ *sufficiently large, we have the decomposition* $\mathcal{T} = \mathcal{T}_1 + \mathcal{T}_2$, *with*

$$\mathcal{T}_1 v := -\sum_{|n|>M_0} W_n\hat{v}_n \exp(i\alpha_n x_1) \quad \textit{satisfying} \quad \mathrm{Re}\left\{-\int_{\Gamma_b} \mathcal{T}_1 v\cdot \overline{v}\, ds\right\} \geq 0,$$

and $\mathcal{T}_2 v := -\sum_{|n|\leq M_0} W_n\hat{v}_n \exp(i\alpha_n x_1)$ *for all* $v \in H_\alpha^{1/2}(\Gamma_b)^2$.

The first assertion of Lemma 3.2 can be deduced directly from the definition of $\mathcal{T}$, whereas the second and third assertions follow from the explicit representation of $\mathcal{T}$.

Next we introduce the sesquilinear form $B(u,\varphi)$ defined by

$$B(u,\varphi) := \int_{\Omega_b} \left(a(u,\overline{\varphi}) - \omega^2 u\cdot\overline{\varphi}\right) dx - \int_{\Gamma_b} \overline{\varphi}\cdot \mathcal{T} u\, ds, \ \forall\, u,\phi \in V_\alpha, \tag{3.16}$$

with $\mathcal{T} u := \mathcal{T}(u|_{\Gamma_b})$. Applying Betti's identity (3.12) to a solution $u = u^{sc} + u^{in}$ of (DP) and using the fact that

$$Tu = T(u^{sc} + u^{in}) = \mathcal{T} u^{sc} + Tu^{in} = \mathcal{T} u + f_0, \quad \text{with} \quad f_0 := Tu^{in} - \mathcal{T} u^{in},$$

we obtain the following variational formulation of (DP): Find $u \in V_\alpha$ such that

$$B(u,\varphi) = \int_{\Gamma_b} f_0\cdot\overline{\varphi}\, ds, \quad \forall\, \varphi \in V_\alpha, \tag{3.17}$$

where

$$f_0 = \begin{cases} 2i\beta_0 k_p(\lambda+2\mu)\, d_0^{-1} \exp(i\alpha x_1 - i\beta_0 b)\,(-\alpha,\gamma_0)^\top & \text{if } u^{in} = u_p^{in}, \\ -2i\gamma_0 k_s \mu\, d_0^{-1} \exp(i\alpha x_1 - i\gamma_0 b)\,(\beta_0,\alpha)^\top & \text{if } u^{in} = u_s^{in}. \end{cases} \tag{3.18}$$

The problems (DP) and (3.17) are equivalent in the following sense. If $u \in H_{loc}^1(\Omega_\Lambda)^2$ is a solution of the boundary value problem (DP), then $u|_{\Omega_b}$ satisfies the variational problem (3.17). Conversely, a solution $u \in V_\alpha$ of (3.17) can be extended to a solution $u = u^{in} + u^{sc}$ of the Navier equation (3.3) for $x_2 \geq b$, where the Rayleigh coefficients $A_{p,n}$ and $A_{s,n}$ of u^{sc} are uniquely determined by the Fourier coefficients $\hat{u}_n$ of $\exp(-i\alpha x_1)\,(u - u^{in})(x_1, b)$ via the relation

$$\hat{u}_n = \begin{pmatrix} \alpha_n & \gamma_n \\ \beta_n & -\alpha_n \end{pmatrix} \begin{pmatrix} A_{p,n}\exp(i\beta_n b) \\ A_{s,n}\exp(i\gamma_n b) \end{pmatrix}. \tag{3.19}$$

The sesquilinear form B obviously generates a continuous linear operator $\mathcal{B} : V_\alpha \to V'_\alpha$ such that

$$B(u, \varphi) = (\mathcal{B}u, \varphi)_{\Omega_b} \quad \forall\, u,\, \varphi \in V_\alpha. \tag{3.20}$$

Here V'_α denotes the dual of the space V_α with respect to the duality $(\cdot, \cdot)_{\Omega_b}$ extending the scalar product in $L^2(\Omega_b)^2$.

Theorem 3.3 ([34]). *Assume that the grating profile Λ is a Lipschitz curve. Then the sesquilinear form B defined in* (3.16) *is strongly elliptic over V_α. Moreover, the operator $\mathcal{B}$ defined by* (3.20) *is always a Fredholm operator with index zero.*

Proof. Using the well-known Korn's inequality (see [31, Chapter 3], [63, Chapter 10]), we obtain

$$\int_{\Omega_b} a(u, \overline{u})dx \geq C_1 \|u\|^2_{H^1(\Omega_b)^2} - C_2\|u\|^2_{L^2(\Omega_b)^2}, \quad \forall\, u \in H^1(\Omega_b)^2,$$

for some constants $C_1, C_2 > 0$ independent of u. On the other hand, it follows from Lemma 3.2 (iii) that $-\mathcal{T}$ can be decomposed into the sum of the positive definite operator $-\mathcal{T}_1$ and the finite rank operator $-\mathcal{T}_2$. Therefore, the sesquilinear form B is strongly elliptic and the operator $\mathcal{B}$ is a Fredholm operator with index zero. □

In contrast to properties of the DtN map for the periodic scalar Helmholtz equation, the operator $-\mathcal{T}$ for the Navier equation is no longer positive definite (cf. Lemma 3.2 (iii) and [44]). Thanks to the periodicity of the domain, we can still justify the Fredholm property of $\mathcal{B}$. Concerning the DtN map for the biperiodic Maxwell equations, we refer to [1, 19, 70].

3.2 Uniqueness and existence for direct scattering problems

We rewrite problem (3.17) in the form

$$\mathcal{B}u = \mathcal{F}_0, \quad \mathcal{F}_0 \in V'_\alpha, \tag{3.21}$$

where $\mathcal{F}_0$ is given by the right-hand side of (3.17). We have the following solvability results for (DP).

Theorem 3.4 ([34]).

(i) *If the grating profile* Λ *is a Lipschitz curve, then there always exists a solution of (DP) under the boundary conditions of the first, second, third or fourth kind. Moreover, uniqueness holds for small frequencies, and for all frequencies excluding a discrete set with the only accumulation point at infinity.*

(ii) *If* Λ *is the graph of a Lipschitz function, then for any frequency* $\omega > 0$ *there exists a unique solution of (DP) under the Dirichlet boundary condition.*

Proof. (i) By the Fredholm alternative and Theorem 3.3, equation (3.21) is solvable if the right-hand side $\mathcal{F}_0$ is orthogonal to all solutions v of the homogeneous adjoint equation $\mathcal{B}^* v = 0$. Note that such v can always be extended to a solution of (3.3) in the unbounded domain Ω_Λ by setting

$$\begin{aligned} v(x) = &\sum_{n\in\mathbb{Z}} A_{p,n} \left(\alpha_n, -\overline{\beta}_n\right)^\top \exp\left(i\,\alpha_n x_1 - i\overline{\beta}_n x_2\right) \\ &+ \sum_{n\in\mathbb{Z}} A_{s,n} \left(-\overline{\gamma}_n, -\alpha_n\right)^\top \exp\left(i\,\alpha_n x_1 - i\,\overline{\gamma}_n x_2\right), \end{aligned}$$

for $x_2 \geq b$, where the Rayleigh coefficients $A_{p,n}, A_{s,n}$ are determined by the nth Fourier coefficient $\hat{v}_n$ of $e^{-i\alpha x_1} v|_{\Gamma_b}$ via an analogous relation to (3.19) with β_n, γ_n replaced by $\overline{\beta}_n, \overline{\gamma}_n$, respectively. On the other hand, it can be derived from

$$0 = (\mathcal{B}^* v, \psi)_{\Omega_b} = (v, \mathcal{B}\psi)_{\Omega_b} = \overline{B(\psi, v)}, \;\; \forall\, \psi \in V_\alpha$$

that v has vanishing Rayleigh coefficients of the incoming modes (see [34]):

$$A_{p,n} = 0 \quad \text{for} \quad |\alpha_n| < k_p \quad \text{and} \quad A_{s,n} = 0 \quad \text{for} \quad |\alpha_n| < k_s. \tag{3.22}$$

Using (3.22), we arrive at

$$(F_0, v)_{\Omega_b} = \int_{\Gamma_b} f_0 \cdot \overline{v} ds = 0,$$

with f_0 given in (3.18). Applying the Fredholm alternative yields the existence of a solution to (DP).

We next prove uniqueness for small frequencies. If $B(u, u) = 0$ for some $u \in V_\alpha$, then it follows from [34, Lemma 4] that the Rayleigh coefficient of order zero of u vanishes. This together with the asymptotic behavior of the matrix W_n (see [34, Lemma 2]) as $\omega \to 0^+$ leads to the estimate

$$\mathrm{Re}\Big\{ - \int_{\Gamma_b} \overline{u} \cdot \mathcal{T} u ds \Big\} \geq C\, \|u\|^2_{H_\alpha^{1/2}(\Gamma_b)^2},$$

for some constant $C > 0$ independent of u and ω. Then, one can prove the coercivity of the sesquilinear form B via the estimate (cf. [34, Remark 2])

$$B(u, u) \geq \int_{\Omega_b} a(u, \overline{u}) dx + C\, \|u\|^2_{H_\alpha^{1/2}(\Gamma_b)^2} - \omega^2 \|u\|^2_{L^2(\Omega_b)^2} \geq C_1 \|u\|^2_{H^1_\alpha(\Omega_b)^2},$$

which implies that $u = 0$ if ω is sufficiently small. In view of the analytic Fredholm theory (see e.g. [27, Theorem 8.26] or [50, Theorem I. 5. 1]), we obtain uniqueness and existence for all frequencies $\omega \in \mathbb{R}^+ \backslash \mathcal{D}$, where $\mathcal{D}$ is a discrete set including the Rayleigh frequencies. Moreover, we conclude from the arguments in [34, Theorem 6] or [42, Theorem 3.3] that $\mathcal{D}$ cannot have a finite accumulation point.
(ii) If Λ is the graph of a smooth function f and $u = 0$ on Λ, by a periodic Rellich identity we have (see [34])

$$\begin{aligned}0 &= -2\mathrm{Re} \int_{\Omega_b} (\Delta^* + \omega) u \cdot \partial_2 \overline{u}\, dx \\ &= \left(\int_{\Lambda} + \int_{\Gamma_b} \right) \left(2\mathrm{Re}(Tu \cdot \partial_2 \overline{u}) - a(u, \overline{u}) n_2 + \omega^2 |u|^2 n_2 \right) ds \qquad (3.23) \\ &= \int_{\Lambda} \left(\mu\, |\partial_{\mathbf{n}} u|^2 + (\lambda + \mu)\, |\operatorname{div} u|^2 \right) n_2\, ds.\end{aligned}$$

Since $n_2 = -1/\sqrt{1 + |f'|^2} < 0$, there holds $\partial_{\mathbf{n}} u = u = 0$ on Λ. Applying Holmgren's theorem leads to uniqueness at arbitrary frequency. By Theorem 3.3, existence follows directly from uniqueness via the Fredholm alternative. Note that the integral in (3.23) over Λ does not make sense for Lipschitz graphs. To deal with the Lipschitz boundary, one can adapt Nečas' method [66, Chap. 5] of approximating the grating profile by smooth graphs. For the details we refer to [44] and [34]. □

Theorem 3.4 (ii) generalizes the results of [44, 54] for the scalar quasiperiodic Helmholtz equation to the case of the Navier equation. More general Rellich identities for the Navier equation (on bounded domains) can be found in [28]. The well-posedness for the boundary value problem with mixed Dirichlet and impedance boundary conditions can also be established for all frequencies, provided Λ is a Lipschitz curve [34]. The uniqueness to (DP) under the second, third, and fourth-kind boundary conditions is not true in general. Nonuniqueness examples can be constructed for a flat grating in the resonance case (see [39]).

3.3 Uniqueness and existence for transmission gratings

Suppose the whole (x_1, x_2)-plane is filled with elastic materials that are homogeneous above and below a certain interface Λ of period 2π. Let $\Omega^\pm$ be the unbounded domains above and below Λ, respectively. We assume that the Lamé coefficients $\mu^\pm, \lambda^\pm$ in $\Omega^\pm$ are constants satisfying $\mu^\pm > 0$, $\lambda^\pm + \mu^\pm > 0$, and that the mass densities $\rho^\pm$ are positive constants in these subdomains. Let $k_p^\pm := \omega \sqrt{\rho^\pm/(2\mu^\pm + \lambda^\pm)}$, $k_s^\pm := \omega \sqrt{\rho^\pm/\mu^\pm}$ be the corresponding compressional and shear wave numbers, respectively. In Section 3, we assume that a time-harmonic plane elastic wave u^{in} with incident angle θ is incident on Λ from Ω^+, which is either an incident pressure wave of the form (3.1), or an incident shear wave of the form (3.2), with k_p, k_s replaced by k_p^+, k_s^+. We then look for the total displacement field u

$$u = u^{in} + u^+ \text{ in } \Omega^+, \quad u = u^- \text{ in } \Omega^-, \tag{3.24}$$

where the scattered fields $u^\pm$ satisfy the corresponding Navier equations

$$(\Delta^* + \omega^2 \rho^\pm) u^\pm = 0 \text{ in } \Omega^\pm, \tag{3.25}$$

with the α-quasiperiodicity condition

$$u^\pm(x_1 + 2\pi, x_2) = \exp(2i\alpha\pi)\, u^\pm(x_1, x_2). \tag{3.26}$$

On the interface the continuity of the displacement and the stress lead to the transmission conditions

$$u^{in} + u^+ = u^-, \quad T^+(u^{in} + u^+) = T^- u^- \text{ on } \Lambda, \tag{3.27}$$

where the corresponding stress operators are defined as in (3.8), with μ, λ replaced by $\mu^\pm$, $\lambda^\pm$. Finally, we need to impose appropriate radiation conditions on the scattered fields $u^\pm$ as $x_2 \to \pm\infty$. Introduce the notation

$$\Lambda^+ := \max_{(x_1,x_2)\in\Lambda} x_2, \quad \Lambda^- := \min_{(x_1,x_2)\in\Lambda} x_2,$$

and define $\beta_n^\pm$ and $\gamma_n^\pm$ as in (3.11) with k_p, k_s replaced by $k_p^\pm$, $k_s^\pm$. Then we insist that the scattered fields $u^\pm$ admit the following Rayleigh expansions (cf. (3.10)), for $x_2 \gtrless \Lambda^\pm$:

$$\begin{aligned} u^\pm(x) = \sum_{n\in\mathbb{Z}} \Big\{ & A_{p,n}^\pm (\alpha_n, \pm\beta_n^\pm)^\top \exp(i\alpha_n x_1 \pm i\beta_n^\pm x_2) \\ & + A_{s,n}^\pm (\pm\gamma_n^\pm, -\alpha_n)^\top \exp(i\alpha_n x_1 \pm i\gamma_n^\pm x_2) \Big\}. \end{aligned} \tag{3.28}$$

Transmission problem (TP): Given a grating profile curve $\Lambda \subset \mathbb{R}^2$ (which is 2π-periodic in x_1) and an incident plane pressure or shear wave u^{in}, find a vector function $u \in H^1_{loc}(\mathbb{R}^2)^2$ that satisfies (3.24)–(3.28).

Introduce artificial boundaries

$$\Gamma^\pm := \{(x_1, b^\pm) : 0 \le x_1 \le 2\pi\}, \quad b^+ > \Lambda^+, \quad b^- < \Lambda^-$$

and the bounded domains

$$\Omega_b := (0, 2\pi) \times (b^-, b^+), \quad \Omega_b^\pm := \Omega^\pm \cap \Omega_b.$$

The DtN maps $\mathcal{T}^\pm$ on the artificial boundaries $\Gamma^\pm$ have the Fourier series representations (cf. Lemma 3.2 (i))

$$\mathcal{T}^\pm u^\pm := -\sum_{n\in\mathbb{Z}} W_n^\pm \hat{u}_n^\pm \exp(i\alpha_n x_1), \quad u^\pm = \sum_{n\in\mathbb{Z}} \hat{u}_n^\pm \exp(i\alpha_n x_1) \in H_\alpha^{1/2}(\Gamma^\pm)^2,$$

where the matrices $W_n^\pm = W_n^\pm(\omega,\alpha)$ take the form (cf. (3.15))

$$W_n^\pm := \frac{1}{i}\begin{pmatrix} \omega^2\rho^\pm\beta_n^\pm/d_n^\pm & 2\mu^\pm\alpha_n - \omega^2\rho^\pm\alpha_n/d_n^\pm \\ -2\mu^\pm\alpha_n + \omega^2\rho^\pm\alpha_n/d_n^\pm & \omega^2\rho^\pm\gamma_n^\pm/d_n^\pm \end{pmatrix},$$

with $d_n^\pm := \alpha_n^2 + \beta_n^\pm\gamma_n^\pm$. Applying the first Betti formula on each subdomain $\Omega^\pm$ to a solution of (TP), and using the transmission conditions (3.27), we obtain the following variational formulation of (TP) on the bounded domain Ω: Find $u \in H_\alpha^1(\Omega)^2$ such that

$$\begin{aligned} B(u,\varphi) &:= \int_{\Omega_b} (a(u,\overline{\varphi}) - \omega^2\rho\, u\cdot\overline{\varphi})\,dx - \int_{\Gamma^+} \overline{\varphi}\cdot\mathcal{T}^+u\,ds - \int_{\Gamma^-}\overline{\varphi}\cdot\mathcal{T}^-u\,ds \\ &= \int_{\Gamma^+} f_0\cdot\overline{\varphi}\,ds, \quad \forall\,\varphi\in H_\alpha^1(\Omega)^2. \end{aligned} \tag{3.29}$$

Here the domain integral is understood as the sum of the integrals

$$\int_{\Omega_b^\pm} \left(a^\pm(u,\overline{\varphi}) - \omega^2\rho^\pm\, u\cdot\overline{\varphi}\right)\,dx,$$

where the bilinear forms $a^\pm$ are defined as in (3.13), with μ, λ replaced by $\mu^\pm$, $\lambda^\pm$, and the function f_0 on the right-hand side is defined analogously as in (3.18), with $\beta_0, \gamma_0, k_p, k_s, \lambda, \mu$ replaced by $\beta_0^+, \gamma_0^+, k_p^+, k_s^+, \lambda^+, \mu^+$, respectively. As in (3.20), the sesquilinear form B defined in (3.29) generates a continuous linear operator $\mathcal{B}$ from $H_\alpha^1(\Omega_b)^2$ into its dual $(H_\alpha^1(\Omega_b)^2)'$, with respect to the pairing $(u,\varphi)_{\Omega_b} = \int_{\Omega_b} u\cdot\overline{\varphi}$, via

$$B(u,\varphi) = (\mathcal{B}u,\varphi)_{\Omega_b}, \quad \forall\,u,\varphi\in H_\alpha^1(\Omega_b)^2. \tag{3.30}$$

The following result extends Theorems 3.3 and 3.4 (i) to problem (TP).

Theorem 3.5 ([34]).
(i) *The sesquilinear form B defined by* (3.29) *is strongly elliptic over $H_\alpha^1(\Omega)^2$, and the operator $\mathcal{B}$ defined in* (3.30) *is Fredholm with index zero.*
(ii) *For an incident plane pressure or shear wave, there always exists a solution to the variational problem* (3.29) *and hence to problem (TP).*
(iii) *Assume that u^{in} is an incident pressure wave of the form* (3.1) *(with $k_p = k_p^+$). Then, there exists a sufficiently small frequency $\omega_0 > 0$ such that the variational problem* (3.29) *admits a unique solution $u \in H_\alpha^1(\Omega_b)^2$ for all incident angles and for all frequencies $\omega \in (0,\omega_0]$. Moreover, for all but a sequence of countable frequencies $\omega_j \to \infty$, the variational problem* (3.29) *admits a unique solution.*

(iv) *For an incident shear wave of the form* (3.2), *the results of the third assertion remain valid under one of the following additional assumptions:*
   (1) $k_p^+ > k_s^+ \sin\theta$, *or equivalently,* $\mu^+/(2\mu^+ + \lambda^+) > \sin^2\theta$,
   (2) $k_p^- > k_s^+$, *or equivalently,* $\rho^-/(2\mu^- + \lambda^-) > \rho^+/\mu^+$.

Remark 3.6. Assume that the elastic material is homogeneous above some periodic Lipschitz interface and below another periodic Lipschitz interface, whereas the elastic medium between the two interfaces may be inhomogeneous with piecewise constant Lamé parameters λ, μ and density ρ having jumps at certain (finitely many) disjoint periodic Lipschitz interfaces. Then Theorem 3.5 can be easily extended to these more general periodic diffractive structures.

In general, the uniqueness result of Theorem 3.4 (ii) does not hold for the transmission problem (TP). Even in the special case of two half-planes with certain elastic parameters $\lambda^\pm$, $\mu^\pm$, $\rho^\pm$ and the transmission conditions (3.27) on the line $\{x_2 = 0\}$, there may exist nontrivial solutions of the homogeneous problem (Rayleigh surface waves) that decay exponentially as $x_2 \to \pm\infty$ [3]. Hence additional conditions must be imposed on the elastic parameters to guarantee the uniqueness. As an example, we present a new uniqueness result for (TP) at any frequency of the incident wave.

Theorem 3.7. *Assume that the interface* Λ *is given by the graph of a* 2π*-periodic Lipschitz function* $x_2 = f(x_1)$. *Suppose further that* $\rho^+ \neq \rho^-$ *and that the Lamé constants* $\lambda^\pm, \mu^\pm$ *in* $\Omega_b^\pm$ *satisfy* $\lambda^+ = \lambda^-, \mu^+ = \mu^-$. *Then there exists a unique solution* u *to problem (TP).*

Proof. By Theorem 3.5 (i), we only need to prove $u = 0$ provided that $u^{in} = 0$. Suppose first that the grating profile is given by a C^∞-smooth function f. Applying the Rellich identity (3.23) to $u^\pm$ in $\Omega_b^\pm$ yields (cf. (3.23))

$$\begin{aligned} 0 &= -2\mathrm{Re} \int_{\Omega_b^\pm} (\Delta^* + \omega) u^\pm \cdot \partial_2 \overline{u}^\pm dx \\ &= \left(\pm \int_\Lambda + \int_{\Gamma^\pm} \right) 2\mathrm{Re}(T^\pm u^\pm \cdot \partial_2 \overline{u}^\pm) - a^\pm(u^\pm, \overline{u}^\pm) n_2 + \omega^2 \rho^\pm |u^\pm|^2 n_2 \, ds, \qquad (3.31) \\ &= \pm \int_\Lambda 2\mathrm{Re}(T^\pm u^\pm \cdot \partial_2 \overline{u}^\pm) - a^\pm(u^\pm, \overline{u}^\pm) n_2 + \omega^2 \rho^\pm |u^\pm|^2 n_2 \, ds. \end{aligned}$$

Note that the normal to Λ is directed into Ω^-. Making use of the transmission conditions (3.27) with $u^{in} = 0, \lambda^+ = \lambda^-, \mu^+ = \mu^-$, we deduce that $\nabla u^+ = \nabla u^-$ on Λ, leading to

$$T^+ u^+ \cdot \partial_2 \overline{u}^+ = T^- u^- \cdot \partial_2 \overline{u}^-, \quad a^+(u^+, \overline{u}^+) = a^-(u^-, \overline{u}^-) \quad \text{on } \Lambda.$$

Thus, from (3.31) we see that

$$0 = \omega^2(\rho^+ - \rho^-) \int_\Lambda |u^+|^2 n_2 \, ds.$$

This relation combined with $n_2 = -1/\sqrt{1 + |f'|^2} < 0$ and the transmission condition (3.27) gives $u^+ = u^- = 0$ on Λ. By the uniqueness to the forward scattering problem with the Dirichlet boundary condition (see Theorem 3.4 (ii)), we conclude that $u^\pm = 0$ in $\Omega_b^\pm$. Thus $u = 0$ in Ω_b. In the general case we proceed analogously to Theorem 3.4 (ii) by using Nečas' method [66, Chap. 5] of approximating the Lipschitz grating profile by smooth graphs. □

Remark 3.8. The solvability results in Theorems 3.4, 3.5 and 3.7 can be generalized to biperiodic diffraction gratings in $\mathbb{R}^3$. However, to prove an analog of Lemma 3.2 for the DtN map, a more sophisticated analysis than for plane elasticity is needed. Note that the 2×2 matrix W_n in (3.14) must be replaced by a 3×3 matrix in the biperiodic case. For further details we refer to [39].

4 Uniqueness for inverse scattering problems

In this section, we will review recent uniqueness results for inverse elastic scattering by periodic structures. Suppose that the two gratings Λ_1, Λ_2 generate the total fields $u_j(x;\theta_m)$ $(j = 1,2)$ for the incident plane wave of the form (3.1) or (3.2) with the incident angle $\theta = \theta_m$. We are interested in the following uniqueness questions about the inverse problem (IP). Does the relation

$$u_1(x_1, b; \theta_m) = u_2(x_1, b; \theta_m), \quad \forall\, x_1 \in (0,2\pi), \text{ for } m = 1,2,\dots M,$$

with some sufficiently large b imply $\Lambda_1 = \Lambda_2$? If yes, what is the minimal number of incident elastic waves needed to assure uniqueness? Moreover, can we describe the exceptional classes of grating profiles that generate the same near field data on $x_2 = b$? A first answer related to these questions is given in [6] for the Dirichlet boundary condition with several incident frequencies.

Investigating the eigenvalues and quasiperiodic eigenfunctions of Lamé operator in a periodic layer, it is proved in [6] that a Dirichlet C^2-smooth surface can be uniquely determined from the scattered fields corresponding to incident pressure waves for one incident angle and an interval of wave numbers. Moreover, a finite number of wave numbers is enough if some a priori information about the height of the grating is available. In particular, uniqueness with one incident pressure wave holds if the height is sufficiently small. This extends the paper by Hettlich and Kirsch on the periodic analog of Schiffer's theorem (see [52]) to the case of elastic scattering. We further refer to [6, Theorem 5] for the minimal number of compressional wave numbers needed to guarantee uniqueness. See also [5, 14, 18, 55] for other uniqueness results within smooth periodic profiles in inverse scattering of acoustic and electromagnetic waves.

As already mentioned in the introduction, there are recent uniqueness results on inverse acoustic and electromagnetic scattering by polygonal and polyhedral obstacles and diffraction gratings [4, 16, 17, 26, 35, 41, 45, 46, 59, 61]. These results are essentially based on reflection principles for the Helmholtz and Maxwell equations. In this section, we present corresponding uniqueness theorems for polygonal elastic diffraction gratings under the third or fourth-kind boundary conditions, where a reflection principle is available. There seems to be no reflection principle for the Navier equation under the physically more relevant boundary conditions of the first (Dirichlet) or second (Neumann) kind. Thus it is still a challenging problem to study uniqueness in the inverse problem under these boundary conditions.

Let $\mathcal{A}$ be our admissible class of periodic structures defined by

$$\begin{aligned}\mathcal{A} := \{\Lambda :\ & \Lambda = (x_1, f(x_1)), \text{ where } f \text{ is a continuous piecewise linear} \\ & \text{function of period } 2\pi \text{ satisfying } \max_{x_1\in\mathbb{R}}\{f(x_1)\} < b, \text{ and} \\ & \text{the graph of } f \text{ restricted to } [0,2\pi] \text{ consists of finitely many} \\ & \text{line segments and is not a straight line parallel to the } x_1\text{-axis}\}.\end{aligned}$$

Note that flat gratings are excluded from our admissible class $\mathcal{A}$, because such gratings cannot be uniquely determined by a fixed number of incident pressure or shear waves [36, Section 4].

4.1 Inverse scattering of incident pressure waves

In this section, we make the following assumptions.

(A1) $u^{in} = u_p^{in} := \hat{\theta} \exp(ik_p x \cdot \hat{\theta})$ with some fixed incident angle $\theta \in (-\frac{\pi}{2}, \frac{\pi}{2})$.

(A2) Without loss of generality, one of the profiles $\Lambda_1, \Lambda_2 \in \mathcal{A}$ has a corner point at the origin $O = (0,0)$.

(A3) The corresponding total fields u_1 and u_2 both satisfy the third (or fourth) kind boundary conditions on Λ_1 and Λ_2, and they coincide on Γ_b:

$$u_1(x_1, b; \theta) = u_2(x_1, b; \theta), \quad x_1 \in (0, 2\pi). \tag{3.32}$$

Introduce the (exceptional) class of polygonal grating profiles $\mathcal{D}_2(\theta, k_p)$ by

$$\begin{aligned}\mathcal{D}_2(\theta, k_p) := \{\Lambda \in \mathcal{A} :\ & \text{each line segment of } \Lambda \text{ lies on a straight line} \\ & \text{defined by } x_2 = x_1 \tan\varphi + \frac{2\pi}{k_p \cos\theta} n \text{ for} \\ & \text{some } n \in \mathbb{Z} \text{ with } \varphi \in \{\tfrac{\theta}{2} + \tfrac{\pi}{4}, \tfrac{\theta}{2} - \tfrac{\pi}{4}\}\}\end{aligned}$$

if $k_p(1 \pm \sin\theta) \in \mathbb{Z}$, and by $\mathcal{D}_2(\theta, k_p) := \emptyset$ otherwise. See Figure 3.1 for two different grating profiles from the class $\mathcal{D}_2(\theta, k_p)$ with $\theta = -\frac{\pi}{6}, k_p = 2$. In fact, the set $\mathcal{D}_2(\theta, k_p)$ contains grating profiles which only depend on the incident pressure

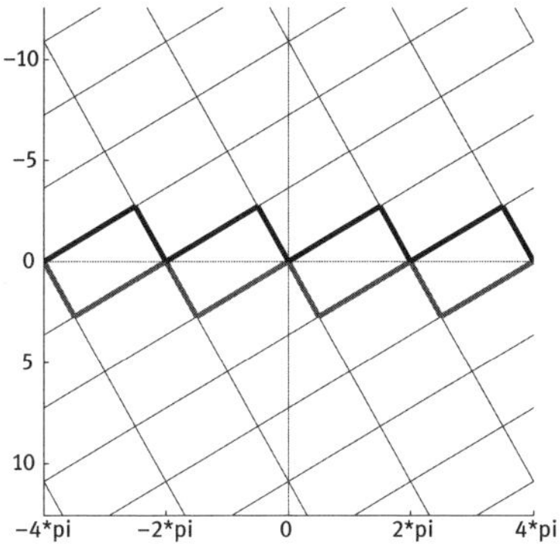

Figure 3.1: (See Color Plates for a color version.) $\mathcal{D}_2(\theta, k_p)$ with $\theta = -\frac{\pi}{6}, k_p = 2$.

wave. One can prove that $\mathcal{D}_2(\theta, k_p) \neq \emptyset$ for all k_p and θ satisfying $k_p(1 \pm \sin\theta) \in \mathbb{Z}$ [36, Lemma 11].

Given a fixed incident angle θ, define

$$\pi_p := \{n \in \mathbb{Z} : \beta_n(\theta) = 0\}, \qquad \pi_s := \{n \in \mathbb{Z} : \gamma_n(\theta) = 0\}.$$

We say that a Rayleigh frequency occurs if either $\pi_p \neq \emptyset$ or $\pi_s \neq \emptyset$, and that Rayleigh frequencies of the compressional (shear) part are excluded if $\pi_p = \emptyset$ ($\pi_s = \emptyset$). Let us now give the main theorem for the fourth-kind boundary conditions.

Theorem 3.9 ([36]). *Assume the boundary conditions of the fourth kind* (3.7) *are imposed on* Λ_1 *and* Λ_2. *Then the relation* (3.32) *implies that one of the following cases must occur:*

(i) $\Lambda_1 = \Lambda_2$.

(ii) $\Lambda_1, \Lambda_2 \in \mathcal{D}_2(\theta, k_p)$, $\pi_p = \emptyset$, *and the total field takes the form*

$$u = \hat{\theta} \exp(ik_p x \cdot \hat{\theta}) - \hat{\theta} \exp(-ik_p x \cdot \hat{\theta}) - e_1 \exp(ik_p x_1) + e_1 \exp(-ik_p x_1)$$

with $u = u_j$ $(j = 1,2)$, $e_1 = (1,0)^\top$.

It follows from Theorem 3.9 (ii) that the class $\mathcal{D}_2(\theta, k_p)$ contains the exceptional grating profiles which generate the same total field, with a finite number of propagating modes only involved in the compressional part of the scattered field. Nonuniqueness examples to (IP) with one incident pressure wave can be easily constructed from the set $\mathcal{D}_2(\theta, k_p)$ (see e.g. [36, Section 6.2]).

The following reflection principle for the Navier equation is the main tool for proving Theorem 3.9 and the other theorems below. We denote by R_l the reflection with respect to a line l in $\mathbb{R}^2$, and by R_l' the reflection with respect to the line l' that passes through the origin O and is parallel to l.

Lemma 3.10 ([36]). *(Reflection principle for the Navier equation) Let Ω be a symmetric domain with respect to a line l, and let $\tilde{l} \subset \Omega$ be a subset of another line such that $R_l(\tilde{l}) \subset \Omega$. Assume $u \in H^1(\Omega)^2$ satisfies the Navier equation $\Delta^* u + \omega^2 u = 0$ in Ω.*

(i) *If u satisfies the boundary conditions of the fourth kind (3.7) on $l \cap \Omega$, then*

$$u(x) + R_l'(u(R_l(x))) = 0 \quad in \quad \Omega. \tag{3.33}$$

(ii) *If u satisfies the boundary conditions of the third kind (3.6) on $l \cap \Omega$, then*

$$u(x) - R_l'(u(R_l(x))) = 0 \quad in \quad \Omega. \tag{3.34}$$

(iii) *In particular, if u satisfies the fourth respectively third-kind boundary conditions on both l and $\tilde{l}$ in Ω, then the same boundary conditions hold on $R_l(\tilde{l})$.*

The original version of the reflection principle for the Navier equation can be found in [47], where it was proved in the three-dimensional case when the domain Ω is symmetric with respect to the (x_1, x_2)-plane. The proof readily carries over to a two-dimensional domain Ω in the above lemma. Note that the relations (3.33) and (3.34) are similar to those given in [60, 61] for the Maxwell equations.

Proof of Theorem 3.9. From the assumption (3.32) and the unique continuation of solutions to the Navier equation, it follows that $u_1(x) = u_2(x)$ for all x lying above the profiles Λ_1 and Λ_2. Relying on the analyticity of u_j and the reflection principle in Lemma 3.10, one can always find an "exit" ray l extending to infinity in $x_2 > b$ such that u_j $(j = 1,2)$ satisfies the fourth-kind boundary conditions on l. Applying properties of almost periodic functions to $u_j|_l$ gives the vanishing Rayleigh coefficients $A_{p,n} = 0$ for $|\alpha_n| > k_p$ and $A_{s,n} = 0$ for $|\alpha_n| > k_s$. Therefore, both of the total fields $u_j = u^{in} + u_j^{sc}$, $j = 1,2$, can be reduced to a finite sum of propagating waves, i.e.

$$\begin{aligned} u_1 = u_2 = u^{in} &+ \sum_{|\alpha_n| \le k_p} A_{p,n}(\alpha_n, \beta_n)^\top e^{i(\alpha_n x_1 + \beta_n x_2)} \\ &+ \sum_{|\alpha_n| \le k_s} A_{s,n}(\gamma_n, -\alpha_n)^\top e^{i(\alpha_n x_1 + \gamma_n x_2)} \end{aligned}$$

for $x_2 > b$, which can be analytically extended to the whole plane. Further, it can be derived from the reflection relation (3.33) that the solution u and its finite number of propagating directions exhibit some rotational and reflection invariance (see [36, Lemmas 6 and 7]) around the origin. This symmetry implies further that $A_{s,n} = 0$ for all $|\alpha_n| \le k_s$, due to the incident pressure wave u_p^{in} and the assumption that Λ_j $(j = 1,2)$ is the graph of a function [36, Lemma 8]. Therefore, the total field only consists of the compressional part with a finite number of propagating directions lying on the circle $|x| = k_p$. Among these directions, only the incident wave of direction $(\alpha, -\beta)^\top$ is propagating downward, whereas the other waves of directions $(\alpha_n, \beta_n)^\top$ are propagating upward when $\beta_n > 0$ or along the x_1-axis when $\beta_n = 0$.

From the reflectional and rotational invariance of these directions, one can describe each propagating direction of the compressional part and its corresponding Rayleigh coefficient in terms of the incident compressional wave number k_p and the incident angle θ. This implies the representation of the total field in Theorem 3.9 (ii) and leads to the definition of the unidentifiable set $\mathcal{D}_2(\theta, k_p)$ (see [36]). □

The following results can be obtained directly from Theorem 3.9.

Remark 3.11.

(i) Let $\theta \in (-\frac{\pi}{2}, \frac{\pi}{2})$. If $\mathcal{D}_2(\theta, k_p) = \emptyset$ or $\Lambda \notin \mathcal{D}_2(\theta, k_p)$, then one incident pressure wave with the incident angle θ uniquely determines $\Lambda \in \mathcal{A}$. Since $\mathcal{D}_2(\theta, k_p) = \emptyset$ if $\pi_p = \emptyset$, we have uniqueness with one incident pressure wave provided the Rayleigh frequencies of the compressional part are excluded.

(ii) Two incident pressure waves are always sufficient to uniquely determine $\Lambda \in \mathcal{A}$ since

$$\mathcal{D}_2(\theta_1, k_p) \cap \mathcal{D}_2(\theta_2, k_p) = \emptyset, \qquad \theta_1 \neq \theta_2.$$

Before stating the main theorem for the third-kind boundary conditions, we introduce the following three (exceptional) classes of polygonal periodic structures.

For $2k_p \sin\theta \in \mathbb{Z}$, let

$$\begin{aligned}\mathcal{N}_2(\theta, k_p) := \{\Lambda \in \mathcal{A} :\ & \text{each line segment of } \Lambda \text{ lies on a straight line defined} \\ & \text{by } x_2 = x_1 \tan\theta + \frac{\pi}{k_p \cos\theta} n \text{ for some } n \in \mathbb{Z}, \text{ or} \\ & x_2 = x_1 \tan(\theta + \frac{\pi}{2}) + c \text{ for some } c \in \mathbb{R}\},\end{aligned}$$

and if $2k_p \sin\theta \notin \mathbb{Z}$, set $\mathcal{N}_2(\theta, k_p) := \emptyset$. Moreover,

$$\begin{aligned}\mathcal{N}_3(\theta, k_p) := \{\Lambda \in \mathcal{A} :\ & \text{each line segment of } \Lambda \text{ lies on a straight line defined} \\ & \text{by } x_2 = x_1 \tan\varphi + \frac{2\pi}{k_p \sqrt{3} \cos\varphi} n \text{ for some } n \in \mathbb{Z}, \text{ with} \\ & \varphi \in \{\theta + \frac{5\pi}{6}, \theta + \frac{\pi}{6}, \theta - \frac{\pi}{2}\}\}\end{aligned}$$

if $\theta \in [-\frac{\pi}{6}, \frac{\pi}{6}]$ and $k_p\sqrt{3}\sin(\frac{\pi}{6} \pm \theta) \in \mathbb{Z}$, and $\mathcal{N}_3(\theta, k_p) := \emptyset$ otherwise. Finally,

$$\begin{aligned}\mathcal{N}_4(0, k_p) := \{\Lambda \in \mathcal{A} :\ & \text{each line segment of } \Lambda \text{ lies on a straight line defined} \\ & \text{by } x_2 = \pm x_1 + \frac{2\pi}{k_p} n \text{ for some } n \in \mathbb{Z}, \text{ or} \\ & x_2 = \frac{\pi}{k_p} m \text{ for some } m \in \mathbb{Z}\}\end{aligned}$$

if $\theta = 0$ and $k_p \in \mathbb{Z}$, and $\mathcal{N}_4(0, k_p) := \emptyset$ otherwise. See Figure 3.2 for examples of the triangular grids on which the profiles of $\mathcal{N}_3(\theta, k_p)$ and $\mathcal{N}_4(0, k_p)$ are located.

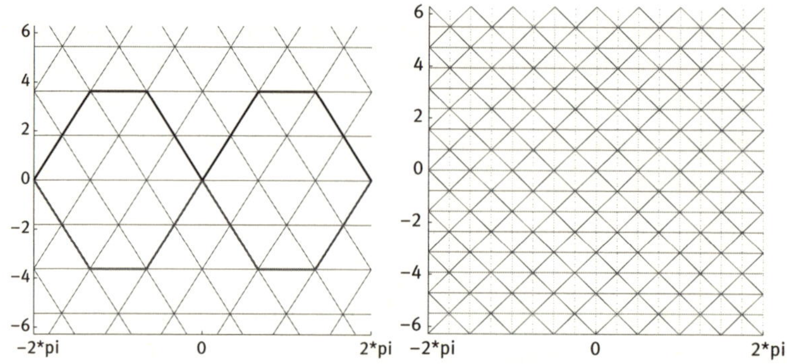

Figure 3.2: (See Color Plates for a color version.) Left: $\mathcal{N}_3(\theta, k_p)$ with $\theta = -\frac{\pi}{6}$, $k_p = 2$. Right: $\mathcal{N}_4(0, k_p)$ with $k_p = 4$.

Theorem 3.12 ([36]). *Assume the third-kind boundary conditions (3.6) are imposed on Λ_1 and Λ_2. Then the relation (3.32) implies one of the following cases:*
(i) $\Lambda_1 = \Lambda_2$.
(ii) (a) $\Lambda_1, \Lambda_2 \in \mathcal{N}_2(\theta, k_p)$, $\pi_p = \emptyset$, *and the total field takes the form*

$$u(x) = \hat{\theta} \exp(ik_p x \cdot \hat{\theta}) - \hat{\theta} \exp(-ik_p x \cdot \hat{\theta}).$$

(b) $\Lambda_1, \Lambda_2 \in \mathcal{D}_2(\theta, k_p)$, $\pi_p \neq \emptyset$, *and the total field can be written as*

$$u(x) = \hat{\theta} \exp(ik_p x \cdot \hat{\theta}) - \hat{\theta} \exp(-ik_p x \cdot \hat{\theta}) + e_1 \exp(ik_p x_1) - e_1 \exp(-ik_p x_1).$$

(iii) $\Lambda_1, \Lambda_2 \in \mathcal{N}_3(\theta, k_p)$ *with* $\theta \in [-\frac{\pi}{6}, \frac{\pi}{6}]$. *In this case,* $\pi_p \neq \emptyset$ *if* $\theta = \frac{\pi}{6}$ *or* $\theta = -\frac{\pi}{6}$, *and the total field takes the form*

$$u(x) = \hat{\theta} \exp(ik_p x \cdot \hat{\theta}) + \mathrm{Rot}_{\frac{2\pi}{3}}(\hat{\theta}) \exp(ik_p x \cdot \mathrm{Rot}_{\frac{2\pi}{3}}(\hat{\theta})) + \mathrm{Rot}_{\frac{4\pi}{3}}(\hat{\theta}) \exp(ik_p x \cdot \mathrm{Rot}_{\frac{4\pi}{3}}(\hat{\theta})),$$

where Rot_φ *denotes the rotation around the origin O by the angle* φ.
(iv) $\Lambda_1, \Lambda_2 \in \mathcal{N}_4(0, k_p)$, $\theta = 0$, $\pi_p \neq \emptyset$, *and the total field can be written as*

$$u(x) = -e_2 \exp(-ik_p x_2) + e_2 \exp(ik_p x_2) + e_1 \exp(ik_p x_1) - e_1 \exp(-ik_p x_1).$$

Here $u = u_j$ $(j = 1,2)$, $e_1 = (1,0)^\top$ *and* $e_2 = (0,1)^\top$.

Corollary 3.13. *Suppose that the assumptions of Theorem 3.12 are satisfied. If (3.32) holds for four incident pressure waves with distinct incident angles, then Λ_1 and Λ_2 must be identical, while three incident waves are always enough to imply $\Lambda_1 = \Lambda_2$ if the Rayleigh frequencies of the compressional part for each incident angle are excluded.*

We refer to [35, Corollary 5] for additional conditions on the incident angles and the compressional wave number k_p guaranteeing that each grating profile $\Lambda \in \mathcal{A}$ can be uniquely determined by three incident pressure waves under the third-kind boundary conditions.

4.2 Inverse scattering of incident shear waves

In this section, we assume that the profiles $\Lambda_1, \Lambda_2 \in \mathcal{A}$ satisfy the conditions (A2) and (A3) again, whereas u^{in} in (A1) has to be replaced by the incident shear wave $u_s^{in}(x) = \hat{\theta}^\perp \exp(ik_s x \cdot \hat{\theta})$. Then the uniqueness results of [36] for incident plane shear waves may be stated as follows.

Theorem 3.14. *Assume the boundary conditions of the third kind* (3.6) *are imposed on* Λ_1 *and* Λ_2. *Then the relation* (3.32) *implies one of the following cases:*
(i) $\Lambda_1 = \Lambda_2$.
(ii) $\Lambda_1, \Lambda_2 \in \mathcal{D}_2(\theta, k_s)$, $\pi_s = \emptyset$, *and the total field takes the form*

$$u(x) = \hat{\theta}^\perp \exp(ik_s x \cdot \hat{\theta}) - \hat{\theta}^\perp \exp(-ik_s x \cdot \hat{\theta}) + e_2 \exp(ik_s x_1) - e_2 \exp(-ik_s x_1)$$

with $u = u_j$ $(j = 1,2)$, $e_2 = (0,1)^\top$.

Theorem 3.15. *Let the fourth-kind boundary conditions* (3.7) *be satisfied on* Λ_1 *and* Λ_2, *and assume that* (3.32) *holds. Then one of the following cases must occur:*
(i) $\Lambda_1 = \Lambda_2$.
(ii) (a) $\Lambda_1, \Lambda_2 \in \mathcal{N}_2(\theta, k_s)$, $\pi_s = \emptyset$, *and the total field takes the form*

$$u(x) = \hat{\theta}^\perp \exp(ik_s x \cdot \hat{\theta}) - \hat{\theta}^\perp \exp(-ik_s x \cdot \hat{\theta}).$$

(b) $\Lambda_1, \Lambda_2 \in \mathcal{D}_2(\theta, k_s)$, $\pi_s \neq \emptyset$, *and the total field takes can be written as*

$$u = \hat{\theta}^\perp \exp(ik_s x \cdot \hat{\theta}) - \hat{\theta}^\perp \exp(-ik_s x \cdot \hat{\theta}) - e_2 \exp(ik_s x_1) + e_2 \exp(-ik_s x_1).$$

(iii) $\Lambda_1, \Lambda_2 \in \mathcal{N}_3(\theta, k_s)$ *with* $\theta \in [-\frac{\pi}{6}, \frac{\pi}{6}]$. *In this case,* $\pi_s \neq \emptyset$ *if* $\theta = \frac{\pi}{6}$ *or* $\theta = -\frac{\pi}{6}$, *and the total field takes the form*

$$\begin{aligned} u(x) &= \hat{\theta}^\perp \exp(ik_s x \cdot \hat{\theta}) + (\mathrm{Rot}_{\frac{2\pi}{3}}(\hat{\theta}))^\perp \exp(ik_s x \cdot \mathrm{Rot}_{\frac{2\pi}{3}}(\hat{\theta})) \\ &\quad + (\mathrm{Rot}_{\frac{4\pi}{3}}(\hat{\theta}))^\perp \exp(ik_s x \cdot \mathrm{Rot}_{\frac{4\pi}{3}}(\hat{\theta})). \end{aligned}$$

(iv) $\Lambda_1, \Lambda_2 \in \mathcal{N}_4(0, k_s)$, $\theta = 0$, $\pi_s \neq \emptyset$, *and the total field can be written as*

$$u(x) = e_1 \exp(-ik_s x_2) - e_1 \exp(ik_s x_2) + e_2 \exp(ik_s x_1) - e_2 \exp(-ik_p x_1).$$

As the counterpart of Remark 3.11 and Corollary 3.13, we have the following uniqueness results with the minimal number of incident shear waves.

Corollary 3.16. *Two (four) incident shear waves with distinct incidence angles are sufficient to uniquely determine a grating profile $\Lambda \in \mathcal{A}$ under the boundary conditions of the third (fourth) kind. If the Rayleigh frequencies of the shear part are excluded, then the minimal number is one (three).*

Remark 3.17. The uniqueness results in Section 4 generalize those for acoustic and electromagnetic gratings [35] to the elastic case. As one would expect, the derivation of the unidentifiable classes in 3D is more complicated than in 2D, since we must take into account biperiodic structures which vary in both x_1 and x_2 and where the incident wave is not perpendicular to the x_3-axis. In general, there exist five (two) classes of unidentifiable polyhedral grating profiles in the case of plane shear (pressure) wave incidence under the boundary conditions of the third (fourth) kind [37].

5 Numerical solution of direct and inverse scattering problems

5.1 A discrete Galerkin method for (DP)

In this section, we discuss the numerical treatment of the direct scattering problem for the Dirichlet boundary condition. The approach is based on the discrete Galerkin method proposed by Atkinson [13] for solving an equivalent integral equation of first kind. A similar method is used in [56] for solving the forward problem of elastic scattering from an open arc in $\mathbb{R}^2$. As mentioned in the introduction, the implementation of this method is easier than the integral equation method with a second-kind integral equation. This approach also leads to the unique solvability of the forward scattering problem for the Dirichlet boundary condition (see [38, Lemma 4.1]). In the sequel, we assume that the incident wave is an incident pressure wave, and that Λ is the graph of some C^2-smooth periodic function f. For piecewise linear gratings where the scattered field may be singular at corner points, one can adopt a mesh grading transformation to parameterize the grating profile (see [38, 43]). In this section, it is supposed that $\pi_p = \pi_s = \emptyset$, that is, the Rayleigh frequencies are excluded.

We first recall the free space fundamental solution $\Phi_k(x, y)$ to the Helmholtz equation $(\Delta + k^2)u = 0$ given by

$$\Phi_k(x,y) = \frac{i}{4} H_0^{(1)}(k|x-y|), \quad x \neq y,\ x = (x_1, x_2), y = (y_1, y_2) \in \mathbb{R}^2,$$

with $H_0^{(1)}(t)$ being the first-kind Hankel function of order zero, and then recall the α-quasiperiodic fundamental solution to the Helmholtz equation defined by (see e.g. [58])

$$G_k(x,y) = \sum_{n\in\mathbb{Z}} \exp(-i\alpha 2\pi n)\,\Phi_k(x+n(2\pi,0),y)$$
$$= \frac{i}{4\pi}\sum_{n\in\mathbb{Z}} \frac{1}{\beta_n}\exp(i\alpha_n(x_1-y_1)+i\beta_n|x_2-y_2|), \tag{3.35}$$

for $x-y\neq n(2\pi,0)$, $n\in\mathbb{Z}$, where β_n are defined as in (3.11) with k_p replaced by k. The free space fundamental solution to the Navier equation (3.3) is given by (see [56])

$$\Gamma(x,y) = \frac{1}{\mu}\Phi_{k_s}(x,y)\,I + \frac{1}{\omega^2}\operatorname{grad}_x \operatorname{grad}_x^T\left[\Phi_{k_s}(x,y) - \Phi_{k_p}(x,y)\right],$$

where I stands for the 2×2 unit matrix. Then, the α-quasiperiodic fundamental solution (Green's tensor) to the Navier equation (3.3) takes the form

$$\Pi(x,y) := \sum_{n\in\mathbb{Z}} \exp(-i\alpha 2\pi n)\,\Gamma(x+n(2\pi,0),y)$$

for $x-y\neq n(2\pi,0), n\in\mathbb{Z}$. The convergence of the above series for $\Pi(x,y)$ is discussed in [7, Section 6]. Define

$$\begin{pmatrix} P_{11}^{(n)} & P_{12}^{(n)} \\ P_{21}^{(n)} & P_{22}^{(n)} \end{pmatrix} = \frac{i}{4\pi\omega^2\beta_n}\begin{pmatrix} \alpha_n^2 & \alpha_n\beta_n \\ \alpha_n\beta_n & \beta_n^2 \end{pmatrix} =: P^{(n)}, \tag{3.36}$$

$$\begin{pmatrix} S_{11}^{(n)} & S_{12}^{(n)} \\ S_{21}^{(n)} & S_{22}^{(n)} \end{pmatrix} = \frac{i}{4\pi\mu\gamma_n} I - \frac{i}{4\pi\omega^2\gamma_n}\begin{pmatrix} \alpha_n^2 & \alpha_n\gamma_n \\ \alpha_n\gamma_n & \gamma_n^2 \end{pmatrix} =: S^{(n)}. \tag{3.37}$$

We can derive from (3.35) that $\Pi(x,y)$ takes the form

$$\Pi(x,y) = \sum_{n\in\mathbb{Z}} P^{(n)}\exp(i[\alpha_n(x_1-y_1)+\beta_n|x_2-y_2|\,])$$
$$+ \sum_{n\in\mathbb{Z}} S^{(n)}\exp(i[\alpha_n(x_1-y_1)+\gamma_n|x_2-y_2|\,]). \tag{3.38}$$

Our approach is based on the following decomposition of the α-quasiperiodic fundamental solution $\Pi(x,y)$ and the periodicity of the grating surface.

Lemma 3.18 ([38]). *On the profile Λ, the α-quasiperiodic Green's tensor $\Pi(x,y)$ has a logarithmic singularity of the form*

$$\Pi(x,y) = -\eta_1\frac{1}{\pi}\ln(|x-y|)\,I + \Pi^*(x,y), \quad \eta_1 = \frac{1}{\omega^2}(k_s^2+k_p^2), \tag{3.39}$$

where $\Pi^*(x,y)$ is a continuously differentiable matrix in the variables $x,y\in\Lambda$.

To solve problem (DP) under the boundary condition (3.4), we make the ansatz for the scattered field u^{sc} in the form

$$u^{sc}(x) = \int_\Lambda \Pi(x,y)\,\phi(y)ds(y), \quad x\in\Omega_\Lambda$$

with some unknown α-quasiperiodic function $\phi(y) \in L^2(\Lambda)^2$. Then we only need to solve the linear first-kind integral equation

$$\int_{\Lambda} \Pi(x,y)\,\phi(y)ds(y) = -u^{in}(x), \quad x \in \Lambda. \tag{3.40}$$

Set

$$\begin{aligned}
&x = (t, f(t)),\ y = (s, f(s)),\\
&g(t) := -u^{in}(t, f(t)) \exp(-i\alpha t),\\
&\rho(s) := \phi(s, f(s)) \exp(-i\alpha s)\sqrt{1 + f'(s)^2},\\
&K(t,s) := \Pi(t, f(t); s, f(s)) \exp(i\alpha(s - t)).
\end{aligned}$$

Multiplying (3.40) by $\exp(-i\alpha t)$ gives the equivalent form

$$\int_0^{2\pi} K(t,s)\rho(s)ds = g(t), \quad 0 \le t \le 2\pi. \tag{3.41}$$

Note that $\rho(t), g(t)$ are both 2π-periodic with respect to t. It follows from the decomposition in (3.39) that

$$K(t,s) = -\eta_1 \frac{1}{\pi} \ln \left| 2e^{-1/2} \sin\left(\frac{t-s}{2}\right)\right| I + H(t,s),$$

where $H(t,s)$ is a continuously differentiable function on $\mathbb{R} \times \mathbb{R}$. Define the integral operators

$$\begin{aligned}
A\rho(t) &:= -\eta_1 \frac{1}{\pi} \int_0^{2\pi} \ln \left| 2e^{-1/2} \sin\left(\frac{t-s}{2}\right)\right| I\rho(s)ds,\\
B\rho(t) &:= \int_0^{2\pi} H(t,s)\,\rho(s)ds.
\end{aligned}$$

Let $H_p^1(0,2\pi)$ denote the Sobolev space of 2π-periodic functions on $(0,2\pi)$. Then, solving the first-kind integral equation (3.41) can be transformed into:

$$\text{Given } g \in H_p^1(0,2\pi)^2, \text{ find } \rho(t) \in L^2(0,2\pi)^2 \text{ such that } A\rho + B\rho = g. \tag{3.42}$$

Let further $\mathcal{T}_n$ denote the $(2n+1)$-dimensional space of trigonometric polynomials of degree not greater than n, with the basis given by $\{\varphi_m(t) := e^{imt},\ -n \le m \le n\}$. Then the orthogonal projection P_n of $L^2(0,2\pi)^2$ onto $\mathcal{T}_n^2$ takes the form

$$(P_n\rho)(t) = \frac{1}{\sqrt{2\pi}} \sum_{m=-n}^{n} \rho_m e^{imt}, \quad \rho_m = \frac{1}{\sqrt{2\pi}} \int_0^{2\pi} \rho(s) e^{-ims} ds \in \mathbb{C}^2,$$

and the corresponding Galerkin method for (3.42) consists of solving the projection equation

$$(A + P_nB)\rho_n = P_n g, \quad \text{for } \rho_n = \sum_{j=-n}^{n} c_j\varphi_j(t) \in \mathcal{J}_n^2, \quad c_j \in \mathbb{C}^2. \tag{3.43}$$

Let $C_p[0,2\pi]$ denote the continuous complex-valued 2π-periodic functions in t. The basic idea of the discrete Galerkin method proposed in [13] is to replace the orthogonal projection P_n by the interpolatory projection $Q_n : C_p[0,2\pi]^2 \to \mathcal{J}_n^2$ at the equidistant grid points $t_j = jh, h = 2\pi/(2n+1)$, and to approximate (3.43) by

$$(A + Q_nB_n)\rho_n = Q_n g, \quad \rho_n \in \mathcal{J}_n^2, \tag{3.44}$$

where the integral operator B is approximated by a finite-dimensional operator using the trapezoidal rule

$$B_n\rho_n(t) = h\sum_{j=0}^{2n} H(t,t_j)\rho_n(t_j). \tag{3.45}$$

To avoid the computation of $H(t,s)$ for $t = s$ (i.e. the diagonal terms), we introduce the collocation points $s_k = kh + h/2, k = 0,1,\ldots,2n$, which is a shift of the equidistant grid points t_j. Then, problems (3.44) and (3.45) are equivalent to

$$\sum_{j=-n}^{n}\left[\frac{\eta_1\varphi_j(s_k)}{\max\{1,|j|\}}I + B_n\varphi_j(s_k)\right]c_j = g(s_k), \quad k = 0,1,\ldots,2n.$$

Using (3.45) and the orthogonality of φ_m, the previous finite linear system becomes (see also [43, section 3])

$$\sum_{j=0}^{2n}\left[\eta_1\sigma_{kj} + hH(s_k,t_j)\right]\rho_n(t_j) = g(s_k), \quad k = 0,1,\ldots,2n, \tag{3.46}$$

in terms of the unknown density ρ, where

$$\sigma_{kj} = \frac{1}{2\pi}h\sum_{m=-n}^{n}\varphi_m(s_k)\overline{\varphi_j(t_m)}/\max\{1,|m|\}.$$

Note that $s_k \neq t_j$ for all $k, j = 0,1,\ldots,2n$, and that the σ_{kj} can be readily computed employing the fast Fourier transform. Solving the linear system (3.46) in the unknowns $\rho_n(t_j)$ then gives an approximate solution $\rho_n \in \mathcal{J}_n$ of the integral equation (3.40). We refer to [13] for a convergence analysis of this approximation method.

5.2 A two-step algorithm for (IP)

In this section, we present a numerical method for the profile reconstruction problem (IP) in the case that the Dirichlet condition is imposed on the unknown profile Λ.

We assume that Λ is the graph of a sufficiently smooth 2π-periodic function f, and suppose that we have the a priori information that Λ lies between the horizontal lines $\{x_2 = 0\}$ and $\{x_2 = b\}$ for some $b > 0$. For simplicity we restrict ourselves to the case of one incident wave with incidence angle θ. We can rewrite the corresponding near-field data $u_b^{sc}(x_1) = u^{sc}(x_1, b)$ and far-field data $u_b^{\infty}(x_1)$ as

$$u_b^{sc}(x_1) = \sum_{n\in\mathbb{Z}} \mathbf{A}_n(\theta)\exp(i\alpha_n x_1), \quad u_b^{\infty}(x_1) = \sum_{n\in\mathcal{U}_s} \mathbf{A}_n^{\infty}(\theta)\exp(i\alpha_n x_1), \tag{3.47}$$

where

$$\mathbf{A}_n(\theta) = A_{p,n}(\alpha_n, \beta_n)^T e^{i\beta_n b} + A_{s,n}(-\gamma_n, \alpha_n)^T e^{i\gamma_n b},$$

$$\mathbf{A}_n^{\infty}(\theta) = \begin{cases} \mathbf{A}_n(\theta), & \text{if } n \in \mathcal{U}_p, \\ A_{p,n}(\alpha_n, \beta_n)^T e^{i\beta_n b}, & \text{if } n \in \mathcal{U}_s\backslash\mathcal{U}_p. \end{cases}$$

We consider the following inverse problem, which is a slightly more general version of problem (IP*) formulated at the end of Section 2 (in the case of a single incidence angle).

(IP′): Given a finite number of Rayleigh coefficients $\mathbf{A}_n(\theta)$, $|n| < K$, for some $K \in \mathbb{N}$, determine the unknown grating profile Λ.

Note that the far-field data $\mathbf{A}_n^{\infty}(\theta)$, $n \in \mathcal{U}_s$, are covered if K is sufficiently large. Since (IP′) is nonlinear and severely ill-posed, it is quite natural to apply regularization and optimization techniques. Introduce the Hilbert space $X = L^2(0,2\pi)^2$ with the scalar product

$$(u(t), v(t)) := \frac{1}{2\pi}\int_0^{2\pi} u(t)\cdot\overline{v(t)}dt,$$

and the norm $\|u\| := \sqrt{(u,u)}$. Define the linear operators $J, S_f : X \to X$ by

$$J\varphi(x_1) := \frac{1}{2\pi}\int_0^{2\pi} \Pi(x_1, b; t, 0)\varphi(t)dt,$$

$$S_f\varphi(x_1) := \frac{1}{2\pi}\int_0^{2\pi} \Pi(x_1, f(x_1); t, 0)\varphi(t)dt.$$

The Kirsch–Kress method adapted to our diffraction problem consists of solving the optimization problem

$$\|T\varphi - u_b\|^2 + \gamma\|\varphi\|^2 + \eta\left\|u^{in}\circ f + S_f\varphi\right\|^2 \to \inf_{f\in\mathcal{M},\varphi\in X}, \tag{3.48}$$

where $\gamma > 0$ denotes the regularization parameter, $\eta > 0$ is a coupling parameter and $\mathcal{M}$ is an admissible set of profile functions with uniformly bounded $C^{0,1}$-norm.

The convergence analysis for problem (3.48) is presented in [44] in the case of the quasiperiodic Helmholtz equation, which we think can carry over to the quasiperiodic Navier equation. Since the combined optimization scheme (3.48) requires the determination of two unknown functions f and φ, to reduce the computational effort, we extend the two-step inversion algorithm proposed in [22] for electromagnetic gratings to the inverse elastic scattering problem (IP′).

Step 1: Reconstruct the scattered field from the near-field data $u_b^{sc}(x_1)$.

We try to represent the scattered field u^{sc} as the single layer potential

$$u^{sc}(x) = \frac{1}{2\pi}\int_0^{2\pi} \Pi(x_1,x_2;t,0)\varphi(t)dt, \quad x \in \Omega_\Lambda$$

with some unknown α-quasiperiodic function $\varphi(t) \in X$. Then we only need to solve the first-kind integral equation

$$J\varphi(x_1) = u_b^{sc}(x_1), \qquad x_1 \in (0,2\pi). \tag{3.49}$$

Expand $\varphi(t) \in X$ into the Fourier series

$$\varphi(t) = \sum_{n\in\mathbb{Z}} \varphi_n \exp(i\alpha_n t), \quad \varphi_n := \left(\varphi_n^{(1)}, \varphi_n^{(2)}\right) \in \mathbb{C}^2.$$

Then, it follows from (3.36)–(3.38) that the operator J takes the form

$$J\varphi(x_1) = \sum_{n\in\mathbb{Z}} M^{(n)}\varphi_n \exp(i\alpha_n x_1), \quad M^{(n)} := \left(P^{(n)}\exp(i\beta_n b) + S^{(n)}\exp(i\gamma_n b)\right).$$

Since the operator J is compact on X, we consider the Tikhonov regularized version $\gamma\varphi + J^*J\varphi = J^*u_b^{sc}$ of equation (3.49) with the regularization parameter $\gamma > 0$, where J^* denotes the adjoint operator of J. Let the singular value decomposition of $M^{(n)}$ be given by $M^{(n)} = U^{(n)}\Sigma^{(n)}(V^{(n)})^*$, where

$$U^{(n)} = (U_1^{(n)}, U_2^{(n)}), \quad V^{(n)} = (V_1^{(n)}, V_2^{(n)}), \quad \Sigma^{(n)} = \mathrm{diag}(\sigma_1^{(n)}, \sigma_2^{(n)}).$$

with $U_j^{(n)}, V_j^{(n)} \in \mathbb{C}^2$ being column vectors and $\sigma_j^{(n)} \in \mathbb{R}^+$ for $n \in \mathbb{Z}$, $j = 1,2$. Then, the regularized solution φ_γ is given by (see [27, Chapter 4])

$$\begin{aligned}\varphi_\gamma &= \sum_{n\in\mathbb{Z}}\sum_{j=1}^{2} \frac{\sigma_j^{(n)}}{(\sigma_j^{(n)})^2+\gamma}\left(u_b^{sc}, U_j^{(n)}\exp(i\alpha_n t)\right) V_j^{(n)}\exp(i\alpha_n t)\\ &\approx \sum_{|n|\le K}\sum_{j=1}^{2} \frac{\sigma_j^{(n)}}{(\sigma_j^{(n)})^2+\gamma}\left(\mathbf{A}_n\cdot\overline{U_j^{(n)}}\right) V_j^{(n)}\exp(i\alpha_n t)\end{aligned}$$

in the case that near-field data are given, where $\mathbf{A}_n \in \mathbb{C}^2$ are defined in (3.47). Now we can represent φ_γ as

$$\varphi_\gamma = \sum_{|n|\le K} \varphi_\gamma^{(n)}\exp(i\alpha_n t), \quad \varphi_\gamma^{(n)} := \sum_{j=1}^{2} \frac{\sigma_j^{(n)}}{(\sigma_j^{(n)})^2+\gamma}\left(\mathbf{A}_n\cdot\overline{U_j^{(n)}}\right) V_j^{(n)}. \tag{3.50}$$

When far-field data are given, we only need to replace $u_b^{sc}, \mathbf{A}_n$ by $u_\infty^{sc}, \mathbf{A}_n^\infty$, respectively.

Step 2: Determine the grating profile function f by minimizing the defect

$$\left\| u^{in} \circ f + S_f \varphi \right\| \to \inf_{f \in \mathcal{M}} \tag{3.51}$$

over some admissible set $\mathcal{M}$ of profile functions.

Having computed φ_γ from the first step, we may consider $S_f(\varphi_\gamma)$ as an approximation of the scattered field on the grating profile. We now turn to investigating the nonlinear least-squares minimization problem (3.51). Using (3.36)–(3.38) and (3.50), we see that the integral operator S_f takes the form

$$S_f \varphi_\gamma(x_1) = \sum_{|n| \le K} P^{(n)} \varphi_\gamma^{(n)} e^{i\alpha_n x_1 + i\beta_n f(x_1)} + S^{(n)} \varphi_\gamma^{(n)} e^{i\alpha_n x_1 + i\gamma_n f(x_1)}.$$

Hence, problem (3.51) is equivalent to

$$\left\| \hat{\theta} e^{-i\beta f(t)} + \sum_{|n| \le K} \left(P^{(n)} \varphi_\gamma^{(n)} e^{i\beta_n f(t)} + S^{(n)} \varphi_\gamma^{(n)} e^{i\gamma_n f(t)} \right) e^{int} \right\|^2 \to \inf_{f \in \mathcal{M}}, \tag{3.52}$$

where, for C^2-smooth grating profiles, an admissible set $\mathcal{M}$ is given by

$$\mathcal{M} = \left\{ f(t) = a_0 + \sum_{m=1}^{M} a_m \cos(mt) + a_{M+m} \sin(mt) \right\}, \tag{3.53}$$

with some fixed number $M \in \mathbb{N}$ and bounded Fourier coefficients a_j. We discretize the objective functional in (3.52) by the trapezoidal rule and then solve the resulting minimization problem in a finite-dimensional space.

Here we present two numerical examples using the two-step algorithm. Take $k_s = 4.45$, $\omega = 5$, and probe the unknown grating profile by a single incident pressure wave with $\theta = 0$ and $k_p = 4.2$. With these settings we have $\mathcal{U}_p = \mathcal{U}_s = \{n \in \mathbb{Z} : |n| \le 4\}$. In the following Examples 3.19 and 3.20, unless otherwise stated we always set $K = 7$ in (3.50) and (3.52) . This implies that all propagating modes of the compressional and shear parts corresponding to $|n| \le 4$ are used, while six additional evanescent modes corresponding to $5 \le |n| \le 7$ are also taken into account.

Example 3.19 (Fourier gratings). Suppose that the grating profile function is given by the finite Fourier series

$$f(t) = 2 + \zeta(\cos(t) + \cos(2t) + \cos(3t)), \quad \zeta = 0.05\pi,$$

where ζ characterizes the steepness of the profile. We used exact near-field data to reconstruct this profile function, which has the form (3.53) with $M = 3$. To generate

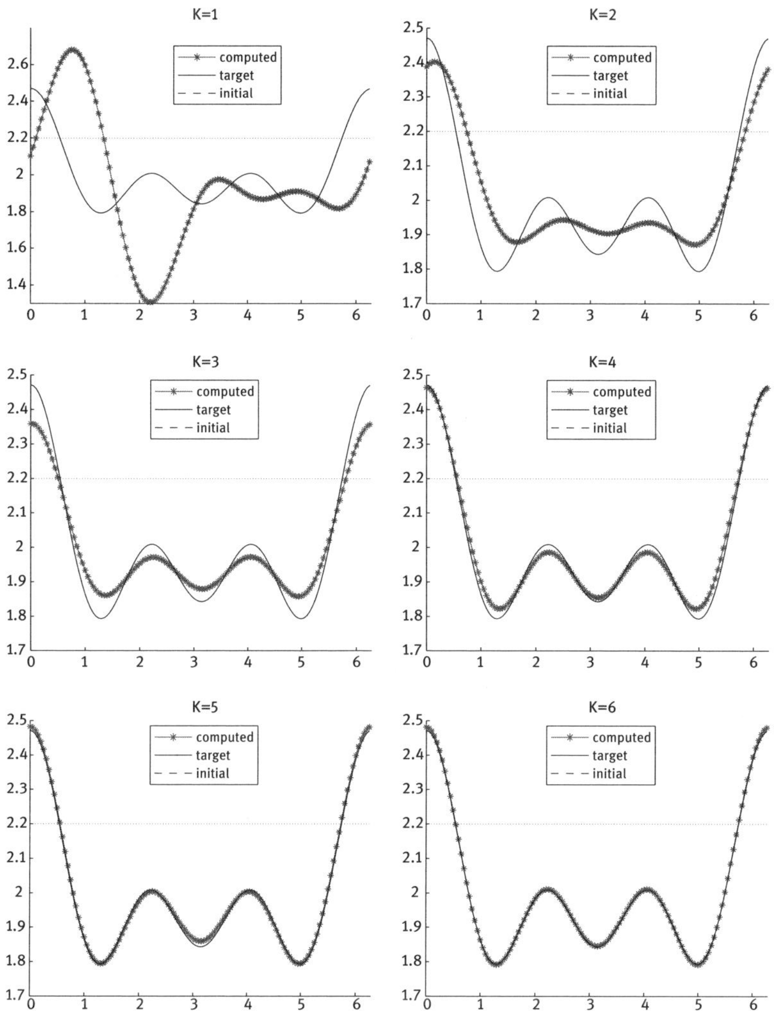

Figure 3.3: (See Color Plates for a color version.) Sensitivity of the method to the parameter K in Example 3.19.

the synthetic data, the discrete trigonometric Galerkin method of Section 5.1 was employed. Figure 3.3 illustrates the sensitivity of the method to the parameter K, from which we see that the propagating modes corresponding to $K = 4$ (i.e., only far-field data are used) lead to satisfactory results. The reconstruction from noisy data even with noise level $\delta = 10\%$ can still produce good results in the case $K = 7$. However, the results are not acceptable if we increase the steepness to 0.1π.

Example 3.20 (General smooth gratings). Suppose that Λ is the graph given by the function

$$f(t) = 1.5 + 0.2\exp(\sin(3t)) + 0.3\exp(\sin(3t)),$$

which can be approximated by the truncated Fourier series

$$f^*(t) = 2.133 - 0.0543\cos(6t) - 0.0814\cos(8t) + 0.22606\sin(3t) + 0.339\sin(4t)$$

with four interior local minima. In this case we chose $M = 8$ in (3.53), with a total of 17 parameters to be determined. The computational results are presented in Figures 3.4 and 3.5. Since the steepness of Λ is relatively large, the downward convex part of the grating surface is not well-reconstructed. One can see from Figure 3.4 that data with 10% noise lead to a larger deviation in contrast to the profile reconstructed from data corresponding to a noise of 5%.

Our experiments have shown that the reconstruction with $K < 7$ also leads to a large deviation and that the one with $K \geq 8$ only slightly improves the computational results in Figure 3.4. Figure 3.5 demonstrates the computational results from different initial guesses using unperturbed propagating modes.

The two-step algorithm is easily implemented, and satisfactory reconstructions can be obtained with a low computational effort for suitable initial values of grating parameters. This is mainly because the singular value decomposition of the derived first-kind integral equation can be readily achieved, and only the unknown grating

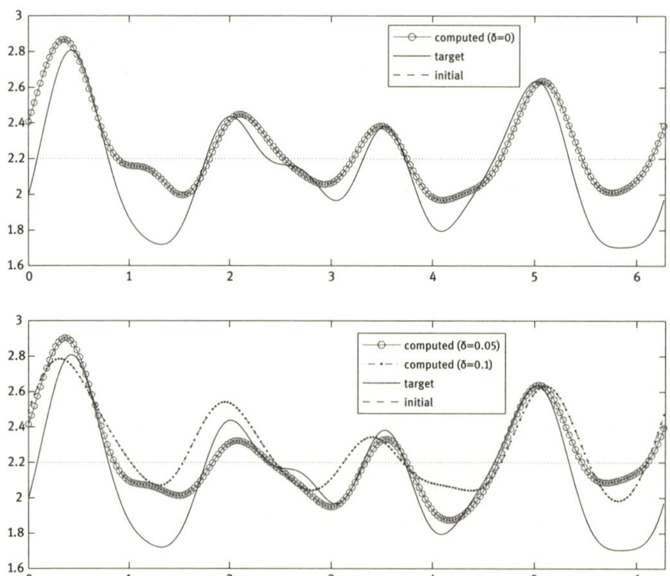

Figure 3.4: (See Color Plates for a color version.) Example 3.20. $K = 7, M = 8$.

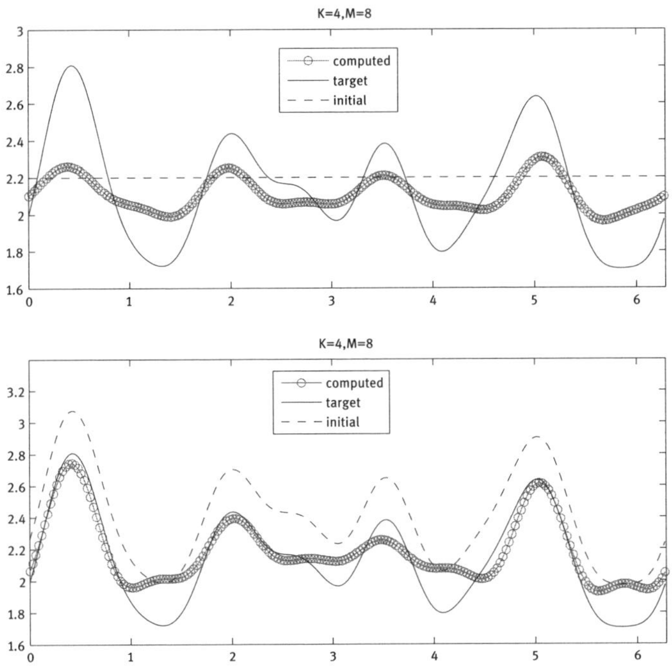

Figure 3.5: (See Color Plates for a color version.) Example 3.20. $K = 4, M = 8, \delta = 0$.

profile function needs to be determined in the second step. Moreover, no direct scattering problems need to be solved in the process of the inversion algorithm. As our experience has shown in [22–24], the two step algorithm is in general faster and more accurate than the minimization of the combined cost functional. The reconstruction scheme can also be applied to piecewise linear grating profiles with a finite number of corners (see [38]), and it can be readily adapted to the case of several incident angles or a finite number of incident frequencies. However, the convergence of the two-step algorithm is still open.

References

[1] T. Abboud, Formulation variationnelle des équations de Maxwell dans un réseau bipériodique de $\mathbb{R}^3$, *C.R. Acad. Sci. Paris* **317** (1993), 245–248.

[2] I. Abubakar, Scattering of plane elastic waves at rough surface I, *Proc. Cambridge Philos. Soc.* **58** (1962), 136–157.

[3] J. D. Achenbach, *Wave Propagation in Elastic Solids* (Amsterdam: North Holland) (1973).

[4] G. Alessandrini and L. Rondi, 2005 Determining a sound-soft polyhedral scatterer by a single far-field measurement, *Proc. Amer. Math. Soc.* **133** (2005), 1685–1691 (Corrigendum: http://arxiv.org/abs/math.AP/0601406).

[5] H. Ammari, Uniqueness theorems for an inverse problem in a doubly periodic structure, *Inverse Problems* **11** (1995), 823–833.

[6] C. Antonios, G. Drossos, and K. Kiriakie, On the uniqueness of the inverse elastic scattering problem for periodic structures, *Inverse Problems* **17** (2001), 1923–1935.

[7] T. Arens, The scattering of plane elastic waves by a one-dimensional periodic surface, *Math. Meth. Appl. Sci.* **22** (1999), 55–72.

[8] T. Arens, A new integral equation formulation for the scattering of plane elastic waves by diffraction gratings, *J. Integral Equations Appl.* **11** (1999), 275–297.

[9] T. Arens, Uniqueness for elastic wave scattering by rough surfaces, *SIAM J. Math. Anal.* **33** (2001), 461–471.

[10] T. Arens, Existence of solution in elastic wave scattering by unbounded rough surfaces *Math. Meth. Appl. Sci.* **25** (2002), 507–528.

[11] T. Arens and N. Grinberg, A complete factorization method for scattering by periodic surface, *Computing* **75** (2005), 111–132.

[12] T. Arens and A. Kirsch, The factorization method in inverse scattering from periodic structures, *Inverse Problems* **19** (2003), 1195–1211.

[13] K. E. Atkinson, A discrete Galerkin method for first kind integral equations with a logarithmic kernel, *J. Integral Equations Appl.* **1** (1988), 343–363.

[14] G. Bao, A uniqueness theorem for an inverse problem in periodic diffractive optics, *Inverse Problems* **10** (1994), 335–340.

[15] G. Bao, L. Cowsar, and W. Masters, eds. *Mathematical Modeling in Optical Science* (Philadelphia: SIAM) (2001).

[16] G. Bao, H. Zhang, and J. Zou, Unique determination of periodic polyhedral structures by scattered electromagnetic fields, *Trans. Amer. Math. Soc.* **363** (2011), 4527–4551.

[17] G. Bao, H. Zhang, and J. Zou, Unique determination of periodic polyhedral structures by scattered electromagnetic fields II: the resonance case, *to appear in: Trans. Amer. Math. Soc.*

[18] G. Bao and Z. Zhou, An inverse problem for scattering by a doubly periodic structure, *Trans. Amer. Math. Soc.* **350** (1998), 4089–4103.

[19] G. Bao, Finite element approximation of time harmonic waves in periodic structures, *SIAM J. Numer. Anal.* **32** (1995), 1155–1169.

[20] G. Bao, Z. Chen, and H. Wu, Adaptive finite-element method for diffraction gratings, *J. Opt. Soc. Amer. A* **22** (2005), 1106–1114.

[21] A. S. Bonnet-Bendhia and F. Starling, Guided waves by electromagnetic gratings and non-uniqueness examples for the diffraction problem, *Math. Meth. Appl. Sci.* **17** (1994), 305–338.

[22] G. Bruckner and J. Elschner, A two-step algorithm for the reconstruction of perfectly reflecting periodic profiles, *Inverse Problems* **19** (2003), 315–329.

[23] G. Bruckner and J. Elschner, The numerical solution of an inverse periodic transmission problem, *Math. Meth. Appl. Sci.* **28** (2005), 757–778.

[24] G. Bruckner, J. Elschner, and M. Yamamoto, An optimization method for profile reconstruction, In: *Progress in Analysis, Proceed. 3rd ISAAC congress* (Singapore: World Scientific) (2003), 1391–1404.

[25] X. Chen and A. Friedman, Maxwell's equations in a periodic structure, *Trans. Amer. Math. Soc.* **323** (1991), 465–507.

[26] J. Cheng and M. Yamamoto, Uniqueness in an inverse scattering problem with non-trapping polygonal obstacles with at most two incoming waves, *Inverse Problems* **19** (2003), 1361–1384 (Corrigendum: *Inverse Problems* **21**, 1193).

[27] D. Colton and R. Kress, *Inverse Acoustic and Electromagnetic Scattering Theory* (Berlin: Springer) (1998).

[28] P. Cummings and X. Feng, Sharp regularity coefficient estimates for complex-valued acoustic and elastic Helmholtz equations, *Mathematical Models and Methods in Applied Sciences* **16** (2006), 139–160.
[29] Y. Dermenjian, J. C. Guillot, Scattering of elastic waves in a perturbed isotropic half space with a free boundary. The limiting absorption principle, *Math. Meth. Appl. Sci.* **10** (1988), 87–124.
[30] D. C. Dobson, A variational method for electromagnetic diffraction in biperiodic structures, *Model. Math. Anal. Numer.* **28** (1994), 419–439.
[31] G. Duvaut and J. L. Lions *Inequalities in Mechanics and Physics* (Berlin: Springer) (1976).
[32] D. Dobson and A. Friedman, The time-harmonic Maxwell equations in a doubly periodic structure, *J. Math. Anal. Appl.* **166** (1992), 507–528.
[33] J. Elschner, R. Hinder, F. Penzel, and G. Schmidt, Existence, uniqueness and regularity for solutions of the conical diffraction problem, *Mathematical Models and Methods in Applied Sciences* **10** (2000), 317–341.
[34] J. Elschner and G. Hu, Variational approach to scattering of plane elastic waves by diffraction gratings, *Math. Meth. Appl. Sci.* **33** (2010), 1924–1941.
[35] J. Elschner and G. Hu, Global uniqueness in determining polygonal periodic structures with a minimal number of incident plane waves, *Inverse Problems* **26** (2010), 115002.
[36] J. Elschner and G. Hu, Inverse scattering of elastic waves by periodic structures: Uniqueness under the third or fourth kind boundary conditions, *Methods and Applications of Analysis* **18** (2011), 215–244.
[37] J. Elschner and G. Hu, Uniqueness in inverse scattering of elastic waves by three-dimensional polyhedral diffraction gratings, *J. Inverse Ill-Posed Probl.* **19** (2011), 717–768.
[38] J. Elschner and G. Hu, An optimization method in inverse elastic scattering for one-dimensional grating profiles, *Communications in Computational Physics* **16** (2012), 1434–1460.
[39] J. Elschner and G. Hu, Scattering of plane elastic waves by three-dimensional diffraction gratings, *Mathematical Models and Methods in Applied Sciences* **22** (2012), 1150019.
[40] J. Elschner and G. Hu, Elastic scattering by unbounded rough surfaces, *SIAM J. Math. Anal.* **44** (2012), 4101–4127.
[41] J. Elschner, G. Schmidt, and M. Yamamoto, An inverse problem in periodic diffractive optics: global uniqueness with a single wavenumber, *Inverse Problems* **19** (2003), 779–787.
[42] J. Elschner and G. Schmidt, Diffraction in periodic structures and optimal design of binary gratings I. Direct problems and gradient formulas, *Math. Meth. Appl. Sci.* **21** (1998), 1297–1342.
[43] J. Elschner and E. P. Stephan, A discrete collocation method for Symm's integral equation on curves with corners, *Journal of Computational and Applied Mathematics* **75** (1996), 131–146.
[44] J. Elschner and M. Yamamoto, An inverse problem in periodic diffractive optics: reconstruction of Lipschitz grating profiles, *Applicable Analysis* **81** (2002), 1307–1328.
[45] J. Elschner and M. Yamamoto, Uniqueness in determining polygonal periodic structures, *Z. Anal. Anwend.* **26** (2007), 165–177.
[46] J. Elschner and M. Yamamoto, Uniqueness in determining polyhedral sound-hard obstacles with a single incoming wave, *Inverse Problems* **24** (2008), 035004.
[47] J. Elschner and M. Yamamoto, Uniqueness in inverse elastic scattering with finitely many incident waves, *Inverse Problems* **26** (2010), 045005.
[48] J. T. Fokkema, Reflection and transmission of elastic waves by the spatially periodic interface between two solids (theory of the integral-equation method), *Wave Motion* **2** (1980), 375–393.
[49] J. T. Fokkema and P. M. van den Berg, Elastodynamic diffraction by a periodic rough surface (stress-free boundary), *J. Acoust. Soc. Am.* **62** (1977), 1095–1101.

[50] I. C. Gohberg and M. G. Krein, *Introduction to the Theory of Linear Nonselfadjoint Operators in Hilbert Space*, Translations of Mathematical Monographs Vol.18 (Providence: AMS) (1969).
[51] F. Hettlich, Iterative regularization schemes in inverse scattering by periodic structures, *Inverse Problems* **18** (2002),701–714.
[52] F. Hettlich and A. Kirsch, Schiffer's theorem in inverse scattering for periodic structures, *Inverse Problems* **13** (1997), 351–361.
[53] K. Ito and F. Reitich, A high-order perturbation approach to profile reconstruction: I. Perfectly conducting grating, *Inverse Problems* **15** (1999), 1067–1085.
[54] A. Kirsch, Diffraction by periodic structures, In: *Proc. Lapland Conf. Inverse Problems* eds. L. Päivärinta et al (Berlin: Springer) (1993), 87–102.
[55] A. Kirsch, Uniqueness theorems in inverse scattering theory for periodic structures, *Inverse Problems* **10** (1994), 145–152.
[56] R. Kress, Inverse elastic scattering from a crack, *Inverse Problems* **12** (1996), 667–684.
[57] V. D. Kupradze et al *Three-dimensional Problems of the Mathematical Theory of Elasticity and Thermoelasticity* (Amsterdam: North-Holland) (1979).
[58] C. M. Linton, The Green's function for the two-dimensional Helmholtz equation in periodic domains, *Journal of Engineering Mathematics* **33** (1998), 377–401.
[59] H. Liu and J. Zou, Uniqueness in an inverse acoustic obstacle scattering problem for both sound-hard and sound-soft polyhedral scatterers, *Inverse Problems* **23** (2006), 515–524.
[60] H. Liu, M. Yamamoto, and J. Zou, Reflection principle for the Maxwell equations and its application to inverse electromagnetic scattering, *Inverse Problems* **23** (2007), 2357–2366.
[61] H. Liu, M. Yamamoto, and J. Zou, New reflection principles for Maxwell's equations and their applications, *Numer. Math. Theory Methods Appl.* **2** (2009), 1–17.
[62] J. W. S. Lord Rayleigh, On the dynamical theory of gratings, *Proc. Roy. Soc. Lon. A.* **79** (1907), 399–416.
[63] W. McLean, *Strongly Elliptic Systems and Boundary Integral Equations* (Cambridge: Cambridge University Press) (2000).
[64] A. Meier, T. Arens, S. N. Chandler-Wilde, and A. Kirsch, A Nyström method for a class of integral equations on the real line with applications to scattering by diffraction gratings and rough surfaces, *J. Integral Equations Appl.* **12** (2000), 281–321.
[65] S. Nakagawa, K. T. Nihei, L. R. Myer, and E. L. Majer, Three-dimensional elastic wave scattering by a layer containing vertical periodic fractures, *J. Acoust. Soc. Am.* **113** (2003), 3012–3023.
[66] J. Nečas, *Les Méthodes Directes en Théorie des Équations Elliptiques* (Paris: Masson) (1967).
[67] J. C. Nedelec and F. Starling, Integral equation methods in a quasi-periodic diffraction problem for the time-harmonic Maxwell equation, *SIAM J. Math. Anal.* **22** (1991), 1679–1701.
[68] A. Rathsfeld, G. Schmidt, and B. H. Kleemann, On a fast integral equation method for diffraction gratings, *Communications in Computational Physics* **1** (2006), 984–1009.
[69] J. H. Rose, Elastic wave inverse scattering in nondestructive evaluation, *Pure and Applied Geophysics* **131** (1989), 715–739.
[70] G. Schmidt, On the diffraction by biperiodic anisotropic structures, *Applicable Analysis* **82** (2003), 75–92.
[71] J. W. C. Sherwood, Elastic wave propagation in a semi-infinite solid medium, *Proc. Phys. Soc.* **71** (1958), 207–219.
[72] C. H. Wilcox, *Scattering Theory for Diffraction Gratings* (New York: Springer) (1984).
[73] T. T. Wu and P. L. Liu, Advancement on the nondestructive evaluation of concrete using transient elastic waves, *Ultrasonics* **36** (1998), 197–204.

Oliver G. Ernst and Martin J. Gander

Multigrid methods for Helmholtz problems: A convergent scheme in 1D using standard components

Abstract: We analyze in detail two-grid methods for solving the 1D Helmholtz equation discretized by a standard finite-difference scheme. We explain why both basic components, smoothing and coarse-grid correction, fail for high wave numbers, and show how these components can be modified to obtain a convergent iteration. We show how the parameters of a two-step Jacobi method can be chosen to yield a stable and convergent smoother for the Helmholtz equation. We also stabilize the coarse-grid correction by using a modified wave number determined by dispersion analysis on the coarse grid. Using these modified components we obtain a convergent multigrid iteration for a large range of wave numbers. We also present a complexity analysis which shows that the work scales favorably with the wave number.

Keywords: multigrid methods, Helmholtz equation, two-step smoother, Krylov smoother, dispersion correction

2010 Mathematics Subject Classification: 65N55, 65F10, 65N22

Oliver G. Ernst: Fakultät Mathematik und Informatik, TU Bergakademie Freiberg, Akademiestr. 6, 09595 Freiberg, Germany, ernst@math.tu-freiberg.de
Martin J. Gander: Section de Mathématiques, Université de Genève, 2–4 rue du Lièvre, CP 64, CH-1211 Genève, Switzerland, Martin.Gander@unige.ch

1 Introduction

The development of multigrid methods in the 1960s represents a milestone for solving linear systems of equations obtained by discretizing elliptic PDEs, allowing such a system to be solved with optimal work and storage complexity, i.e. proportional to the number of unknowns. While enormously effective when applied to elliptic equations as well as a multitude of more difficult problems (see [45] and the references therein), extending the multigrid method for solving the Helmholtz equation with the same efficiency remains an open problem.

The Helmholtz equation

$$\Delta u + k^2 u = f \tag{4.1}$$

is satisfied by the spatial part $u = u(x)$ of a time-harmonic solution $w(x,t) = u(x)e^{-i\omega t}$ with angular frequency $\omega \in \mathbb{R}$ of the linear wave equation $w_{tt} = c^2 \Delta w - \tilde{f}$ with propagation speed c. For a homogeneous medium, which we shall assume throughout, the *wave number* k is given by the constant $k = \omega/c$. The wave number is real for models of purely propagating waves, which is the case of primary interest and highest difficulty. When the underlying medium is absorbing, k is complex and can be normalized to have a nonnegative real part (cf. [14]).

Multigrid methods consist of an ingenious combination of the two complementary iterative solution techniques *smoothing* and *coarse-grid correction*. Smoothing, in its simplest form, refers to classical so-called *relaxation methods*, a term introduced by Southwell in the 1930s [43], for successively updating the solution values of a discretized problem, see also Richardson's early paper [41] from 1910[1]. The comprehensive theory of relaxation methods for elliptic PDEs is summarized in the monograph of Varga [49]. Smoothers are characterized algorithmically by updating each solution value by local averages of neighboring values, resulting in linear complexity for each smoothing step. For elliptic PDEs this typically results in strong damping of oscillatory components of the solution error in each step, while smooth components remain essentially unaffected. These smooth error components remaining after smoothing are reduced by the coarse-grid correction, in which an approximation of the original elliptic problem on a coarser grid is solved, in the simplest case by solving the residual equation for the error. This simple combination of smoothing and coarse-grid correction is known as a *two-grid iteration*; a *multigrid* method is obtained by solving the coarse-grid equation in the same way, descending recursively through increasingly coarser grids and ending ultimately at a grid for which the discrete equation can be solved with trivial effort. Classical multigrid theory is described in [10, 12, 33, 45].

It was realized early on [4, 6, 8, 42, 50] that the basic multigrid approach will indeed work for the Helmholtz equation if the wave number is small relative to the mesh size. Essentially, this means that the coarsest grids employed in the multigrid hierarchy must resolve the waves of wave number k. This is however not very useful in practical situations, particularly for large-scale calculations, since then the finest grid is precisely chosen to resolve the waves of wave number k, and one could not use any coarsening at all in a multigrid context.

Thus, it is still the case that classical multigrid methods do not work when applied to discretizations of the Helmholtz equation, see for example [20, 22] and references therein. Textbooks on multigrid often comment on the difficulty of the case of indefinite problems, for example the early multigrid guide by Brandt and Livne from 1984,

---

1 This chapter is exemplary for a modern paper in numerical analysis: there are several physical problems modeled by PDEs, which are then discretized, and solved by iteration (Richardson iteration, which is almost the Chebyshev semi-iterative method in its original form, albeit constructing an approximate Chebyshev polynomial by trial and error, and Richardson extrapolation), with realistic wallclock times (flops Richardson and his boy calculators were able to attain), computing costs ($\frac{n}{18}$ pence per coordinate relaxation of Δ_h, n being the number of digits), and finally there is a real application, a masonry dam.

which has recently been republished in the SIAM Classics in Applied Mathematics series [10, page 72], contains the quote:

> "If σ is negative, the situation is much worse, whatever the boundary conditions."

In the formative 1985 monograph by Hackbusch [33, page 212], we find

> "We have restricted our interest to weakly indefinite problems since severe difficulties arise for very small ε."

And in the more recent multigrid book by Trottenberg, Oosterlee and Schüller [45, page 400], the authors state

> "The multigrid treatment of the case $c < 0$ can become rather involved and will be a real difficulty if $-c$ gets very large."

In these references, the authors then give some brief indications of why standard multigrid methods fail to work effectively for Helmholtz type problems, and many other textbooks simply do not mention the difficult Helmholtz case at all. In the research literature, in an early paper [9], Brandt and Ta'asan investigated why a slight indefiniteness already poses problems for the classical multigrid algorithm, and proposed to treat the small subspace where the algorithm fails explicitly by deflation. This subspace, however, becomes unmanageably large for increasingly large wave number, and therefore one can no longer treat it explicitly when the goal is an algorithm with optimal complexity. A new technique for treating the subspace, namely the ray cycle, was introduced in the PhD thesis of Livshits [39], with a detailed study of the one-dimensional case, and a higher dimensional variant appeared in Brandt and Livshits [11], see also [40]. The so called wave-ray algorithm has, however, a considerable number of technical and algorithmic complications, which make both the implementation and the rigorous convergence analysis challenging, and this has prevented the algorithm from being widely used. A different approach at first sight is to write the Helmholtz equation as a first-order least squares system, see Lee et al. [38]. This does, however, in no way remove the specific problems of the multigrid algorithm in the case of the Helmholtz equation, and again ray components need to be introduced, based on several heuristics. While numerical experiments on the unit square illustrate the performance of this method, there is no convergence analysis. Vaněk et al. [46] present a two-level algebraic multigrid method for the Helmholtz problem, where they use an explicit plane wave representation on the coarse grid, localized within an aggregate of nodes, and zero outside, smoothed by a Chebyshev iteration. Numerical experiments show good computational complexity and scalability of the method. Since multigrid works well in the Helmholtz case provided the mesh is fine enough, and fine meshes are needed because of the pollution effect [3, 35, 36],

Kim and Kim [37] propose to use just standard multigrid for the first few finest levels, and then to switch to a domain decomposition solver of optimized Schwarz type (see below), when multigrid ceases to be effective. Proceeding in a different direction, Elman, Ernst and O'Leary studied the convergence problems of the classical multigrid algorithm applied to the Helmholtz equation in more detail by spectral analysis in [17], and used Krylov methods, both as smoothers and as outer convergence accelerator for multigrid, in order to obtain a practical and more robust method. One can also use the Laplacian, as proposed by Bayliss et al. [5], or the shifted Laplacian, as advocated by Erlangga et al. [21], as a preconditioner for the Helmholtz equation, and then apply the preconditioner using multigrid, see also [32], and for an algebraic version [2], where it is shown numerically that, asymptotically, the number of iterations grows linearly with the wave number for this approach, see also [15, 48]. This illustrates well that in the shifted Laplacian preconditioner, there are two conflicting requirements: the shift should be large enough in order for multigrid to work, and not too large, in order to still have a good preconditioner of the original problem, and one cannot, in general, satisfy both; see [22] for a simple Fourier analysis argument, and [31] for a complete theoretical treatment, and the preprint [15] for a detailed local Fourier analysis and a thorough numerical study.

But what happens precisely in the multigrid algorithm when one applies it to the Helmholtz equation? Why and how do standard components of a multigrid algorithm fail? Is it even possible to obtain a convergent algorithm using just the standard components? This chapter is for people interested in understanding the details on how multigrid fails for the Helmholtz equation. We explain all our results on a one-dimensional model problem, and show what is needed in one dimension in order to obtain a convergent method using only the standard multigrid components of smoothing and coarse-grid correction. Some of the ideas presented in this chapter, such as the smoothing components, can easily be extended to higher dimensions. Others are more difficult to extend, in particular the shifting of the coarse problem, which is a perfect remedy in one spatial dimension, but will not cure all problems of the coarse-grid correction in higher dimensions. With this text, we hope to share some of our passions we developed over the years analyzing iterative methods for the Helmholtz equation, which has also led to a certain success; for example there are now quite successful incomplete factorization preconditioners, based on the analytic factorization of the operator [26], for the Helmholtz equation see [30] for an analytic incomplete LU (AILU) preconditioner. The newest variant in this class of preconditioners is called the sweeping preconditioner [18, 19]. Also in domain decomposition there are successful preconditioners, with the first fundamental contribution [16], which then led to optimized Schwarz methods for the Helmholtz equation [28, 29]. There are also specialized FETI methods, such as FETI-H [25] and FETI-DPH [23], with a convergence analysis in [24]. For a thorough numerical comparison of all domain decomposition methods specialized for the Helmholtz equation applied to a realistic test equation [27]. For multigrid methods, we are still far from being able to give definitive answers in this chapter, except maybe for one spatial dimension.

The remainder of this chapter is organized as follows. In Section 2, we recall the ineffectiveness and instability of the classical relaxation methods when used as smoothers for the Helmholtz equation using a spectral analysis of the damped Jacobi smoother for a 1D model problem. We propose a two-step Jacobi smoother, a second-order relaxation method, as a possible working alternative and, again using spectral analysis, show how its parameters can be chosen depending on the type of indefinitenes, i.e. the level of discretization within a multigrid hierarchy. We also indicate the fundamental limitations of such a stationary iteration at a certain level of the multigrid hierarchy, the so-called *resonance level*. In Section 3, we perform a detailed spectral analysis of the coarse-grid correction operator using standard linear interpolation and full weighting restriction. We give closed-form expressions of its eigenvalues and show how, with decreasing resolution, the correction operator changes from a projection to an increasingly unstable operation, which can be interpreted as a resonance phenomenon. Guided by dispersion analysis of the finite difference scheme, we explore a variety of modified wave numbers on the coarse grids, resulting in stable correction operators on much coarser grids. Section 4 considers the combination of these remedies in a two-grid iteration and proposes a number of additional modifications yielding, for our model problem, a uniformly convergent two-grid algorithm for essentially all wave numbers that can occur within a full multigrid cycle. Section 5 then gives the results of numerical experiments for solving one-dimensional Helmholtz boundary value problems using multigrid methods constructed from the new techniques.

2 Smoothing

Nearly all multigrid algorithms for solving a linear system of equations $\boldsymbol{A}\boldsymbol{u} = \boldsymbol{f}$ employ for their smoothing component *stationary linear iterations* of the form

$$\boldsymbol{u}_{m+1} = \boldsymbol{u}_m + \boldsymbol{M}^{-1}(\boldsymbol{f} - \boldsymbol{A}\boldsymbol{u}_m), \quad m = 0, 1, \ldots$$

based on a splitting $\boldsymbol{A} = \boldsymbol{M} - \boldsymbol{N}$ of the coefficient matrix $\boldsymbol{A}$. Denoting the (algebraic) error of the mth iterate by $\boldsymbol{e}_m := \boldsymbol{u} - \boldsymbol{u}_m$ and the associated residual $\boldsymbol{r}_m := \boldsymbol{f} - \boldsymbol{A}\boldsymbol{u}_m$, the error propagation operator of such an iteration is given by

$$\boldsymbol{S} = \boldsymbol{I} - \boldsymbol{M}^{-1}\boldsymbol{A}. \tag{4.2}$$

2.1 Smoothing analysis

We give several representations of the spectral properties of smoothers for the 1D model problem

$$\begin{aligned} -u''(x) - k^2 u(x) &= f, \quad x \in (0,1), \\ u(0) = u(1) &= 0. \end{aligned} \tag{4.3}$$

With the standard three-point centered finite difference approximation for the second derivative operator on a uniform mesh with N interior grid points and mesh width $h = 1/(N+1)$, (4.3) is approximated by the linear system $\boldsymbol{A}\boldsymbol{u} = \boldsymbol{f}$ for the function values $u_j \approx u(x_j)$, $j = 1, \dots, N$, at the grid points $x_j = jh$, where

$$\boldsymbol{A} = \frac{1}{h^2} \operatorname{tridiag}(-1, 2 - k^2h^2, -1). \tag{4.4}$$

Spectral analysis

The matrix $\boldsymbol{A}$ is symmetric and therefore has a complete set of orthogonal eigenvectors. These are given by the discrete sine functions

$$\boldsymbol{v}_j = [\sin j\ell\pi h]_{\ell=1}^N, \quad j = 1, \dots, N. \tag{4.5}$$

When it is necessary to rescale these eigenvectors to have unit Euclidean norm this is achieved by the factor $\sqrt{2h}$ (for all j). The associated eigenvalues are given by

$$\lambda_j = \frac{2(1 - \cos j\pi h)}{h^2} - k^2 = \frac{4}{h^2} \sin^2 \frac{j\pi h}{2} - k^2, \quad j = 1, \dots, N. \tag{4.6}$$

The form of the eigenvectors (4.5) reveals that these become more oscillatory with increasing index j. We refer to the eigenpairs associated with the indices $1 \le j \le N/2$ as the *smooth* part of the spectrum I_{sm} and the remainder as the *oscillatory* part I_{osc}, i.e.

$$I_{\text{sm}} := \{\lambda_j : 1 \le j \le N/2\}, \quad I_{\text{osc}} := \{\lambda_j : N/2 < j \le N\}. \tag{4.7}$$

When N is odd the eigenpair with index $j = (N+1)/2$ lies exactly in the middle, and we denote by $n = (N+1)/2 - 1 = 1/(2h) - 1$, so that the middle index is $n + 1$. The associated eigenvector has a wavelength of $4h$.

Fourier analysis

For simple model problems smoothing analysis may be performed based on the spectral decomposition of the error propagation operator (4.2). This usually requires that the eigenvalues of $\boldsymbol{S}$ be accessible analytically, as is the case e.g. with Jacobi smoothing applied to Poisson's equation on a square with uniform meshes. An alternative approach, sometimes known as *local mode analysis* or *normal mode analysis*, is to perform Fourier analysis, replacing the given boundary conditions by periodic ones or assuming the regular mesh to be infinite. In the latter case the Fourier variable is continuous, in the former discrete. For a given uniform mesh size $h > 0$ the infinite grid

$$x_j = jh, \quad j \in \mathbb{Z},$$

supports Fourier modes

$$u : \mathbb{Z} \to \mathbb{C}, \quad u_j = e^{i\xi x_j} = e^{i\xi jh}, \quad j \in \mathbb{Z}, \tag{4.8}$$

in which the Fourier variable (frequency) ξ may be restricted to the fundamental domain $\xi \in [-\pi/h, \pi/h)$. In this case, however, the smooth part of the spectrum is given by $I_{\text{sm}} = [-\frac{\pi}{2h}, \frac{\pi}{2h}]$ and the oscillatory part is $I_{\text{osc}} = \{|\xi| \geq \frac{\pi}{2h}\}$. Applying the discrete Helmholtz operator $A_h = -\Delta_h - k^2$ resulting from the usual centered three-point finite difference stencil to a Fourier mode (4.8) yields

$$\begin{aligned}
\left[(-\Delta_h - k^2)u\right]_j &= \frac{1}{h^2}(-u_{j-1} + 2u_j - u_{j+1}) - k^2 u_j \\
&= \frac{e^{i\xi x_j}}{h^2}(-e^{-i\xi h} + (2 - k^2h^2) - e^{i\xi h}) \\
&= \frac{e^{i\xi x_j}}{h^2}(2 - 2\cos(\xi h) - k^2h^2),
\end{aligned}$$

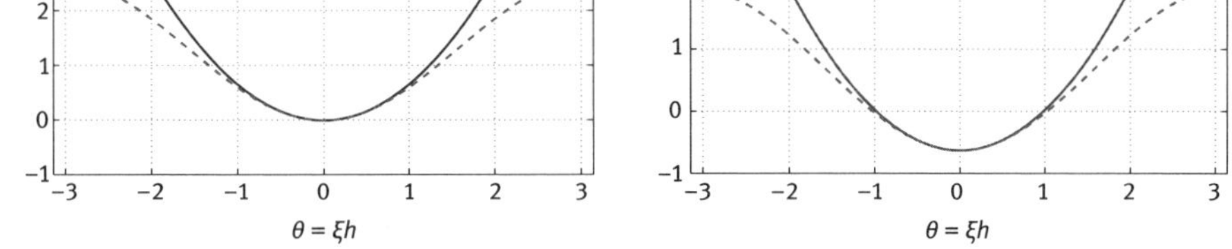

Figure 4.1: Symbols $\hat{A}(\theta)$ and $\hat{A}_h(\theta)$, $\theta = \xi h \in [-\pi, \pi)$ of continuous and discrete 1D Helmholtz operators. The mesh size is $h = 1/256$ with wave numbers $k_1 = 10\pi$ (left) and $k_2 = 80\pi$ (right). The wavelength-to-mesh size ratio λ/h is a measure of resolution, indicating how many mesh intervals of length h correspond to each wave of length $\lambda = 2\pi/k$.

which reveals the so-called *symbol* $\hat{A}_h = \hat{A}_h(\xi)$ of the discrete Helmholtz operator A_h to be

$$\hat{A}_h(\xi) = \frac{2 - 2\cos(\xi h) - k^2h^2}{h^2}, \quad \xi \in \left[-\frac{\pi}{h}, \frac{\pi}{h}\right).$$

Figure 4.1 displays the Fourier symbols of the continuous and discrete Helmholtz operators $A = -\Delta - k^2$ and $A_h = -\Delta_h - k^2$. To facilitate the comparison of symbols associated with different mesh sizes it is sometimes preferable to plot the symbol against the variable $\theta = \xi h$, which then ranges over the fixed interval $[-\pi, \pi)$.

2.2 Jacobi smoothing

The Jacobi smoother is based on the splitting $\boldsymbol{A} = \boldsymbol{D} - (\boldsymbol{D} - \boldsymbol{A})$, where $\boldsymbol{D} = \operatorname{diag}(\boldsymbol{A})$, resulting in the iteration

$$\boldsymbol{u}_{m+1} = \boldsymbol{u}_m + \boldsymbol{D}^{-1}(\boldsymbol{f} - \boldsymbol{A}\boldsymbol{u}_m).$$

For smoothing one usually introduces a damping factor ω for the update, giving

$$\boldsymbol{u}_{m+1} = \boldsymbol{u}_m + \omega\boldsymbol{D}^{-1}(\boldsymbol{f} - \boldsymbol{A}\boldsymbol{u}_m),$$

which corresponds to the splitting $\boldsymbol{A} = \frac{1}{\omega}\boldsymbol{D} - (\frac{1}{\omega}\boldsymbol{D} - \boldsymbol{A})$. The associated error propagation operator is

$$\boldsymbol{S}_\omega = \boldsymbol{I} - \omega\boldsymbol{D}^{-1}\boldsymbol{A}.$$

Note that

$$\boldsymbol{D} = \frac{2 - (kh)^2}{h^2}\boldsymbol{I},$$

we conclude that $\boldsymbol{A}$ and $\boldsymbol{D}$ are simultaneously diagonalizable, and we obtain for the eigenvalues σ_j of $\boldsymbol{S}_\omega$

$$\sigma_j = \sigma_j(\omega) = 1 - \omega\left(1 - \frac{2\cos(j\pi h)}{2 - (kh)^2}\right) =: 1 - \omega\frac{\lambda_j}{\delta}, \quad j = 1, \dots, N, \tag{4.9}$$

where we have introduced $\delta = \delta(k, h) := (2 - k^2h^2)/h^2$ to denote the diagonal entry in the Jacobi splitting, which is constant for this model problem. In multigrid methods the smoothing parameter ω is chosen to maximize damping on the oscillatory half of the spectrum I_{osc}. For the Laplace operator ($k = 0$) the eigenvalues of $\boldsymbol{D}^{-1}\boldsymbol{A}$ are given by $\lambda_j/\delta = 1 - \cos(j\pi h)$, $j = 1, \dots, N$, specifically

$$\frac{\lambda_1}{\delta} = \frac{\pi^2h^2}{4} + O(h^4), \quad \frac{\lambda_N}{\delta} = 2 - \frac{\pi^2h^2}{2} + O(h^4) \quad (h \to 0) \text{ and } \frac{\lambda_{n+1}}{\delta} = 1.$$

As a result, up to order h^2, the spectral interval of $\boldsymbol{D}^{-1}\boldsymbol{A}$ is $[0,2]$, with $I_{\text{osc}} = [1,2]$ containing the eigenvalues belonging to oscillatory modes. Maximal damping on I_{osc} thus translates to the requirement

$$1-\omega = -(1-2\omega), \quad \text{i.e. } \omega = \omega_0 := \frac{2}{3}.$$

For this optimal value of the damping parameter ω, each eigenmode belonging to the oscillatory part of the spectrum $\text{span}\{\boldsymbol{v}_{n+1}^h, \ldots, \boldsymbol{v}_N^h\}$ is reduced by at least a factor of

$$\sigma_{n+1}(\omega_0) = 1-\omega_0 = \frac{1}{3}$$

in each smoothing step, independently of the mesh size h.

Figure 4.2 shows the spectrum of $\boldsymbol{S}_\omega$ for the discrete 1D Laplacian on the unit interval with mesh width $h = 1/50$ for the values $\omega = 0$ (undamped) and the optimal value $\omega = 2/3$, on the left-plotted against the eigenvalues of $\boldsymbol{A}$. Another common way to visualize the spectral properties of a smoother shown on the right is to plot the eigenvalues of $\boldsymbol{S}_\omega$ against the eigenvalue index j or, in a more normalized representation, against jh, in which the eigenvalues of the smoother lie on a curved line.

The same considerations applied to the 1D Helmholtz operator ($k > 0$) result in eigenvalues

$$\frac{\lambda_j}{\delta} = 1 - \frac{2\cos j\pi h}{2-k^2h^2}, \quad j = 1, \ldots, N,$$

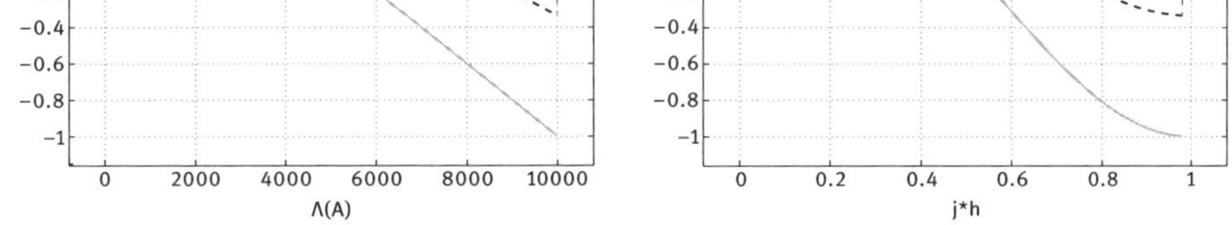

Figure 4.2: (See Color Plates for a color version.) Left: eigenvalues of the undamped and optimally damped Jacobi smoother plotted against those of the associated 1D Laplacian $-\Delta_h$, $h = 1/50$, divided into smooth and oscillatory parts I_{sm} and I_{osc}. The dashed red lines indicate the spectral radius of $\boldsymbol{S}_\omega$ restricted to the space of oscillatory eigenfunctions. Right: same results, but with smoother eigenvalues σ_j plotted against jh.

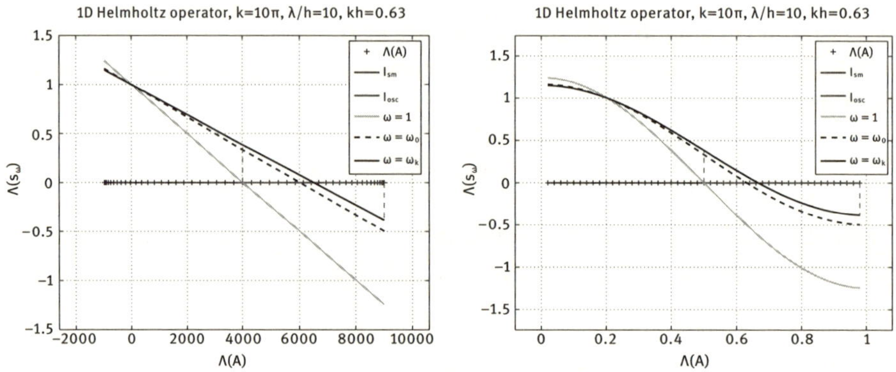

Figure 4.3: (See Color Plates for a color version.) Left: eigenvalues of the Jacobi smoother plotted with $\omega = \omega_k$, $\omega = 1$ and $\omega = \omega_0$ against those of the associated 1D Helmholtz operator $-\Delta_h - k^2$, $h = 1/50$ with wavelength-to-mesh ratio $\lambda/h = 10$. Right: same result in equispaced representation.

of $\boldsymbol{D}^{-1}\boldsymbol{A}$ and therefore, again up to $O(h^2)$, a spectral interval ranging between the extremal eigenvalues

$$\frac{\lambda_1}{\delta} = \frac{-k^2h^2}{2-k^2h^2}, \quad \frac{\lambda_N}{\delta} = \frac{4-k^2h^2}{2-k^2h^2},$$

so that maximal smoothing on the oscillatory half of the spectrum is obtained for ω characterized by

$$1 - \omega\frac{\lambda_N}{\delta} = -\left(1 - \omega\frac{\lambda_1 + \lambda_N}{2\delta}\right),$$

which results in

$$\omega = \omega_k := \frac{2-k^2h^2}{3-k^2h^2}. \tag{4.10}$$

Figure 4.3 shows the spectrum of $\boldsymbol{S}_\omega$ for the discrete 1D Helmholtz operator on the unit interval with mesh width $h = 1/50$ for different values of ω, on the left plotted against the eigenvalues of $\boldsymbol{A}$, and on the right against the eigenvalue index j.

Remark 4.1. In the Laplace case, the optimal choice of the smoothing parameter ω guarantees a reduction of the high-frequency error components by a factor of at least $1/3$ independently of the mesh size. This is not so in the case of the Helmholtz equation, since there we obtain for the reduction factor

$$|1 - \omega_k\frac{\lambda_N}{\delta}| = |\sigma_N(\omega_k)| = \left|\frac{1+2\cos N\pi h}{3-k^2h^2}\right| \approx \frac{1}{3-k^2h^2}.$$

This reduction factor is only smaller than one if $kh < \sqrt{2}$, which means the resolution $r := 2\pi/(kh)$, i.e. the ratio of wavelength $2\pi/k$ to the mesh size h, must be larger than $\sqrt{2}\pi \approx 4.44$ points per wavelength, otherwise the smoother is not working. This was also the assumption under which the parameter ω_k was derived. For given $kh < \sqrt{2}$ one has to perform

$$\nu \geq \frac{-\log \epsilon}{\log(3 - k^2h^2)} \tag{4.11}$$

smoothing steps to achieve the same reduction factor $\epsilon = 1/3$ as a single smoothing step in the Laplace case. Expanding the number of smoothing steps close to the limit $kh = \sqrt{2}$, we find for the number of smoothing steps the estimate

$$\nu = \frac{-\sqrt{2}\log \epsilon}{4\left(\sqrt{2} - kh\right)} + O(1).$$

In addition, the Jacobi smoother is not stable when used for the Helmholtz equation, since low frequency modes are amplified. The strongest amplification occurs for the lowest mode, namely

$$1 - \omega_k \frac{\lambda_1}{\delta} = \sigma_1(\omega_k) = \frac{1 + 2\cos \pi h}{3 - k^2h^2} \approx \frac{3}{3 - k^2h^2} > 1$$

for $kh < \sqrt{2}$, and the worst amplification could be as large as 3.

2.3 Two-step Jacobi smoothing

As the mesh is coarsened beyond a resolution of roughly $\lambda = 8h$ the standard damped Jacobi smoother becomes less efficient because

(a) there is less damping of modes in the oscillatory part I_{osc} of the spectrum (cf. (4.7));

(b) negative eigenvalues in the smooth part I_{sm} are amplified, some severely.

This is illustrated in Figure 4.4, where the eigenvalues of the damped Jacobi smoother with different values of the damping parameter are shown for wavelength-to-mesh ratios of 8 and 5. Even for the wave number-adapted value $\omega = \omega_k$ we observe that damped Jacobi relaxation is becoming less effective as a smoother on I_{osc} while at the same time amplifying the smoothest modes in I_{sm}, for $\lambda/h = 5$ up to a factor of 2.

More flexibility in the construction of the smoother is obtained for higher-order relaxation methods (cf. [49, Chapter 5]). Rather than allowing arbitrarily high order (cf. the preprint [47] or Krylov smoothers [7, 17]) we consider a *two-step Jacobi smoother* consisting of two consecutive smoothing steps, each with its own damping parameter ω_1 and ω_2:

$$\boldsymbol{S}_{\omega_1,\omega_2} = \left(\boldsymbol{I} - \omega_1\boldsymbol{D}^{-1}\boldsymbol{A}\right)\left(\boldsymbol{I} - \omega_2\boldsymbol{D}^{-1}\boldsymbol{A}\right).$$

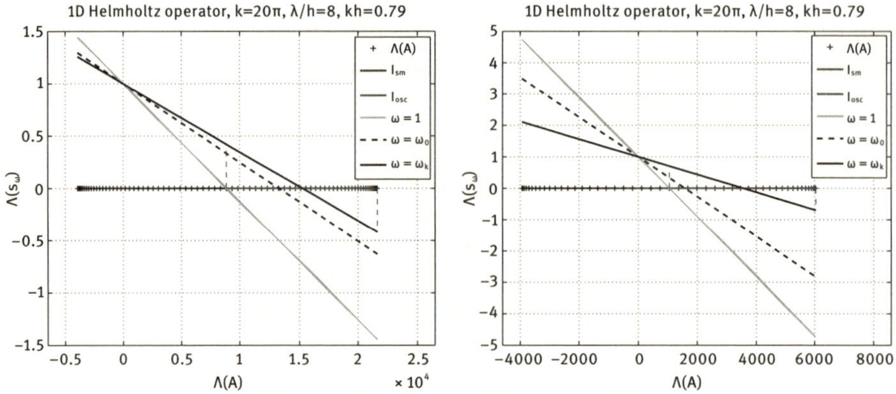

Figure 4.4: (See Color Plates for a color version.) Eigenvalues of the damped Jacobi smoother with the larger wavelength-to-mesh ratio $\lambda/h = 8$ and $\lambda/h = 5$. The damping on I_{osc} is becoming less while smooth modes at the lower end of I_{sm} are amplified significantly.

Such a two-step Jacobi scheme was proposed as a (standalone) solver for Helmholtz problems by Hadley [34].

Denoting again the eigenvalues of $\boldsymbol{D}^{-1}\boldsymbol{A}$ by λ_j/δ, the eigenvalues σ_j of $\boldsymbol{S}_{\omega_1,\omega_2}$ are given by

$$\sigma_j = \sigma(\lambda_j), \quad \sigma(\lambda) = \left(1 - \omega_1 \frac{\lambda}{\delta}\right)\left(1 - \omega_2 \frac{\lambda}{\delta}\right), \tag{4.12}$$

i.e. the eigenvalues of the two-step Jacobi smoother lie on a parabola with vertex

$$\lambda_v = \frac{\delta}{2} \frac{\omega_1 + \omega_2}{\omega_1 \omega_2} \quad \text{with value } \sigma(\lambda_v) = -\frac{1}{4} \frac{(\omega_1 - \omega_2)^2}{\omega_1 \omega_2}.$$

To determine values of the smoothing parameters ω_1 and ω_2 to ensure stability on I_{sm} as well as good damping on I_{osc} we distinguish three cases with regard to the location of the two halves of the spectral interval $[\lambda_1, \lambda_N]$ relative to the origin. Specifically, introducing

$$\lambda_{mid} := \frac{\lambda_1 + \lambda_N}{2},$$

we distinguish as to whether

$$0 < \lambda_{mid}, \quad \lambda_{mid} \le 0 \le \lambda_N, \quad \text{or} \quad \lambda_N < 0,$$

i.e. whether I_{osc} lies to the right, contains, or lies to the left of the origin, respectively. Using the estimates $\lambda_1 \approx -k^2$ and $\lambda_N \approx 4/h^2 - k^2$, these cases are characterized by

$$kh \in [0, \sqrt{2}), \quad kh \in [\sqrt{2}, 2], \quad \text{and} \quad kh > 2. \tag{4.13}$$

Case (1): $0 < \lambda_{\text{mid}}$

When I_{osc} lies to the right of the origin we design the smoother to damp the modes associated with the eigenvalues there while at the same time not amplifying the smooth modes, i.e. those with eigenvalues in I_{sm}, which lie closer to and on the left of the origin. We accomplish this by fixing the maximum value of the parabola $\sigma = \sigma(\lambda)$ at $\lambda = 0$ while requiring maximal damping on the oscillatory interval I_{osc}:

$$\sigma'(0) = 0, \quad \sigma(\lambda_{\text{mid}}) = -\sigma(\lambda_N).$$

The first constraint yields $\omega_2 = -\omega_1$ which, together with the second, results in (note that $\delta > 0$ in this case)

$$\omega_1 = \sqrt{\frac{8\delta^2}{\lambda_1^2 + 2\lambda_1\lambda_N + 5\lambda_N^2}} = \frac{2\sqrt{2}\delta}{\sqrt{(\lambda_1 + \lambda_N)^2 + 4\lambda_N^2}}. \tag{4.14}$$

Figure 4.5 illustrates this choice of parameters. On the left we see the eigenvalue plot for the two-step Jacobi smoother with this choice of relaxation parameters for a well-resolved situation $\lambda = 20h$. Note that damping on I_{osc} is reasonable, although weaker than for the one-step Jacobi smoother at this resolution. On the right the same plot is shown for a lower resolution of $\lambda = 5h$. Note that damping has become weaker with decreasing resolution, but there is no amplification.

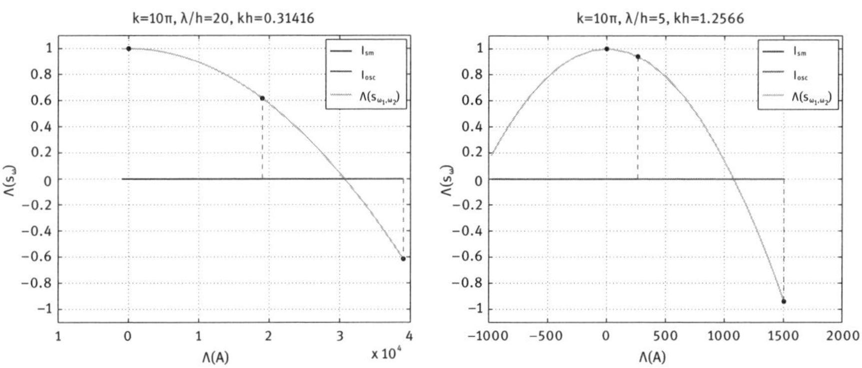

Figure 4.5: (See Color Plates for a color version.) Two-step Jacobi smoothing with parameter choice (4.14).

Remark 4.2. Once more, this smoother does not lead to a fixed mesh-independent reduction factor of the high-frequency modes. Specifically, the reduction factor is

$$\left|\left(1-\omega_1\frac{\lambda_N}{\delta}\right)\left(1-\omega_2\frac{\lambda_N}{\delta}\right)\right| = \left|1-\omega_1^2\frac{\lambda_N^2}{\delta^2}\right| = \left|1-\frac{8\lambda_N^2}{(\lambda_1+\lambda_N)^2+4\lambda_N^2}\right|$$
$$\approx \frac{6-2k^2h^2}{10-6k^2h^2+k^4h^4},$$

where we have used the estimates $\lambda_1 \approx -k^2$ and $\lambda_N \approx 4/h^2 - k^2$. This reduction factor is again only strictly smaller than one if $kh < \sqrt{2}$, for which it was constructed. For given $kh < \sqrt{2}$, one now has to perform

$$\nu \geq \frac{\log \epsilon}{\log((6 - 2k^2h^2)/(10 - 6k^2h^2 + k^4h^4))} \tag{4.15}$$

smoothing steps to achieve the same reduction factor $\epsilon = 1/3$ as one smoothing step for the Laplace case, but this smoother is stable. Expanding the number of smoothing steps close to the critical value $kh = \sqrt{2}$, we find

$$\nu = \frac{-\log \epsilon}{4\left(\sqrt{2} - kh\right)^2} + O\left(\frac{1}{kh - \sqrt{2}}\right),$$

which shows that the price for stability one has to pay is the squaring of the number of smoothing steps close to the resolution limit, compared to the classical single-step Jacobi method.

Case (2): $\lambda_{\text{mid}} < 0 < \lambda_N$

When the center of the spectral interval moves to the left of the origin, requiring equal damping on different sides of the origin can cause the smoother to become unstable for the smooth modes. This can be seen to occur in Figure 4.6 on the left for a resolution of $\lambda = 4h$. Instead, we require for this situation $\sigma'(0) = 0$ as before and, to guarantee stability for the smooth modes, $\sigma(\lambda_1) = -1$. This is achieved for the parameters

$$\omega_1 = \frac{\sqrt{2}|\delta|}{|\lambda_1|}, \quad \omega_2 = -\omega_1. \tag{4.16}$$

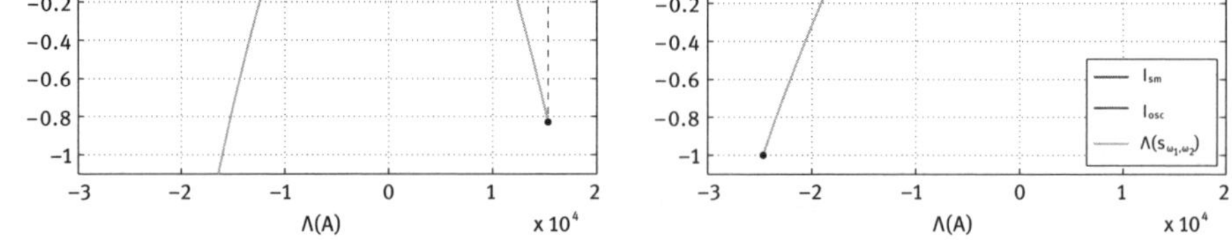

Figure 4.6: (See Color Plates for a color version.) Left: two-step Jacobi smoother with parameter choice (4.14) from Case (1) at resolution $\lambda/h = 4$, for which $\lambda_{\text{mid}} < 0 < \lambda_N$. The smooth modes are seen to be strongly amplified. Right: two-step Jacobi smoother with parameter choice (4.16) from Case (2) at resolution $\lambda/h = 4$.

Figure 4.6 (right) shows the smoother for this parameter choice for the same resolution. The amplification of the smooth modes is seen to have been eliminated. However, we see that no damping occurs for modes associated with eigenvalues near the origin. This is an inherent problem for the indefinite Helmholtz operator.

Remark 4.3. We call this most difficult level for the smoothing operation the resonance level. The reduction factor we obtain with the two-step Jacobi smoother for each $j = n+1, \dots, N$ is

$$\left|\left(1-\omega_1\frac{\lambda_j}{\delta}\right)\left(1-\omega_2\frac{\lambda_j}{\delta}\right)\right| = \left|1-\omega_1^2\frac{\lambda_j^2}{\delta^2}\right| = \left|1-2\left(\frac{4\sin^2\frac{j\pi h}{2}-k^2h^2}{4\sin^2\frac{\pi h}{2}-k^2h^2}\right)^2\right|. \tag{4.17}$$

This reduction factor can be arbitrarily close to one, since $2\sin\frac{j\pi h}{2}$ can be arbitrarily close to $kh \in (\sqrt{2}, 2)$ on this level, but they cannot be equal, since otherwise the corresponding Dirchlet problem would be singular. For a reduction by a factor ϵ, one has to perform

$$\nu \approx \frac{-\log\epsilon}{2}\left[\frac{k^2h^2}{4\sin^2\frac{j_0\pi h}{2}-k^2h^2}\right]^2 \tag{4.18}$$

smoothing steps, where j_0 is the closest integer to the solution of the equation $4\sin^2\frac{j\pi h}{2} = k^2h^2$, i.e.

$$j_0 = \text{round}\left(\frac{2}{\pi h}\arcsin\frac{kh}{2}\right).$$

We arrive at (4.18) by approximating $\log q$, where q denotes the reduction factor in (4.17), to first order by $\log q = \log(1-\delta) \approx -\delta$ for δ small and bounding $|4\sin^2\frac{\pi h}{2} - k^2h^2|$ by k^2h^2.

In the best case, k would lie exactly between two adjacent values of j, $k = \frac{4}{h^2}\sin^2((j+1/2)\pi h)/2$, and then for a $j > n = \frac{1}{2h} - 1$, e.g. $j = \frac{1}{2h} + c$ for some integer $c \geq 0$, we obtain for h small

$$\begin{aligned}\left[4\sin^2\frac{j\pi h}{2} - k^2h^2\right]^2 &= \left[4\sin^2\frac{\left(\frac{1}{2h}+c\right)\pi h}{2} - 4\sin^2\frac{\left(\frac{1}{2h}+c+\frac{1}{2}\right)\pi h}{2}\right]^2 \\ &= \pi^2h^2 + O(h^4),\end{aligned}$$

independently of the constant c. Since on the resonance level $kh = C$ with $C \in (\sqrt{2}, 2)$, this shows that one has to perform about

$$\nu \approx -\frac{\log\epsilon}{2}\frac{C^2k^2}{\pi^2} = O(k^2)$$

smoothing steps for an error reduction by a factor of ϵ in the high-frequency modes, a truly expensive level for smoothing. It is therefore recommended not to use a simple two-step Jacobi smoother on this level. A better idea would be to use a Chebyshev smoother tuned for the oscillatory part of the spectrum, or even a Krylov method, which can lead to a fixed error reduction with only $O(k)$ iterations.

Case (3): $\lambda_N < 0$

In the remaining case where also the largest eigenvalue has crossed to the left, i.e. the problem has become negative definite, we further distinguish between the cases

$$\lambda_1 < 3\lambda_N \quad \text{and} \quad \lambda_1 > 3\lambda_N.$$

In terms of the approximations $\lambda_1 \approx -k^2$ and $\lambda_N \approx 4/h^2 - k^2$, this corresponds to whether or not $kh < \sqrt{6}$.

In the first case we fix $\sigma(\lambda_1) = 1$, thus making the parabola symmetric to $\lambda = \lambda_1/2$ (since also $\sigma(0) = 1$ by construction) and thus ensuring stability. We further specify equal damping at the right end of the spectrum λ_N and at the vertex of the parabola $\lambda_v = \lambda_1/2$, i.e. $\sigma(\lambda_v) = -\sigma(\lambda_N)$. This results in

$$\omega_1 = \frac{2(2+\sqrt{2})\delta}{\lambda_1 + (2+2\sqrt{2})\lambda_N}, \qquad \omega_2 = \frac{2(2-\sqrt{2})\delta}{\lambda_1 + (2-2\sqrt{2})\lambda_N}. \tag{4.19}$$

For the reduction factor on the oscillatory part I_{osc} of the spectrum we obtain

$$|\sigma(\lambda_N)| = |\sigma(\lambda_v)| = \frac{(\lambda_1 - 2\lambda_N)^2}{\lambda_1 + 4\lambda_1\lambda_N - 4\lambda_N^2}.$$

This situation is depicted on the left of Figure 4.7.

In the final remaining case $\lambda_1 > 3\lambda_N$, we choose the parabola to be the shifted and scaled Chebyshev polynomial of degree 2 which is 1 at $\lambda = 0$ and has minimal

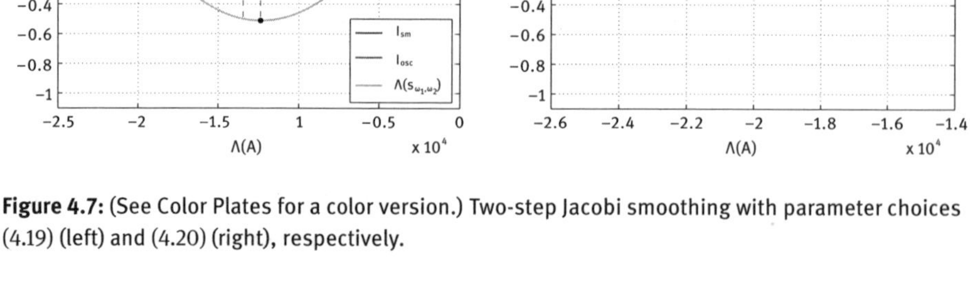

Figure 4.7: (See Color Plates for a color version.) Two-step Jacobi smoothing with parameter choices (4.19) (left) and (4.20) (right), respectively.

supremum on I_{osc}. This is not an option when $\lambda_1 < 3\lambda_N$ since then the Chebyshev polynomial will generally be greater than one on I_{sm}. In this case we obtain the parameters

$$\omega_1 = \frac{4(2+\sqrt{2})\delta}{\lambda_1 + (7+4\sqrt{2})\lambda_N}, \quad \omega_2 = \frac{4(2-\sqrt{2})\delta}{\lambda_1 + (7-4\sqrt{2})\lambda_N}. \tag{4.20}$$

This situation at a resolution of $\lambda/h = 2$ is shown on the right of Figure 4.7. In the limiting case $\lambda_1 = 3\lambda_N$ (4.19) and (4.20) yield the same two parameters.

Remark 4.4. While the smoothing in this last case is again quite effective, there are in general no error components left any more on these levels, if the two-step Jacobi smoother with the $O(k^2)$ smoothing steps required on the previous level was used, as one can see from the error reduction curve shown in Figure 4.6: with so many smoothing steps, all coarse error components will have been damped out as well, even more so than the oscillatory ones. This may, however, not be so if a special smoother is used on the resonance level.

So for completeness, here we give an estimate of the number of smoothing steps needed. In the first case, $2 < kh < \sqrt{6}$, the reduction factor on the high-frequency modes with the parameter values (4.19) is at least

$$\left|\left(1-\omega_1\frac{\lambda_N}{\delta}\right)\left(1-\omega_2\frac{\lambda_N}{\delta}\right)\right| = \frac{(\lambda_1 - 2\lambda_N)^2}{(\lambda_1+2\lambda_N)^2 - 8\lambda_N^2} \approx \frac{(8-k^2h^2)^2}{-64+16k^2h^2+k^4h^4},$$

where we have again used the estimates $\lambda_1 \approx -k^2$ and $\lambda_N \approx 4/h^2 - k^2$. Hence for an error reduction in the high frequencies by a factor ϵ, one has to perform

$$\nu \geq \frac{\log\epsilon}{\log\left((8-k^2h^2)^2/(-64+16k^2h^2+k^4h^4)\right)} \tag{4.21}$$

smoothing steps. In the second case, $kh > \sqrt{6}$, the reduction factor on the high frequency modes is with the choice (4.20)

$$\left|\left(1-\omega_1\frac{\lambda_N}{\delta}\right)\left(1-\omega_2\frac{\lambda_N}{\delta}\right)\right| = \frac{(\lambda_1 - \lambda_N)^2}{(\lambda_1+7\lambda_N)^2 - 32\lambda_N^2} \approx \frac{1}{17-12k^2h^2+2k^4h^4}.$$

For an error reduction in the high frequencies by a factor ϵ, one has to perform in this last case

$$\nu \geq \frac{-\log\epsilon}{\log(17-12k^2h^2+2k^4h^4)} \tag{4.22}$$

smoothing steps.

We finally note that for $kh > \sqrt{6}$, one could also use the simple one-step Jacobi smoother, as it achieves a reduction factor of 1/3 in that case. Using equioscillation to maximize damping on the high frequencies, the parameter should be chosen as

$$\omega = \frac{4\delta}{\lambda_1 + 3\lambda_N} = \frac{2-k^2h^2}{3-k^2h^2-2\sin^2\frac{\pi h}{2}}.$$

With this, one obtains a reduction in the high frequency components by at least a factor of

$$\left|1 - \omega\frac{\lambda_N}{\delta}\right| = \frac{\lambda_1 - \lambda_N}{3\lambda_N + \lambda_1} \approx \frac{1}{k^2h^2 - 3},$$

using again the estimates $\lambda_1 \approx -k^2$ and $\lambda_N \approx 4/h^2 - k^2$. Thus for an error reduction in the high-frequency components by a factor of ϵ, one would have to perform

$$\nu > \frac{-\log \epsilon}{\log(k^2h^2 - 3)}$$

smoothing steps.

To summarize this section, we have shown by way of spectral analysis for our 1D model problem how damped Jacobi relaxation, one of the standard multigrid smoothers, breaks down at high wave numbers. As a possible remedy, we have shown how adapting the parameters of a two-step Jacobi iteration depending on the product kh can yield a stable smoother which, however, on the resonance level, may require up to $O(k^2)$ smoothing steps to reduce the error on the oscillatory part of the spectrum by a fixed amount.

3 Coarse-grid correction

The following spectral analysis of the coarse-grid correction for the 1D case extends that in [33, Chapter 2] and [12, Chapter 5] for the Dirichlet–Laplacian model problem ($k = 0$). We consider problem (4.3) discretized using centered finite differences on a fine grid

$$\Omega^h := \{x_j = jh : j = 0, \dots, N + 1\}$$

with mesh width $h = \frac{1}{N+1}$ as well as a coarser grid[2]

$$\Omega^H := \{x_j = jH : j = 0, \dots, n + 1\}$$

with twice the mesh width $H = 2h$ so that $N = 2n + 1$. We introduce a grid transfer operator I_H^h mapping functions $\boldsymbol{u}^H = [u_0^H, \dots, u_{n+1}^H]$ defined on the coarse-grid Ω^H to a function $I_H^h\boldsymbol{u}^H$ defined on the fine grid Ω^h using linear interpolation. This results in the linear mapping $\boldsymbol{u}^H \mapsto I_H^h\boldsymbol{u}^H$ defined by

$$\left[I_H^h\boldsymbol{u}^H\right]_j = \begin{cases} [\boldsymbol{u}^H]_{j/2} & \text{if } j \text{ is even,} \\ \frac{1}{2}\left([\boldsymbol{u}^H]_{(j-1)/2} + [\boldsymbol{u}^H]_{(j+1)/2}\right) & \text{if } j \text{ is odd,} \end{cases} \qquad j = 0, \dots, N + 1, \tag{4.23}$$

2 We remark that our choice of notation is intended to reflect that, on the fine grid, N is large while h is small and vice versa on the coarse grid.

and has the matrix representation

$$I_H^h = \frac{1}{2}\begin{bmatrix} 2 & & & & \\ 1 & 1 & & & \\ & 2 & & & \\ & 1 & & & \\ & & \ddots & 1 & \\ & & & 2 & \\ & & & 1 & 1 \\ & & & & 2 \end{bmatrix} \in \mathbb{R}^{(N+2)\times(n+2)}$$

with respect to the standard unit coordinate bases in $\mathbb{R}^{n+2}$ and $\mathbb{R}^{N+2}$.

Note that in case of Dirichlet boundary conditions at both interval ends, with which we begin our analysis, the values at the endpoints are omitted, resulting in vectors of length N and n on Ω^h and Ω^H, respectively, as well as omitting of the first and last rows and columns in $\boldsymbol{I}_H^h$.

We analyze the mapping properties of the linear interpolation operator I_H^h on the coarse-grid eigenvectors $\{\boldsymbol{v}_j^H\}_{j=1}^n$ of the discrete 1D Dirichlet–Laplacian, where

$$\left[\boldsymbol{v}_j^H\right]_\ell = \sin(j\ell\pi H), \quad \ell = 1,\dots,n.$$

Proposition 4.5. *The coarse-grid eigenvectors are mapped by the interpolation operator I_H^h according to*

$$I_H^h\boldsymbol{v}_j^H = c_j^2\boldsymbol{v}_j^h - s_j^2\boldsymbol{v}_{N+1-j}^h, \quad j = 1,\dots,n,$$

where we define

$$c_j := \cos\frac{j\pi h}{2}, \quad s_j := \sin\frac{j\pi h}{2}, \quad j = 1,\dots,n. \tag{4.24}$$

In particular, $\boldsymbol{v}_{n+1}^h$ is not in the range of interpolation.

Proof. For any $j \in \{1,\dots,n\}$ we distinguish the cases of odd and even index $\ell \in \{1,\dots,N\}$ corresponding to the two cases in the definition (4.23). In the first case we obtain using elementary trigonometric identities

$$\begin{aligned}
\left[I_H^h\boldsymbol{v}_j^H\right]_\ell &= \frac{1}{2}\left(\left[\boldsymbol{v}_j^H\right]_{(\ell-1)/2} + \left[\boldsymbol{v}_j^H\right]_{(\ell+1)/2}\right)\\
&= \frac{1}{2}\left(\sin\frac{j(\ell-1)\pi H}{2} + \sin\frac{j(\ell+1)\pi H}{2}\right)\\
&= \frac{1}{2}\left(\sin(j(\ell-1)\pi h) + \sin(j(\ell+1)\pi h)\right) = \cos(j\pi h)\sin(j\ell\pi h)\\
&= \left(c_j^2 - s_j^2\right)\sin(j\ell\pi h) = c_j^2\sin(j\ell\pi h) - s_j^2\sin((N+1-j)\ell\pi h)\\
&= c_j^2\left[\boldsymbol{v}_j^h\right]_\ell - s_j^2\left[\boldsymbol{v}_{N+1-j}^h\right]_\ell .
\end{aligned}$$

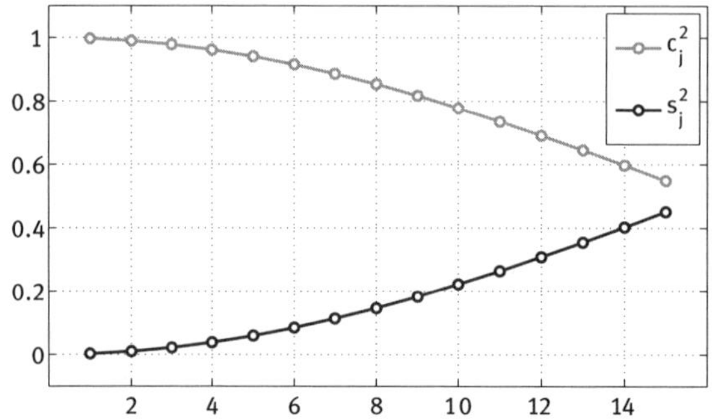

Figure 4.8: Coefficients c_j^2 and s_j^2 of the eigenvectors of the discrete 1D Dirichlet–Laplacian unter the linear interpolation operator for $N = 31$, i.e. $n = 15$.

For even ℓ we obtain

$$\begin{aligned}\left[I_H^h \boldsymbol{v}_j^H\right]_\ell = \left[\boldsymbol{v}_j^H\right]_{\ell/2} = \sin\frac{j\ell\pi H}{2} = \sin(j\ell\pi h) = \left(c_j^2 + s_j^2\right)\sin(j\ell\pi h) \\ = c_j^2\sin(j\ell\pi h) - s_j^2\sin((N+1-j)\ell\pi h) = c_j^2\left[\boldsymbol{v}_j^h\right]_\ell - s_j^2\left[\boldsymbol{v}_{N+1-j}^h\right]_\ell .\end{aligned}$$

□

The coarse-grid modes $\boldsymbol{v}_j^H$ are thus mapped under the interpolation operator to a linear combination of their fine-grid counterparts $\boldsymbol{v}_j^h$ and a *complementary mode* $\boldsymbol{v}_{j'}^h$ with index $j' := N + 1 - j$. Note the relations

$$c_{j'} = s_j \quad s_{j'} = c_j, \quad j = 1, \dots, n, \tag{4.25}$$

between complementary s_j and c_j. Interpolating coarse-grid functions therefore always activates high-frequency modes on the fine grid, with a factor that is less than one but grows with j (cf. Figure 4.8). This can be viewed as an aliasing phenomenon.

To transfer fine-grid functions to the coarse grid we employ the *full weighting* restriction operator

$$I_h^H : \boldsymbol{u}^h \mapsto I_h^H \boldsymbol{u}^h$$

defined by

$$\left[I_h^H \boldsymbol{u}^h\right]_j = \frac{1}{4}\left([\boldsymbol{u}^h]_{2j-1} + 2[\boldsymbol{u}^h]_{2j} + [\boldsymbol{u}^h]_{2j+1}\right), \quad j = 1, \dots, n. \tag{4.26}$$

The associated matrix representation is given by $\boldsymbol{I}_h^H = \frac{1}{2}[\boldsymbol{I}_H^h]^\top$. The restriction operator is thus seen to be the adjoint to the interpolation operator if one introduces on $\mathbb{R}^n$ and $\mathbb{R}^N$ the Euclidean inner product weighted by the mesh size H and h, respectively. Denoting by $\mathcal{N}(\cdot)$ and $\mathcal{R}(\cdot)$ the null space and range of a matrix, the basic relation

$$\mathbb{R}^N = \mathcal{R}\left(\boldsymbol{I}_H^h\right) \oplus \mathcal{N}\left(\left[\boldsymbol{I}_H^h\right]^\top\right) = \mathcal{R}\left(\boldsymbol{I}_H^h\right) \oplus \mathcal{N}\left(\boldsymbol{I}_h^H\right) \tag{4.27}$$

reveals that the range of interpolation and the null space of the restriction are complementary linear subspaces of $\mathbb{R}^N$, which are also orthogonal with respect to the Euclidean inner product. Since the columns of $\boldsymbol{I}_H^h$ are seen to be linearly independent, the interpolation operator has full rank, which together with (4.27) implies

$$\dim \mathcal{R}\left(\boldsymbol{I}_H^h\right) = n, \quad \dim \mathcal{N}\left(\boldsymbol{I}_h^H\right) = N - n = n + 1.$$

Proposition 4.6. *The fine-grid eigenvectors are mapped by the restriction operator I_h^H according to*

$$I_h^H \boldsymbol{v}_j^h = c_j^2 \boldsymbol{v}_j^H, \qquad j = 1, \dots, n, \tag{4.28a}$$

$$I_h^H \boldsymbol{v}_{N+1-j}^h = -s_j^2 \boldsymbol{v}_j^H, \qquad j = 1, \dots, n, \tag{4.28b}$$

$$I_h^H \boldsymbol{v}_{n+1}^h = \boldsymbol{0}. \tag{4.28c}$$

Proof. By (4.26) and elementary trigonometric relations we have for $j, \ell \in \{1, \dots, n\}$

$$\begin{aligned}
\left[I_h^H \boldsymbol{v}_j^h\right]_\ell &= \frac{1}{4}\left(\left[\boldsymbol{v}_j^h\right]_{2\ell-1} + 2\left[\boldsymbol{v}_j^h\right]_{2\ell} + \left[\boldsymbol{v}_j^h\right]_{2\ell+1}\right) \\
&= \frac{1}{4}\left(\sin\left((2\ell-1)j\pi h\right) + 2\sin(2\ell j\pi h) + \sin\left((2\ell+1)j\pi h\right)\right) \\
&= \frac{1}{4}\left(2\sin(2\ell j\pi h) + 2\cos(j\pi h)\sin(2\ell j\pi h)\right) \\
&= \frac{1}{2}\left(1 + \cos(j\pi h)\right)\sin(2\ell j\pi h) \\
&= \cos^2\frac{j\pi h}{2}\sin(\ell j\pi H) = c_j^2\left[\boldsymbol{v}_j^H\right]_\ell,
\end{aligned}$$

which is (4.28a). For $j = n+1$ we have $2jh = 1$, implying $\sin(2\ell j\pi h) = 0 \ \forall \ell$, and thus the fourth equality above results in (4.28c). To show (4.28b) first note that

$$\left[\boldsymbol{v}_{N+1-j}^h\right]_\ell = -(-1)^\ell \sin(j\ell\pi h), \quad j = 1, \dots, n; \quad \ell = 1, \dots, N,$$

and therefore

$$\begin{aligned}
\left[I_h^H \boldsymbol{v}_{N+1-j}^h\right]_\ell &= \frac{1}{4}\left(2\cos(j\pi h)\sin(j\ell\pi H) - (-1)^{2\ell}2\sin(j\ell\pi H)\right) \\
&= \frac{1}{2}\left(\cos(j\pi h) - 1\right)\sin(j\ell\pi H) \\
&= -\sin^2\frac{j\pi h}{2}\sin(j\ell\pi H) = -s_j^2\left[\boldsymbol{v}_j^H\right]_\ell.
\end{aligned}$$

□

In the standard "correction mode" of multigrid schemes for general linear problems the coarse-grid correction of an approximation $\boldsymbol{u}^h$ to the solution of (4.3) on the fine grid Ω^h is obtained by solving the error equation $\boldsymbol{A}_h \boldsymbol{e}^h = \boldsymbol{b} - \boldsymbol{A}_h \boldsymbol{u}^h = \boldsymbol{r}^h$ on the coarse grid. To this end, the residual is first restricted to the coarse grid, and a coarse-grid representation $\boldsymbol{A}_H$ of the differential operator used to obtain the approximation

$A_H^{-1} I_h^H r^h$ of the error $A_h^{-1} r^h$ on Ω^H. The update is then obtained, after interpolating this error approximation to Ω^h, as

$$u^h \leftarrow u^h + I_H^h A_H^{-1} I_h^H \left(b - A_h u^h\right)$$

with associated error propagation operator

$$C := I - I_H^h A_H^{-1} I_h^H A_h. \tag{4.29}$$

When the so-called *Galerkin condition* is satisfied, i.e. when the coarse-grid operator is related to interpolation, restriction and the fine-grid operator by

$$A_H = I_h^H A_h I_H^h, \tag{4.30}$$

then it is easily verified that $C^2 = C$, i.e. C is idempotent and thus a projection operator. More precisely, it is the complementary projection of $I_H^h A_H^{-1} I_h^H A_h$, the range of which is the range of interpolation $\mathcal{R}(I_H^h)$. We conclude that the coarse-grid correction is without effect on error components outside the range of interpolation. In the same vein, coarse-grid correction is without effect for an error component whose associated residual vector, i.e. its image under A_h, lies in the null space of restriction. Moreover, the relation $(A_h C)^\top = A_h C$, which holds, e.g. when restriction is a scalar multiple of the transpose of interpolation[3], implies that C is A_h-self-adjoint, i.e. self-adjoint with respect to the inner product associated with A_h whenever A_h is symmetric and positive definite. In this case C is identified as the A_h-orthogonal projection onto the A_h-orthogonal complement of $\mathcal{R}(I_H^h)$. We may thus write the direct sum (4.27) equivalently as the A_h-orthogonal decomposition

$$\mathbb{R}^N = \mathcal{R}\left(I_H^h\right) \oplus \mathcal{N}\left(I_h^H A_h\right).$$

Returning to the 1D Helmholtz problem: in view of Propositions 4.5 and 4.6 the coarse-grid correction operator C is seen to possess the invariant subspaces

$$\operatorname{span}\{v_{n+1}^h\} \quad \text{and} \quad \operatorname{span}\{v_j^h, v_{j'}^h\}, \quad j' = N+1-j, \quad j = 1, \ldots, n.$$

Note that the eigenvalues of A_h and A_H are given by

$$\lambda_j^h = \frac{4}{h^2} \sin^2 \frac{j\pi h}{2} - k^2, \quad j = 1, \ldots, N, \tag{4.31a}$$

and

$$\lambda_j^H = \frac{4}{H^2} \sin^2 \frac{j\pi H}{2} - k^2, \quad j = 1, \ldots, n, \tag{4.31b}$$

3 More generally, it is true if transpose is replaced by adjoint and restriction is adjoint to interpolation, where both adjoints are taken with respect to the same inner product.

respectively, the action of the coarse-grid correction operator on these invariant subspaces is given by

$$\boldsymbol{C}\begin{bmatrix}\boldsymbol{v}_j^h & \boldsymbol{v}_{j'}^h\end{bmatrix} = \begin{bmatrix}\boldsymbol{v}_j^h & \boldsymbol{v}_{j'}^h\end{bmatrix}\boldsymbol{C}_j, \qquad j = 1,\dots,n,$$

$$\boldsymbol{C}\boldsymbol{v}_{n+1}^h = \boldsymbol{v}_{n+1}^h,$$

where

$$\boldsymbol{C}_j = \begin{bmatrix}1 & 0\\ 0 & 1\end{bmatrix} - \begin{bmatrix}c_j^2\\ -s_j^2\end{bmatrix}\frac{1}{\lambda_j^H}\begin{bmatrix}c_j^2 & -s_j^2\end{bmatrix}\begin{bmatrix}\lambda_j^h & 0\\ 0 & \lambda_{j'}^h\end{bmatrix} = \begin{bmatrix}1 - c_j^4\frac{\lambda_j^h}{\lambda_j^H} & c_j^2 s_j^2\frac{\lambda_{j'}^h}{\lambda_j^H}\\ c_j^2 s_j^2\frac{\lambda_j^h}{\lambda_j^H} & 1 - s_j^4\frac{\lambda_{j'}^h}{\lambda_j^H}\end{bmatrix}. \tag{4.32}$$

Proposition 4.7. *The eigenvalues of the* 2×2 *blocks* (4.32) *representing the coarse-grid correction operator are given by*

$$\Lambda(\boldsymbol{C}_j) = \left\{1 - \frac{c_j^4\lambda_j^h + s_j^4\lambda_{j'}^h}{\lambda_j^H}, 1\right\}, \quad j = 1,\dots,n. \tag{4.33}$$

A set of corresponding eigenvectors is given by

$$\boldsymbol{w}_j^{(1)} = \begin{bmatrix}c_j^2\\ -s_j^2\end{bmatrix} \quad \text{and} \quad \boldsymbol{w}_j^{(2)} = \begin{bmatrix}s_j^2\lambda_{j'}^h\\ c_j^2\lambda_j^h\end{bmatrix} = \frac{4}{h^2}\begin{bmatrix}s_j^2\left(c_j^2 - \left(\frac{hk}{2}\right)^2\right)\\ c_j^2\left(s_j^2 - \left(\frac{hk}{2}\right)^2\right)\end{bmatrix}, \tag{4.34}$$

respectively.

Proof. In view of (4.31a), (4.31b), and (4.25), the quantities appearing in (4.32) may be written as

$$\lambda_j^h = \frac{4}{h^2}s_j^2 - k^2, \quad \lambda_{j'}^h = \frac{4}{h^2}c_j^2 - k^2, \quad \lambda_j^H = \frac{4}{h^2}s_j^2 c_j^2 - k^2, \qquad j = 1,\dots,n.$$

The characteristic polynomial of $\boldsymbol{C}_j$ is then

$$\begin{aligned}\det(\lambda \boldsymbol{I} - \boldsymbol{C}_j) &= \left(\lambda - 1 + c_j^4\frac{\lambda_j^h}{\lambda_j^H}\right)\left(\lambda - 1 + s_j^4\frac{\lambda_{j'}^h}{\lambda_j^H}\right) - c_j^4 s_j^4\frac{\lambda_{j'}^h\lambda_j^h}{\left(\lambda_j^H\right)^2}\\ &= (\lambda - 1)^2 + (\lambda - 1)\frac{c_j^4\lambda_j^h + s_j^4\lambda_{j'}^h}{\lambda_j^H}\\ &= (\lambda - 1)\left(\lambda - 1 + \frac{c_j^4\lambda_j^h + s_j^4\lambda_{j'}^h}{\lambda_j^H}\right)\end{aligned}$$

revealing the eigenvalues. The eigenvectors are easily verified by direct calculation. □

Note that the second eigenvalue and the first eigenvector of each 2×2 block $\boldsymbol{C}_j$ are independent of k.

3.1 The Laplacian

For $k = 0$ we observe

$$\frac{\lambda_j^h}{\lambda_j^H} = \frac{4s_j^2}{(2s_jc_j)^2} = \frac{1}{c_j^2} \quad \text{as well as} \quad \frac{\lambda_{j'}^h}{\lambda_j^H} = \frac{4c_j^2}{(2s_jc_j)^2} = \frac{1}{s_j^2}, \quad j = 1, \ldots, n, \tag{4.35}$$

and therefore directly from (4.32) we obtain

$$C_j = \begin{bmatrix} 1 - c_j^2 & c_j^2 \\ s_j^2 & 1 - s_j^2 \end{bmatrix} = \begin{bmatrix} s_j^2 & c_j^2 \\ s_j^2 & c_j^2 \end{bmatrix}, \quad j = 1, \ldots, n.$$

A matrix of the form $X = \begin{bmatrix} \xi & \eta \\ \xi & \eta \end{bmatrix}$ has eigenvalues and spectral norm given by

$$\Lambda(X) = \{0, \xi + \eta\}, \tag{4.36a}$$

$$\|X\| = \|XX^\top\|^{1/2} = \sqrt{\xi^2 + \eta^2} \left\| \begin{bmatrix} 1 & 1 \\ 1 & 1 \end{bmatrix} \right\|^{1/2} = \sqrt{\xi^2 + \eta^2} \cdot \sqrt{2}. \tag{4.36b}$$

For C_j we thus obtain

$$\Lambda(C_j) = \{0, s_j^2 + c_j^2\} = \{0, 1\}, \quad \|C_j\| = \sqrt{2\left(s_j^4 + c_j^4\right)}, \quad j = 1, \ldots, n.$$

From $s_j^2 \in [0, \frac{1}{2}]$ for $j = 1, \ldots, n$ we obtain the bound

$$\|C_j\| \le \max_{0 \le t \le \frac{1}{2}} \sqrt{2[t^2 + (1 - t)^2]} = \sqrt{2}, \quad j = 1, \ldots, n.$$

By consequence, we also have $\Lambda(C) = \{0, 1\}$, which reflects the fact that C is a projection in case $k = 0$. This is true since the Galerkin condition (4.30) holds if A_h and A_H result from the centered finite difference discretizations of the 1D Laplacian on the uniform meshes of widths h and $H = 2h$. Note that the Galerkin property is lost for this discretization once $k > 0$.

What we observe in the Laplacian case is the usual division of labor of classical multigrid methods. The coarse-grid correction has the task of eliminating the low-frequency components of the error. In view of (4.34) for $k = 0$, a set of eigenvectors of C_j associated with eigenvalues zero and one, i.e. with the null space and range of the coarse-grid correction, is

$$w_j^{(1)} = \begin{bmatrix} c_j^2 \\ -s_j^2 \end{bmatrix} \quad \text{and} \quad w_j^{(2)} = \begin{bmatrix} 1 \\ 1 \end{bmatrix}.$$

Note that these are A_h-orthogonal. For small values of the index j the complementary eigenmodes v_j^h and $v_{j'}^h$ of A_h are very smooth and highly oscillatory, respectively. Moreover, since in this case $c_j^2 \approx 1, s_j^2 \approx 0$, the eigenmode $w_j^{(1)}$ eliminated by the coarse-grid correction is closely aligned with the low-frequency mode v_j^h. This alignment becomes less as j increases.

3.2 The Helmholtz operator

In the Helmholtz case $k > 0$ the eigenvalues of the coarse-grid correction operator are given by (4.33). Elementary manipulations reveal that the nonunit eigenvalues of $\boldsymbol{C}$ are

$$\gamma_j = \gamma_j(kh) = \frac{\frac{(kh)^2}{2}}{\left(\frac{kh}{2s_jc_j}\right)^2 - 1}, \qquad j = 1, \ldots, n.$$

For a fixed value of $kh > 0$ this expression is seen to be a rational function of s_jc_j with a simple pole whenever $2s_jc_j = kh$. Since $2s_jc_j \in (0,1)$ for $j \in \{1, \ldots, n\}$, this pole occurs near one of the indices j whenever $kh \leq 1$.

This is illustrated in Figure 4.9 for a fine mesh containing $N = 31$ interior points. The left plot shows a stem plot of the eigenvalues of the 2×2 blocks $\boldsymbol{C}_j$ for the Laplacian, which consists of ones and zeros. On the right of Figure 4.9 we see the analogous plot for $k = 6.3\pi$. Note that the unit eigenvalues remain, but that the second eigenvalue of each pair is no longer zero. In particular, mode number 13 is amplified by a factor of nearly -4.

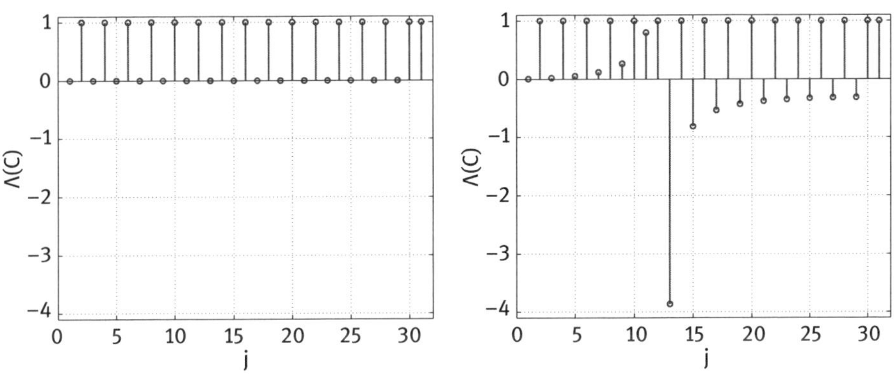

Figure 4.9: Eigenvalues of the coarse-grid correction operator with respect to a fine mesh with $h = 1/32$ for $k = 0$ (left) and $k = 6.3\pi$ (right).

Dispersion correction

A simple device for obtaining a more effective coarse-grid correction for Helmholtz operators results from taking into account the *dispersion properties* of the discretization scheme. Dispersion analysis (cf. [13, 44]) for discretization schemes is concerned with which propagating or standing waves solve a discrete homogeneous differential equation on a uniform grid extending over full space compared to the underlying continuous problem. For the centered finite-difference discretization of the 1D Helmholtz operator with constant k

$$\mathcal{L}u \approx \frac{-u_{j-1} + 2u_j - u_{j+1}}{h^2} - k^2 u_j,$$

plane-wave solutions $e^{ik^h x_j}$ of the *discrete* homogeneous Helmholtz equation possess a discrete wave number k^h characterized by

$$\frac{k^h}{k} = \frac{1}{kh} \arccos\left(1 - \frac{k^2h^2}{2}\right) > 1.$$

As a result, the discrete solution exhibits a *phase lead* with respect to the true solution, and this phase lead grows with h. In the same way, coarse-grid approximations in a multigrid hierarchy will be out of phase with fine-grid approximations. This suggests "slowing down" the waves on coarse grids in order that the coarse-grid correction again be in phase with the fine-grid approximation. For our example, this is achieved by using a modified wave number $\tilde{k}$ in the coarse-grid Helmholtz operator defined by the requirement

$$k^H = \frac{1}{H} \arccos\left(1 - \frac{\tilde{k}^2H^2}{2}\right) = k, \quad \text{i.e.,} \quad \tilde{k} = \sqrt{\frac{2(1 - \cos(kH))}{H^2}}. \tag{4.37}$$

Alternatively, one could adjust the coarse-grid correction by matching the coarse-grid discrete wave number k^H with the fine-grid discrete wave number k^h, which is obtained by choosing the modified wave number $\tilde{k}$ on the coarse grid to satisfy

$$\tilde{k}^H = k^h,$$

i.e.

$$\frac{1}{H} \arccos\left(1 - \frac{\tilde{k}^2H^2}{2}\right) = \frac{1}{h} \arccos\left(1 - \frac{k^2h^2}{2}\right),$$

or

$$1 - \frac{\tilde{k}^2H^2}{2} = \cos\left(\frac{H}{h} \arccos\left(1 - \frac{k^2h^2}{2}\right)\right).$$

Noting $\frac{H}{h} = 2$ and using the relation $\cos(2\alpha) = 2\cos^2(\alpha) - 1$ this becomes

$$1 - \frac{\tilde{k}^2H^2}{2} = 2\left(1 - \frac{k^2h^2}{2}\right)^2 - 1,$$

implying

$$\tilde{k} = k\sqrt{1 - \frac{k^2h^2}{4}}. \tag{4.38}$$

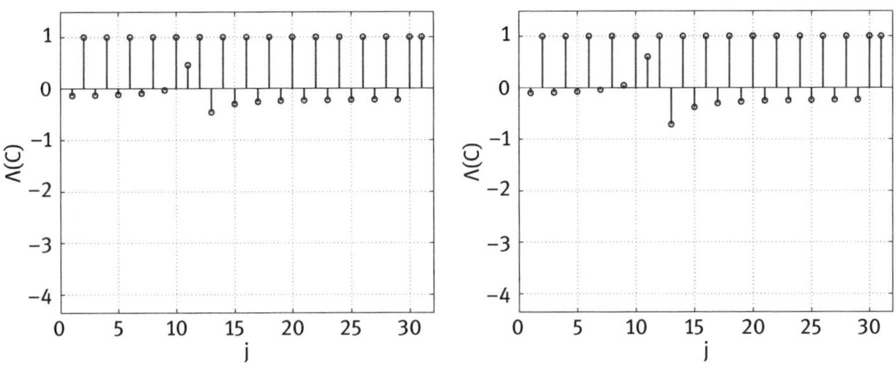

Figure 4.10: Eigenvalues of the coarse-grid correction operator with respect to a fine mesh with $h = 1/32$ for $k = 6.3\pi$ using the modified wave number $\tilde{k}$ given in (4.37) (left) and (4.38) (right) in the coarse-grid Helmholtz operator.

Choosing a modified wave number according to (4.38) is also equivalent to avoiding a possible "singularity" in the term λ_j^h/λ_j^H in (4.32) by forcing the vanishing of λ_j^H as a continuous function of j to occur in the same location as for λ_j^h.

Figure 4.10 shows the eigenvalues of the coarse-grid correction operator depicted on the right of Figure 4.9 with the modified wave number (4.37) (left) and (4.38) (right) used on the coarse grid. The strong amplification of mode number 13 is seen to be much less severe, all nonunit eigenvalues now being less than one in modulus.

A closer analysis of the spectrum of the coarse-grid correction operator (4.32) reveals how the wave number $\tilde{k}$ on the coarse grid can be chosen to stabilize the correction yet further. Denoting the nonunit eigenvalue of $\boldsymbol{C}_j$ in (4.32) again by γ_j, upon changing the coarse-grid wave number from k to $\tilde{k}$ we obtain

$$1 - \gamma_j = \frac{c_j^4 \lambda_j^h + s_j^4 \lambda_{j'}^h}{\lambda_j^H} = \frac{c_j^4 \left[\frac{4}{h^2} s_j^2 - k^2\right] + s_j^2 \left[\frac{4}{h^2} c_j^2 - k^2\right]}{\frac{4}{h^2} s_j^2 c_j^2 - \tilde{k}^2}$$
$$= \frac{s_j^2 c_j^2 - \left(s_j^4 + c_j^4\right)\left(\frac{kh}{2}\right)^2}{s_j^2 c_j^2 - \left(\frac{\tilde{k}h}{2}\right)^2}.$$

Setting $\tau_j := s_j^2 c_j^2$ and noting that $s_j^4 + c_j^4 = 1 - 2\tau_j$, this becomes

$$1 - \gamma_j = \frac{\tau_j - (1 - 2\tau_j)\left(\frac{kh}{2}\right)^2}{\tau_j - \left(\frac{\tilde{k}h}{2}\right)^2} = \left(1 + \frac{k^2 h^2}{2}\right) \frac{\tau_j - \frac{k^2h^2/4}{1 + k^2h^2/2}}{\tau_j - \frac{\tilde{k}^2 h^2}{4}}.$$

We observe that the nonunit eigenvalue of $\boldsymbol{C}_j$ can be made constant for all j by adjusting $\tilde{k}$ to cancel the numerator and denominator on the right, which occurs for

$$\tilde{k} = \frac{k}{\sqrt{1 + \frac{k^2h^2}{2}}}, \tag{4.39}$$

see Figure 4.11. In this case the nonunit eigenvalue is

$$\gamma_j = -\frac{k^2h^2}{2}.$$

Such a dispersion analysis can be carried out for all standard discretization schemes, and it is found that higher-order schemes have much lower phase error (cf., e.g. [1]), making them a favorable choice also from the point of view of multigrid solvers. In higher dimensions higher-order methods also possess nearly isotropic dispersion relations, a necessary requirement for (scalar) dispersion correction.

Eigenvalues and singular values

We compare the eigenvalues and singular values of the coarse-grid correction operator for the same problem as above with wave numbers $k = 0$ and $k = 6.3\pi$ on a mesh with $h = 1/32$ (Figure 4.12). For the positive wave number we compare the unmodified case as well as modified wave numbers according to (4.38) and (4.39), respectively.

Figure 4.12 shows the eigenfunctions and eigenvalues sorted increasingly. As a debugging measure, we computed these both by applying MATLAB's `eig` and `svd` routines to the matrix representing the complete coarse-grid correction operator as well as to each 2×2 block or, in case of the eigenvalues, using (4.33). We observe that modifying the wave number according to (4.39) almost perfectly restores the zero-one spectrum of the coarse-grid correction for the 1D Laplacian. However, this occurs at the price of a large singular value, leading to a large norm for the operator.

Figure 4.13 displays the same eigen- and singular values ordered according to the invariant subspaces. The large singular value is seen to lie in the subspace with index $j = 6$.

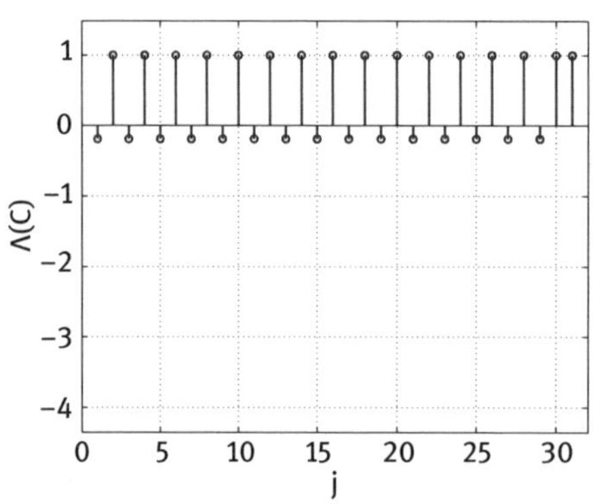

Figure 4.11: Eigenvalues of the coarse-grid correction operator with respect to a fine mesh with $h = 1/32$ for $k = 6.3\pi$ using the modified wave number $\tilde{k}$ given in (4.39) in the coarse-grid Helmholtz operator.

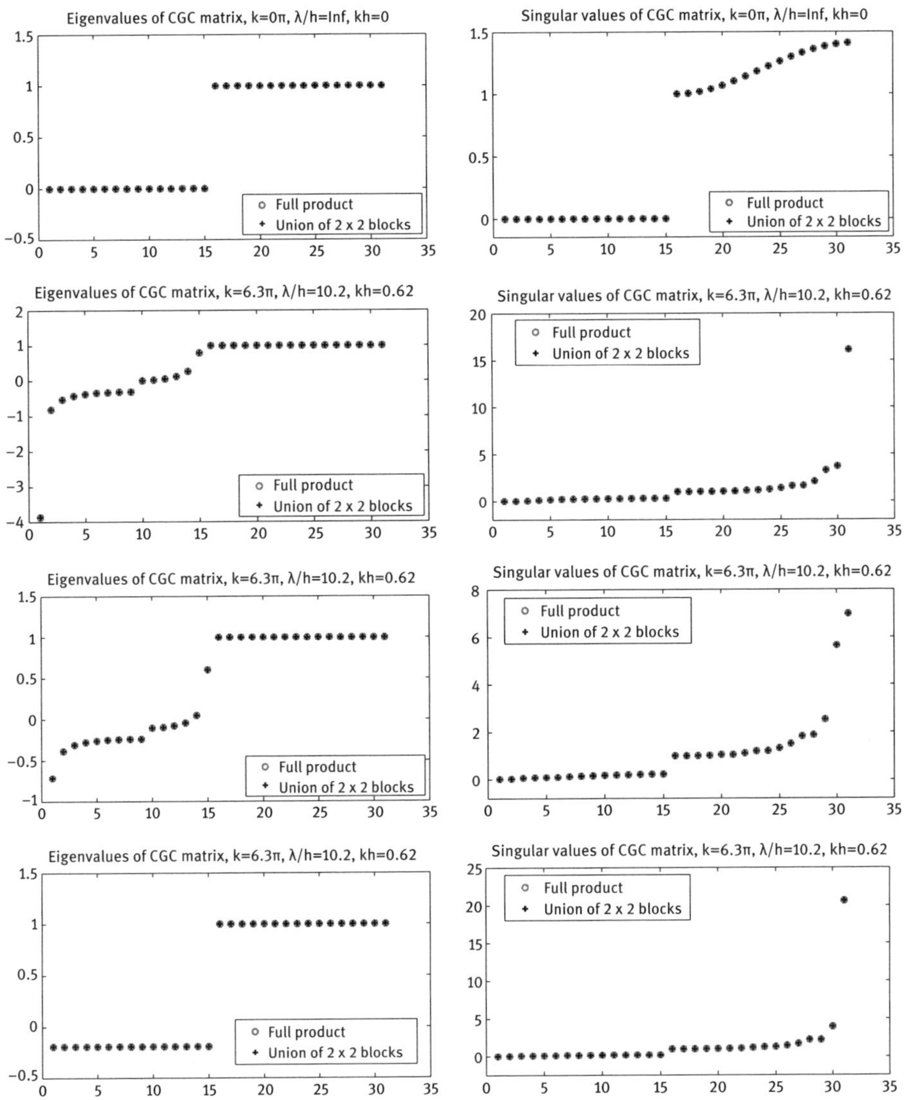

Figure 4.12: Eigenvalues (left) and singular values (right) of the coarse-grid correction operator in the cases $k = 0$ (row 1) and, for wave number $k = 6.3\pi$, the unmodified wave number (row 2), modified wave number according to (4.38) (row 3) and modified wave number according to (4.39) (row 4).

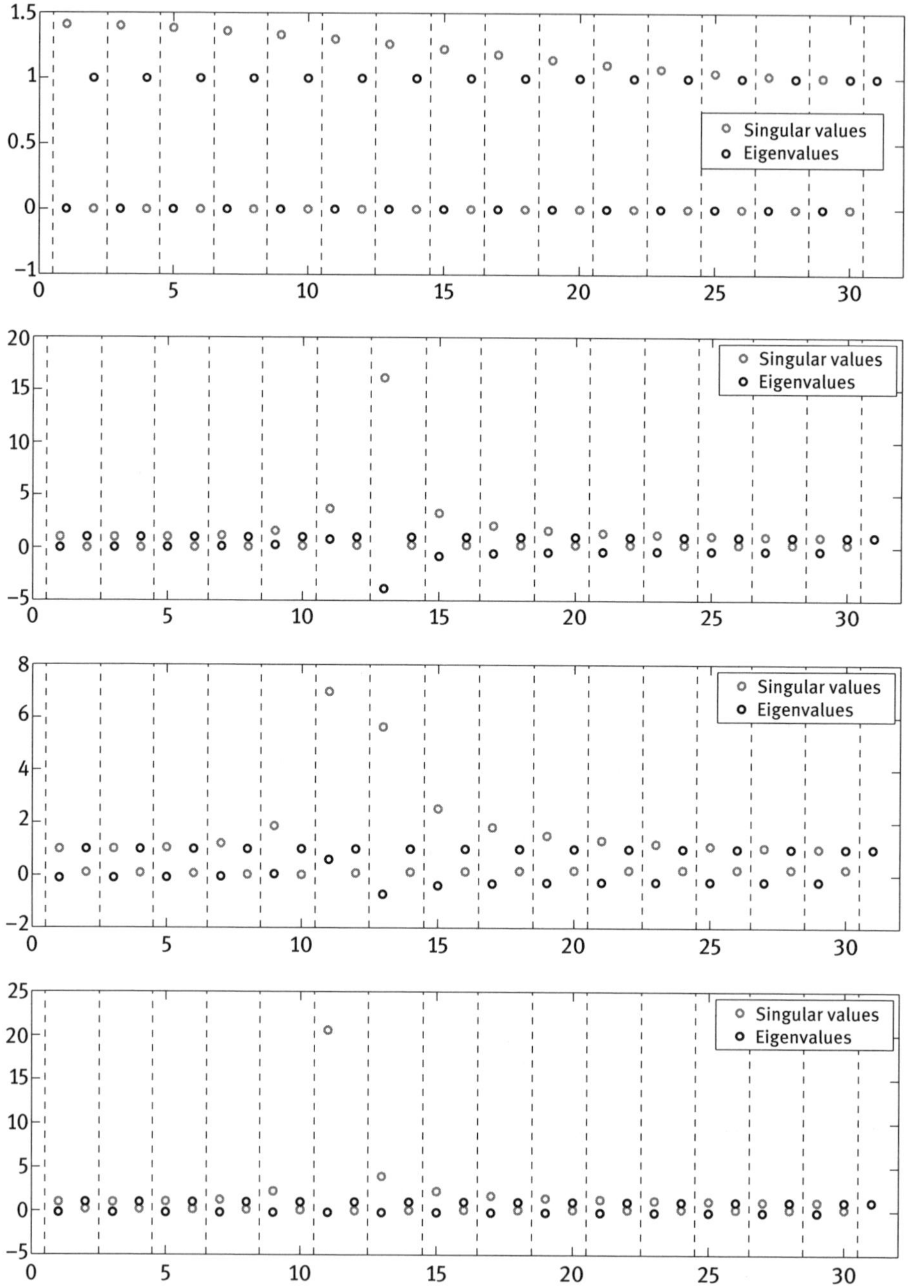

Figure 4.13: Eigenvalues and singular values of the coarse-grid correction operator arranged by invariant two and one-dimensional subspaces in the cases $k = 0$ (row 1) and, for wave number $k = 6.3\pi$, the unmodified wave number (row 2), modified wave number according to (4.38) (row 3) and modified wave number according to (4.39) (row 4).

4 Two-grid iteration

The two-grid iteration combines one or more applications of the smoothing iteration with a coarse-grid correction. If ν_1 and ν_2 denote the number of *pre-* and *postsmoothing* steps carried out before and after the coarse-grid correction, the error propagation operator of the resulting *two-grid operator* is obtained as

$$T = S^{\nu_2} C S^{\nu_1}.$$

Choosing the damped Jacobi iteration with relaxation factor ω as the smoothing operator, the results on the spectral analysis of the damped Jacobi smoother and coarse-grid correction allow us to decompose the analysis of the two-grid operator into the subspaces

$$\operatorname{span}\{\boldsymbol{v}_1^h, \boldsymbol{v}_N^h\}, \operatorname{span}\{\boldsymbol{v}_2^h, \boldsymbol{v}_{N-1}^h\}, \ldots, \operatorname{span}\{\boldsymbol{v}_n^h, \boldsymbol{v}_{n+2}^h\}, \operatorname{span}\{\boldsymbol{v}_{n+1}^h\}$$

of n pairs of complementary modes and the remaining 'middle mode' $\boldsymbol{v}_{n+1}^h$. The action of T on these one- and two-dimensional subspaces is represented by the block diagonal matrix

$$\boldsymbol{T} = \operatorname{diag}(\boldsymbol{T}_1, \ldots, \boldsymbol{T}_n, \boldsymbol{T}_{n+1})$$

with

$$\boldsymbol{T}_j = \begin{bmatrix} \sigma_j & 0 \\ 0 & \sigma_{j'} \end{bmatrix}^{\nu_2} \begin{bmatrix} 1 - c_j^4 \frac{\lambda_j^h}{\lambda_j^H} & c_j^2 s_j^2 \frac{\lambda_{j'}^h}{\lambda_j^H} \\ c_j^2 s_j^2 \frac{\lambda_j^h}{\lambda_j^H} & 1 - s_j^4 \frac{\lambda_{j'}^h}{\lambda_j^H} \end{bmatrix} \begin{bmatrix} \sigma_j & 0 \\ 0 & \sigma_{j'} \end{bmatrix}^{\nu_1} \quad j = 1, \ldots, n, \qquad (4.40)$$

and

$$\boldsymbol{T}_{n+1} = (1 - \omega)^{\nu_1 + \nu_2},$$

the latter resulting from $\sigma_{n+1} = 1 - \omega$ (cf. (4.9)).

As a result, in the spectral basis $\{\boldsymbol{v}_j^h\}_{j=1}^N$ of the discrete 1D Helmholtz equation the complete two-grid operator decouples into diagonal 2×2 blocks in addition to a trivial block for the middle mode $\boldsymbol{v}_{n+1}^h$. We can therefore reduce the following analysis to the action of the two-grid iteration on the associated two-dimensional subspaces.

4.1 The Laplacian

We consider the case $k = 0$, in which, due to (4.35), the 2×2 blocks in (4.40) simplify to

$$\boldsymbol{T}_j = \begin{bmatrix} \sigma_j & 0 \\ 0 & \sigma_{j'} \end{bmatrix}^{\nu_2} \begin{bmatrix} s_j^2 & c_j^2 \\ s_j^2 & c_j^2 \end{bmatrix} \begin{bmatrix} \sigma_j & 0 \\ 0 & \sigma_{j'} \end{bmatrix}^{\nu_1}$$
$$\text{with} \quad \sigma_j = 1 - 2\omega s_j^2, \; \sigma_{j'} = 1 - 2\omega c_j^2.$$

A further simplification, considered in [33, Section 2.2 ff.], results from the (suboptimal) choice of $\omega = \frac{1}{2}$ for the smoothing parameter as well as $\nu_1 = \nu$, $\nu_2 = 0$, in which case

$$T_j = \begin{bmatrix} s_j^2 & c_j^2 \\ s_j^2 & c_j^2 \end{bmatrix} \begin{bmatrix} c_j^2 & 0 \\ 0 & s_j^2 \end{bmatrix}^\nu = \begin{bmatrix} s_j^2 c_j^{2\nu} & c_j^2 s_j^{2\nu} \\ s_j^2 c_j^{2\nu} & c_j^2 s_j^{2\nu} \end{bmatrix}, \quad j = 1, \dots, n, \quad T_{n+1} = \left(\frac{1}{2}\right)^\nu.$$

Using (4.36a) we thus obtain for the spectral radius

$$\rho(T_j) = s_j^2 c_j^{2\nu} + c_j^2 s_j^{2\nu}, \quad j = 1, \dots, n, \quad \rho(T_{n+1}) = 2^{-\nu}.$$

Noting that $c_j^2 = 1 - s_j^2$ and $s_j^2 \in [0, \frac{1}{2}]$ for all j, we obtain the upper bound

$$\rho(T_j) \le R_\nu := \max_{0 \le t \le \frac{1}{2}} R_\nu(t), \quad R_\nu(t) := t(1-t)^\nu + (1-t)t^\nu$$

for $j = 1, \dots, n$. Since $R_\nu(\frac{1}{2}) = (\frac{1}{2})^\nu$ this bound holds also for T_{n+1}.

A common upper bound for the spectral norms $\|T_j\|$ is obtained in an analogous way using (4.36b) as

$$\|T_j\| \le N_\nu := \max_{0 \le t \le \frac{1}{2}} N_\nu(t), \quad N_\nu(t) := \sqrt{2[t^2(1-t)^{2\nu} + (1-t)^2 t^{2\nu}]},$$

which, in view of $N_\nu(\frac{1}{2}) = (\frac{1}{2})^\nu$, is again also valid for T_{n+1}.

Choosing the optimal smoothing parameter $\omega = \frac{2}{3}$ as well as $\nu_1 = \nu$, $\nu_2 = 0$ results in

$$T_j = \begin{bmatrix} s_j^2 \sigma_j^\nu & c_j^2 \sigma_{j'}^\nu \\ s_j^2 \sigma_j^\nu & c_j^2 \sigma_{j'}^\nu \end{bmatrix}, \quad j = 1, \dots, n, \quad T_{n+1} = \left(\frac{1}{3}\right)^\nu$$

where now

$$\sigma_j = \frac{1}{3}\left(3 - 4s_j^2\right), \quad \sigma_{j'} = \frac{1}{3}\left(4s_j^2 - 1\right), \quad j = 1, \dots, n.$$

Proceeding as above, this results in the upper bound

$$\rho(T_j) \le \tilde{R}_\nu := \max_{0 \le t \le \frac{1}{2}} \tilde{R}_\nu(t), \quad \tilde{R}_\nu(t) := t\left(\frac{3-4t}{3}\right)^\nu + (1-t)\left(\frac{4t-1}{3}\right)^\nu$$

for all $j = 1, \dots, n$ and, since $\tilde{R}_\nu(\frac{1}{2}) = (\frac{1}{3})^\nu$, also for T_{n+1}. For the spectral norm we obtain the bound

$$\|T_j\| \le \tilde{N}_\nu := \max_{0 \le t \le \frac{1}{2}} \tilde{N}_\nu(t),$$
$$\tilde{N}_\nu(t) := \sqrt{2\left[t^2\left(\frac{3-4t}{3}\right)^{2\nu} + (1-t)^2\left(\frac{4t-1}{3}\right)^{2\nu}\right]},$$

which holds for all $j = 1 \dots, n+1$ due to $\tilde{N}_\nu(\frac{1}{2}) = (\frac{1}{3})^\nu$.

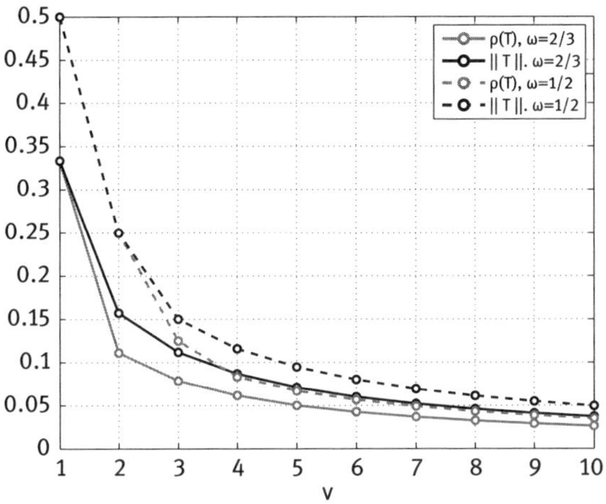

Figure 4.14: Spectral radius and spectral norm of the two-grid operator for the 1D model problem with $k = 0$ using ν steps of presmoothing with optimal relaxation parameter $\omega = 2/3$ (solid lines) and $\omega = 1/2$ (dashed lines).

We show in Figure 4.14 the spectral radii and 2-norms of the two-grid operator as a function of the number of smoothing steps for the case of Laplace's equation with both the optimal damping $\omega = 2/3$ and the suboptimal damping $\omega = 1/2$. Clearly, it is worthwhile to use the optimal damping parameter, especially when the number of smoothing steps is small.

4.2 The Helmholtz operator

Table 4.1 gives the spectral radius of the two-grid operator for the Helmholtz equation with ν steps of presmoothing using damped Jacobi for a range of wave numbers k. We observe that the iteration is divergent for $k = 6.3\pi$, which corresponds to a resolution of roughly ten points per wavelength. Moreover, while additional smoothing steps resulted in a faster convergence rate for k close to zero, this is no longer the case for higher wave numbers.

In order to improve the situation, we note that we have two parameters, the damping parameter ω from the smoothing analysis and the shifted wave number $\tilde{k}$ we can choose in the coarse-grid correction. We thus first investigate numerically whether a better choice is possible for these parameters. This choice needs to guarantee that each block $\boldsymbol{T}_j$ in (4.40) is a contraction, i.e. its eigenvalues must be smaller than one. If, in addition, the eigenvalues are distinct, then each block $\boldsymbol{T}_j$ is diagonalizable and the convergence rate determined by the eigenvalues is not only asymptotic, but visible right from the start, up to a constant given by the condition of the eigenvectors.

Table 4.1: Spectral radius of the two-grid operator for the Helmholtz equation with $h = 1/32$ for varying wave number k and (pre-) smoothing step number ν.

$\nu\backslash\rho(T)$	$k = 0$	$k = 1.3\pi$	$k = 4.3\pi$	$k = 6.3\pi$
1	0.3333	0.3364	0.4093	0.8857
2	0.1111	0.1170	0.2391	1.8530
3	0.0787	0.0779	0.2623	1.6455
4	0.0617	0.0613	0.2481	1.6349
5	0.0501	0.0493	0.2561	1.5832
10	0.0263	0.0256	0.2668	1.3797

After some simplification, we find for the eigenvalues $\mu_{1,2}(\boldsymbol{T}_j)$ the formula

$$\mu_{1,2}(\boldsymbol{T}_j) = \frac{\sigma_j^\nu(\lambda_j^H - c_j^4\lambda_j^h) + \sigma_{j'}^\nu(\lambda_j^H - s_j^4\lambda_{j'}^h) \pm \sqrt{D}}{2\lambda_j^H}, \tag{4.41}$$

$$D = \sigma_j^{2\nu}\left(\lambda_j^H - c_j^4\lambda_j^h\right)^2 + \sigma_{j'}^{2\nu}\left(\lambda_j^H - s_j^4\lambda_{j'}^h\right)^2 + 2\sigma_j^\nu\sigma_{j'}^\nu\left(\lambda_j^H s_j^4\lambda_{j'}^h + \lambda_j^H c_j^4\lambda_j^h + \lambda_{j'}^h\lambda_j^h c_j^4 s_j^4 - \left(\lambda_j^H\right)^2\right),$$

where we set $\nu := \nu_1 + \nu_2$, since only the sum appears in the eigenvalues. We now use the free parameters ω and $\tilde{k}$ to minimize numerically the maximum of $|\mu_{1,2}|$ over all $j = 1, 2, \ldots, n$ in order to test if this leads to a working algorithm. Doing so, we obtain the results shown in Table 4.2. We see that it is indeed possible to obtain a converging multigrid algorithm in 1D using this approach. But what happens at higher wave numbers k?

In order to investigate this, we fix the mesh size to $h = 1/32$, and apply our two-grid analysis for increasing wave number k, reflecting a typical realistic situation where, in a multigrid algorithm for a problem with large wave number k, one would reach such a situation after several coarsening steps. We must, however, be careful not to generate singular problems due to the Dirichlet boundary condition assumption in our analysis. To avoid this, we choose the particular wave number sequence

$$k_j = \sqrt{\frac{2}{h^2}\left(\sin^2\frac{(j-1)\pi h}{2} + \sin^2\frac{j\pi h}{2}\right)}, \quad j = 1, 2, \ldots, N, \tag{4.42}$$

placing k_j^2 exactly between two Dirichlet eigenvalues. This yields N Helmholtz problems, and we apply to each of these the two-grid analysis. In Figure 4.15, we show as a function of k_j what spectral radius one obtains with the various choices of components in the two-grid method.

The first curve corresponds to the classical two-grid method without modifying the wave number, and using ten smoothing steps (which would correspond to five

Table 4.2: Spectral radius of the two-grid operator for the Helmholtz equation with $h = 1/32$ for varying wave number k and (pre-) smoothing step number ν, using optimized coarse wave number and relaxation parameter $(\tilde{k}, \omega)$ shown in parentheses (for comparison, the original values are $k = 1.3\pi = 4.0841$ with $\omega = \frac{2-h^2k^2}{3-h^2k^2} = 0.6648$, $k = 4.3\pi = 13.5088$ with $\omega = \frac{2-h^2k^2}{3-h^2k^2} = 0.6456$, and $k = 6.3\pi = 20.1062$ with $\omega = \frac{2-h^2k^2}{3-h^2k^2} = 0.6179$).

$\nu\backslash\rho(T)$	$k = 1.3\pi$	$k = 4.3\pi$	$k = 6.3\pi$
1	0.3302 (4.4660, 0.6632)	0.3605 (13.6088, 0.6322)	0.4046 (19.1171, 0.5875)
2	0.1129 (4.1825, 0.6637)	0.1358 (13.2518, 0.6294)	0.1602 (18.7491, 0.5959)
3	0.0708 (4.1382, 0.7040)	0.0738 (13.2373, 0.6355)	0.1184 (18.9766, 0.5076)
4	0.0590 (4.1309, 0.7030)	0.0544 (13.2857, 0.6544)	0.1207 (18.9835, 0.4098)
5	0.0394 (4.1093, 0.7583)	0.0475 (13.2537, 0.6135)	0.1239 (18.9881, 0.3552)
10	0.0226 (4.0963, 0.7797)	0.0477 (13.2570, 0.3243)	0.1342 (18.9773, 0.2036)

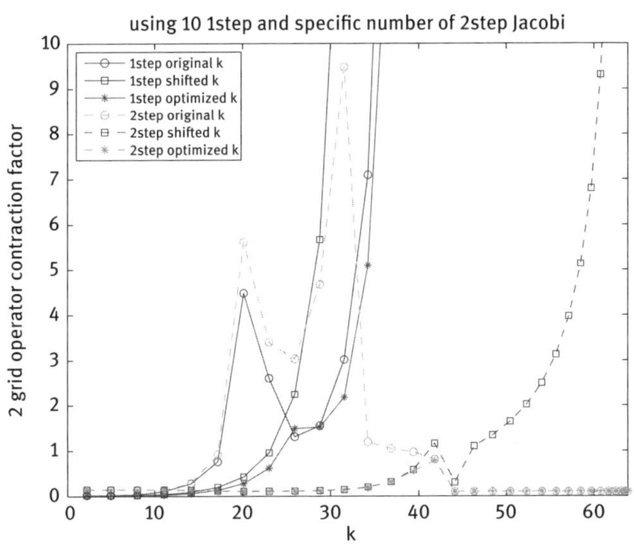

Figure 4.15: (See Color Plates for a color version.) Spectral radii of the two-grid algorithm using various components, for a fixed mesh size h, and increasing wave number k.

pre- and five postsmoothing steps) of the one-step Jacobi smoother with the relaxation parameter appropriate for the Helmholtz equation given in (4.10). We see that this method diverges as soon as the wave number k is larger than about 18, which corresponds to about $2\pi/(kh) \approx 11$ points per wavelength, and is consistent with the observation in Table 4.1.

The second curve shows the spectral radius of the classical two-grid method with modified wave number according to the theoretical shift proposed in (4.38). We see that the two-grid algorithm is already performing somewhat better, a contraction now being obtained up to wave number $k = 23$, corresponding to $2\pi/(kh) \approx 9$ points per wavelength.

Numerically optimizing the shift by minimizing the spectral radius gives a slightly better contraction at this wave number, but does not lead to a convergent method at higher wave numbers, as confirmed by the third, red curve. One needs to modify the smoother.

The next three curves result when using the two-step Jacobi smoother with the number of smoothing steps given in (4.15) from Remark 4.2 and (4.18) from Remark 4.3, requiring a reduction by a factor of 10 in the high-frequency part of the spectrum. In the fourth curve we see that, without the shift, it does not help to use the convergent two step Jacobi smoother: as for the classical one-step Jacobi smoother, the two-grid method diverges as soon as one has fewer than about 11 points per wavelength resolution. This changes fundamentally, however, when one also uses the shifted wave number k according to (4.38): now the two-grid algorithm is convergent for all wave numbers up to about $k = 40$, which corresponds to about $2\pi/(kh) \approx 5$ points per wavelength. Afterward, the shift becomes counterproductive, and the shifted two-grid algorithm diverges. However, in that regime, the unshifted algorithm does converge, and the numerical optimization of the shift indicates that one should indeed not shift as soon as the resolution has become less than about five points per wavelength. This does also make sense intuitively since, at that resolution, the corresponding coarse problem has now become (negative) definite, and is no longer a wave problem.

We now show in Figure 4.16 the various shifts that were used in these experiments. We see that the theoretical shifting formula (4.38) is very good for the two-grid method with the two-step Jacobi smoother, as long as the wave number is smaller than $k \approx 40$, which corresponds to five points per wavelength. For higher values of k the theoretical shift becomes meaningless, and the unshifted problem should be used.

We finally show in Figure 4.17 the number of smoothing steps that were used for the spectral radii shown in Figure 4.15. The graph corresponds to the theoretically required smoothing steps (4.15) and (4.18) to obtain a reduction by a factor of 10 in the high-frequency part of the spectrum.

We see that the number of smoothing steps is moderate for values of the wave number less that about 40, i.e. five points per wavelength, but then increases dramatically to the order of about 1000, on the level where k gets close to or into the resonance level. This is, however, not a surprise if one looks at the smoothing curve in Figure 4.6: clearly the smoother has difficulties to smooth close to and in the resonance level.

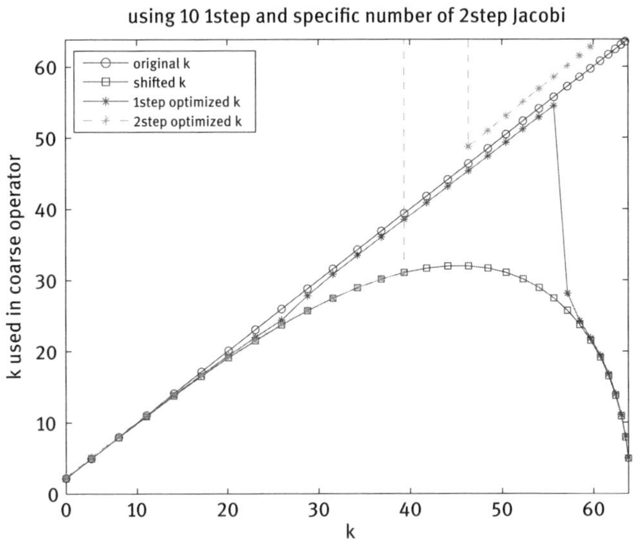

Figure 4.16: (See Color Plates for a color version.) Shifts used in the two-grid algorithm of Figure 4.15.

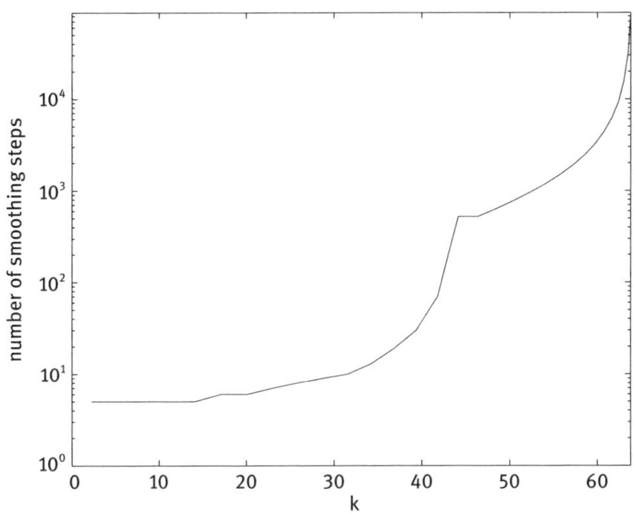

Figure 4.17: Number of smoothing steps used in the two-step Jacobi smoother for the two-grid algorithm of Figure 4.15.

When taking a final look at Figure 4.15 again, we see that the shifted two-grid method with two-step Jacobi smoother seems to have difficulties to contract around $k = 38 - 44$. This is, however, only due to our estimate in Remark 4.2 for the smooth-

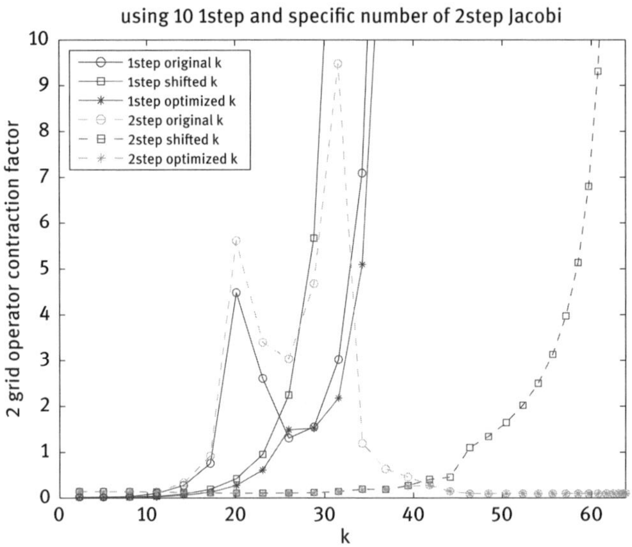

Figure 4.18: (See Color Plates for a color version.) Contraction factors of the two-grid algorithm using various components, for a fixed mesh size h, and increasing wave number k, now using the $O(k^2)$ smoothing steps already a bit before the resonance level.

ing steps needed in the two-step Jacobi smoother, which was based on the smoother alone. For the case k just above 40, this estimate leads to 71 smoothing steps, which is not quite enough, since the shift for the coarse problem starts to lose its effectiveness. We show in Figure 4.18 the same sequence of experiments again, but now using the estimate for the number of smoothing steps in the resonance level already for $kh > \sqrt{2} - 0.3$, to overcome the difficulty of the coarse-grid correction just before the shift becomes completely ineffective. The value 0.3 is just chosen from looking at these numerical experiments, an analysis would need to take into account the estimates for $\mu_{1,2}$ given in (4.41). Using more smoothing steps a bit before the resonance level leads to the results shown in Figure 4.18. The precise number of steps used is shown in Figure 4.19.

We see that it is indeed possible with this strategy to obtain a uniformly convergent two-grid algorithm for all possible values of k with the mesh resolution $h = 1/32$.

We show in Figures 4.20 and 4.21 the same type of experiment, but now for a finer mesh resolution $h = 1/128$. In Figure 4.21, we see that in the range of k between about 130 and 180, both the shifted and unshifted coarse problems lead to convergent methods, i.e., there is a zone of overlap, which is very reassuring.

We finally show in Figure 4.22 the case of $h = 1/512$. From these experiments, we can see that a reasonable strategy for a multigrid algorithm for the Helmholtz equation in 1D is to simply use a classical one-step Jacobi smoother and a shifted coarse

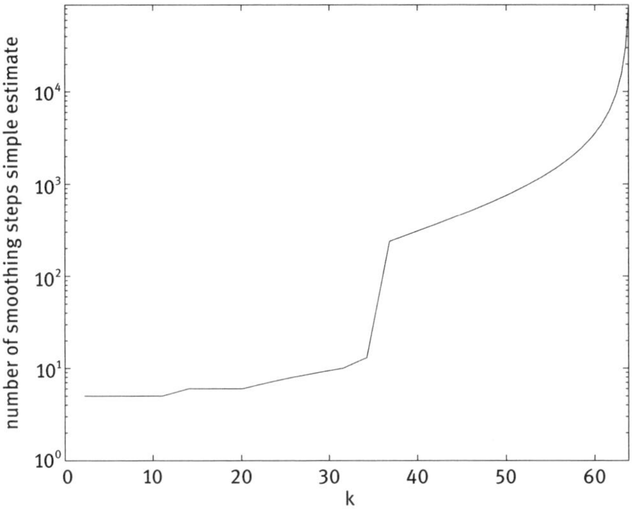

Figure 4.19: Number of smoothing steps used in the two step Jacobi smoother for the two-grid algorithm of Figure 4.18.

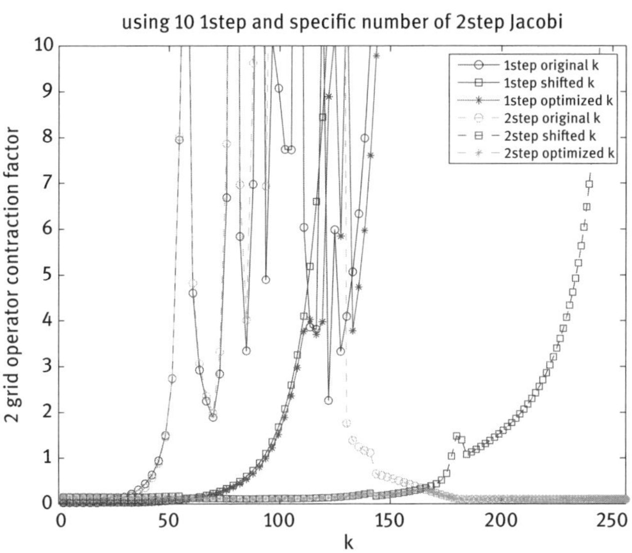

Figure 4.20: (See Color Plates for a color version.) Contraction factors of the two-grid algorithm using various components, for a fixed mesh size $h = 1/128$, and increasing wave number k.

problem, as long as the wave number and mesh size satisfy $hk < 0.5$ for the current level mesh size h, which means we have about 13 points per wavelength or more. For $0.5 \leq kh < 1.25$, one has to switch to the two-step Jacobi smoother, and use

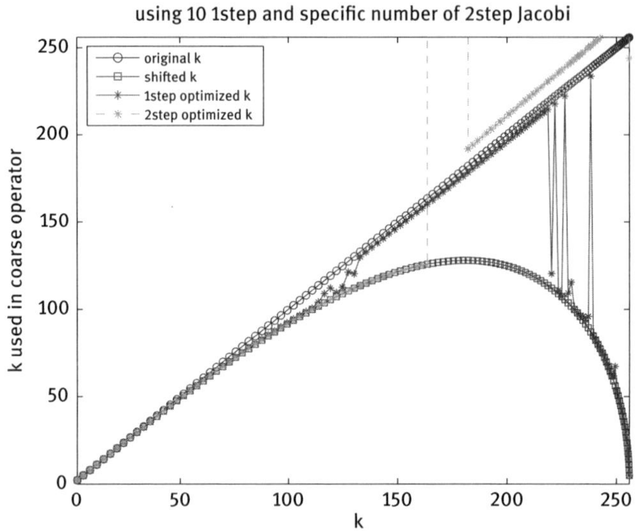

Figure 4.21: (See Color Plates for a color version.) Shifts used in the two-grid algorithm of Figure 4.20.

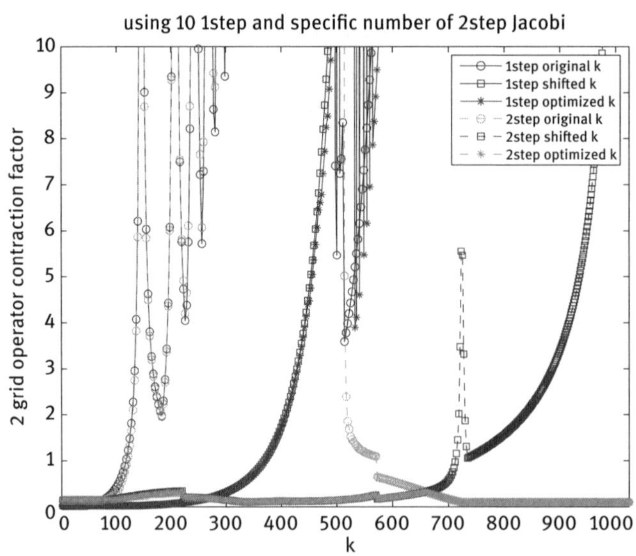

Figure 4.22: (See Color Plates for a color version.) Contraction factors of the two-grid algorithm using various components, for a fixed mesh size $h = 1/512$, and increasing wave number k.

a shifted coarse problem, and as soon as $kh \geq 1.25$, one should not use a shifted coarse problem any more, and needs to use $O(k^2)$ smoothing steps of the two-step Jacobi smoother, or a more effective smoothing method based on Chebyshev polynomials or a Krylov method.

5 Numerical examples

In this section we describe some numerical experiments with two model problems. The first (cf. (4.3)) is the Helmholtz equation

$$\begin{aligned} -u'' - k^2 u &= f \quad \text{on } (0,1), \\ u(0) = u(1) &= 0, \end{aligned} \tag{4.43}$$

with constant wave number $k > 0$, homogeneous Dirichlet boundary conditions and a source term f. This may be viewed as a *cavity problem* in one dimension in view of the fact that the negative 1D Dirichlet–Laplacian has positive eigenvalues leading to a singular operator for the discrete values

$$k = j\pi, \quad j \in \mathbb{N},$$

the so-called *resonance values* of the wave number k. Problem (4.43) is well-posed for all $0 < k \neq j\pi$, and discretizations thereof are well-behaved for sufficiently fine resolution.

As a second 1D model problem we replace the Dirichlet condition at $x = 1$ with the Sommerfeld radiation condition, which in one dimension is a simple local condition of Robin type:

$$\begin{aligned} -u'' - k^2 u &= f \quad \text{on } (0,1), \\ u(0) &= 0, \\ u'(1) - iku(1) &= 0. \end{aligned} \tag{4.44}$$

For the constant source function $f \equiv 1$ we obtain the exact solution

$$\begin{aligned} u(x) &= \frac{1}{2k^2}\left((e^{ik} - 2)e^{ikx} - e^{ik}e^{-ikx} - 2\right) \\ &= \frac{-1 - \cos(kx) - \sin k \sin(kx) - i(1 - \cos k)\sin(kx)}{k^2}. \end{aligned}$$

The same example with $f \equiv -1$ was used in [36] but the expression given there contains a sign error.

We apply the modifications of the basic multigrid V-cycle as developed in the previous sections to the two model problems using the same uniform second-order finite difference discretization considered before. In all cases, the coarse-grid operators were formed not by Galerkin projection but by using the same finite difference scheme on the coarse grid as on the fine grid. Specifically, we use two-step Jacobi smoothing with the parameters chosen depending on the relation of the discrete eigenvalues of the Dirichlet problem as developed in Section 2.3. In all the estimates of the required number of smoothing steps we set the parameter ϵ indicating the desired reduction factor to 0.1. Moreover, we apply the modified wave number as given in (4.38) on the coarse levels. To illustrate the behavior of the modified multigrid scheme we apply it to the model problems over the discrete range of wave numbers given in (4.42).

5.1 Two-grid experiments

We begin with two-grid iterations for the model problem (4.43) with a coarse-grid correction employing a modified wave number on the coarse grid according to (4.38)

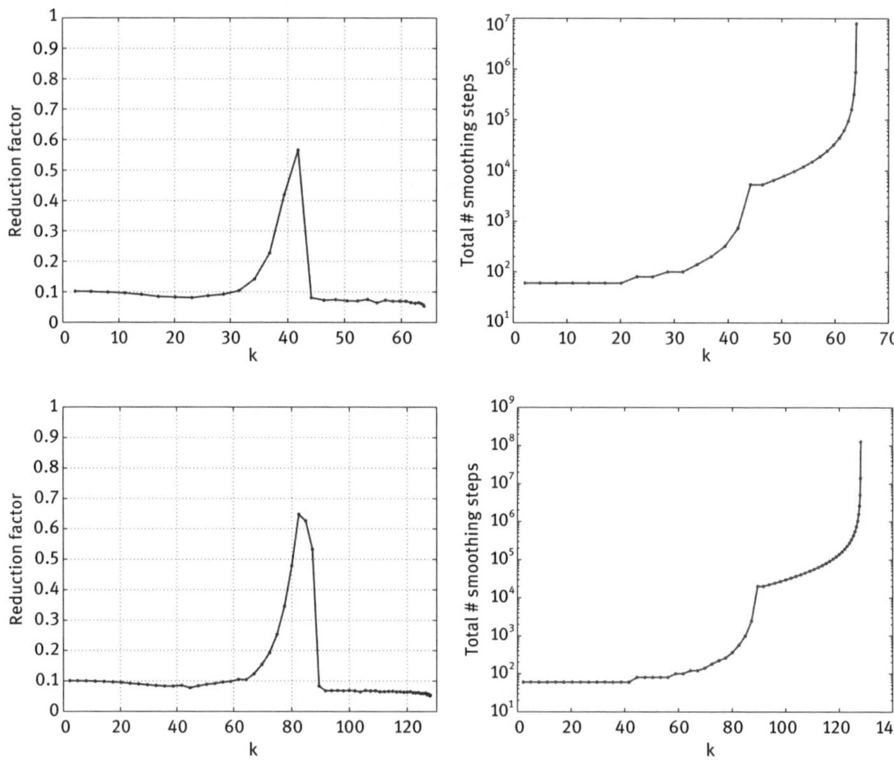

Figure 4.23: Measured error reduction factor for a two-grid method (left) and total number of smoothing steps (right) using two-step Jacobi smoothing and modified wave number for problem (4.43) with mesh width $h = 1/32$ (top) and $h = 1/64$ (bottom) for the sequence (4.42) of wave numbers k.

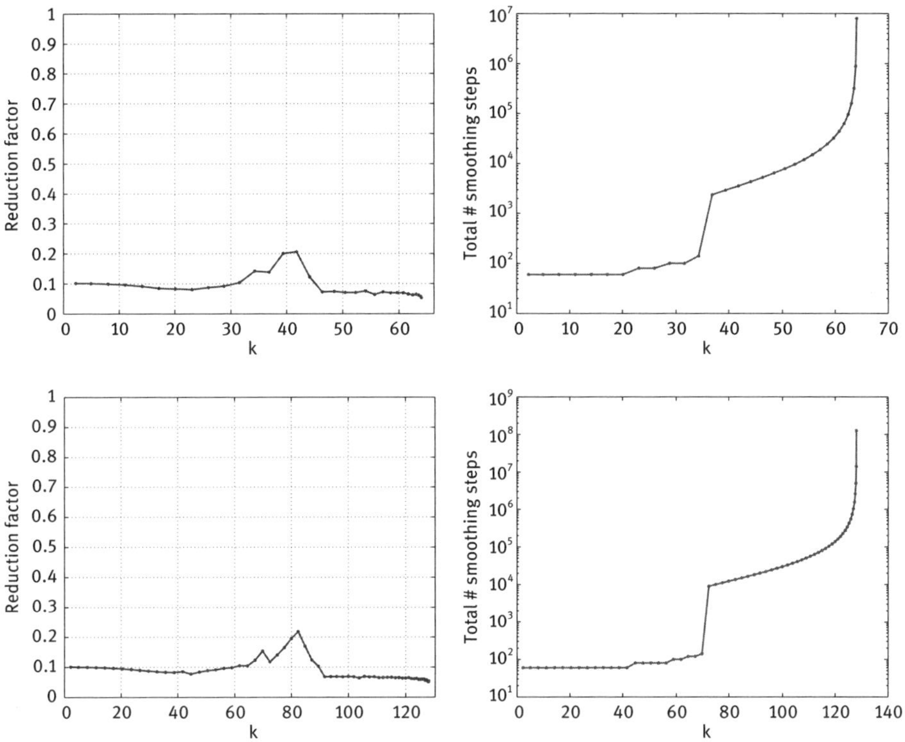

Figure 4.24: Reduction factor of the two-grid method using two-step Jacobi with early transition to the resonance level number of smoothing steps (4.18) and modified wave number for problem (4.43) with mesh width $h = 1/32$ (top) and $h = 1/64$ (bottom).

combined with two-step Jacobi smoothing with the smoothing parameters chosen according to the location of the eigenvalues of the discrete Dirichlet–Helmholtz operator as described in Section 2.3. Following the observations made in Section 4, we employ a modified wave number on all coarse grids which still have a resolution of at least $\lambda/h \geq 5$. Figure 4.23 shows the corresponding measured reduction factors and the total number of smoothing iterations. We note the region just below $kh = \sqrt{2}$ where a reduction factor much larger than the desired $\epsilon = 0.1$ results. As discussed in Section 4.2, this is due to the weakening of the effect of shifting the wave number on the coarse grid, which was not taken into account when estimating the number of smoothing steps required based on the smoother alone. When using the remedy proposed in Section 4.2, which means we switch for the estimate of the number of smoothing steps in the two-step Jacobi smoother from the first to the second case already when $kh \geq \sqrt{2} - 0.3$, leads to the results shown in Figure 4.24. We observe that the two-grid cycle has now a uniform error reduction factor of about 0.2.

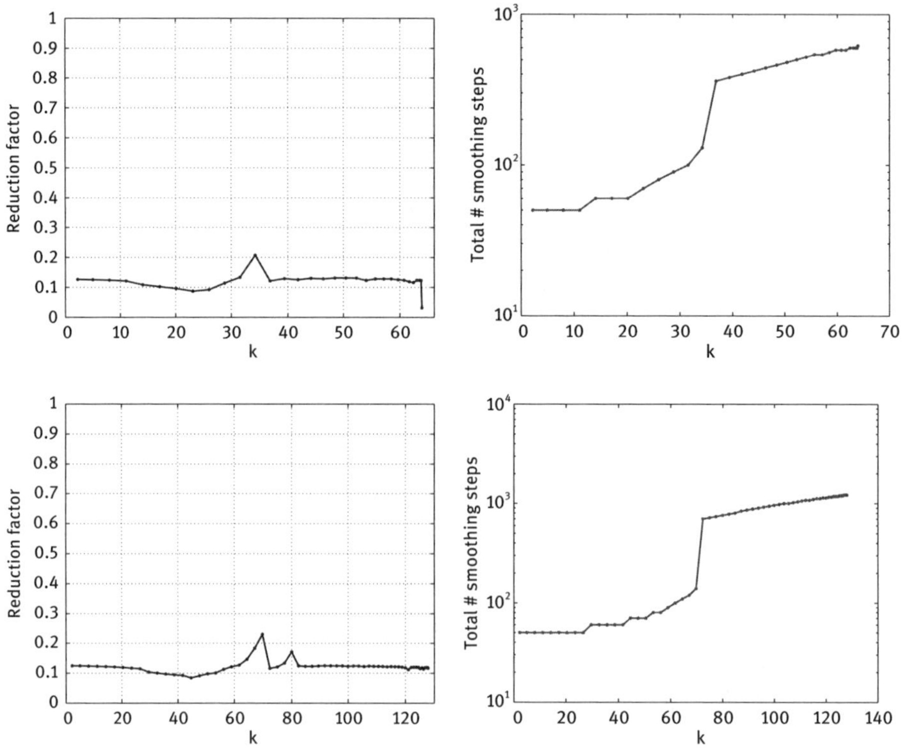

Figure 4.25: Reduction factor of the two-grid method using two-step Jacobi with early transition to resonance level with Krylov smoothing and modified wave number for problem (4.43) with mesh width $h = 1/32$ (top) and $h = 1/64$ (bottom).

We also notice, however, that the number of smoothing steps required by the estimate (4.18) on the resonance level is excessively high. We therefore now replace the two-step Jacobi smoother by GMRES whenever we would have to used the estimate (4.18) for the number of smoothing steps, i.e. on the resonance level and slightly before, $\sqrt{2} - 0.3 < kh < 2$, and we use $\nu = \frac{3k}{2\pi}$ as the number of GMRES smoothing steps, which is linear in k. This leads to the reduction factors and total smoothing counts shown in Figure 4.25 for $h = 1/32$ and $h = 1/64$. We again observe a uniformly small reduction factor across all wave numbers and a much smaller number of smoothing steps.

5.2 Multigrid experiments, complexity

We next show multigrid experiments, beginning with a complexity estimate. For the Laplacian, if we perform a fixed number ν of smoothing steps, and the smoother has a cost per step linear in $\frac{1}{h}$, e.g. $\frac{C_J}{h}$, then the cost of one V-cycle is, for $h = 2^{-m}$,

$$c_\Delta := \nu \frac{C_J}{h} + \nu \frac{C_J}{2h} + \cdots + \nu C_J = \nu \frac{C_J}{h} \sum_{j=0}^{m} \frac{1}{2^j} = \nu \frac{C_J}{h} \left(2 - \frac{1}{2^m}\right)$$
$$= \nu \frac{C_J}{h}(2-h) < \nu \frac{2C_J}{h}.$$

In the Helmholtz case, we have seen that the number of smoothing steps becomes large on one particular level, the resonance level, where $O(k^2)$ two-step Jacobi smoothing steps, or $O(k)$ Krylov smoothing steps are necessary. On the remaining levels the number of smoothing steps remains again bounded by a number $\nu = O(1)$. The cost of our multigrid algorithm therefore contains a component like for the Laplacian, c_Δ, and the cost of the resonance level. If we assume that, to avoid the pollution effect, we have chosen a fine-level resolution satisfying $k^3h^2 = C$, then we have in the multigrid hierarchy, before reaching the resonance level ℓ, further levels, where ℓ satisfies

$$k(2^\ell h) < \sqrt{2}, \quad 2^\ell \approx \frac{\sqrt{2}}{kh}, \tag{4.45}$$

see (4.13). The size of the system on the resonance level is then $\frac{1}{2^{\ell+1}h} = \frac{k}{2^{3/2}}$, where we used estimate (4.45). On this level, we need either $O(k^2)$ two-step Jacobi smoothing steps, which leads to a cost of

$$c_J = k^2 C_J \frac{k}{2^{\frac{3}{2}}} = \frac{C_J k^3}{2^{\frac{3}{2}}} = \frac{C_J C}{2^{\frac{3}{2}} h^2},$$

where we used $k^3h^2 = C$, or $O(k)$ Krylov smoothing steps, which, assuming that also the cost of Krylov steps is linear in the size of the system with constant of proportionality C_K, leads to a cost

$$c_K = kC_K \frac{k}{2^{\frac{3}{2}}} = \frac{C_K k^2}{2^{\frac{3}{2}}} = \frac{C_K C}{2^{\frac{3}{2}} h^{\frac{4}{3}}}.$$

The overall cost of this multigrid algorithm is therefore, for one V-cycle, bounded by

$$c_J^{tot} < c_\Delta + c_J < \nu \frac{2C_J}{h} + \frac{C_J C}{2^{\frac{3}{2}} h^2} \tag{4.46}$$

in the case of the two-step Jacobi smoother on the resonance level, and

$$c_K^{tot} < c_\Delta + c_K < \nu \frac{2C_J}{h} + \frac{C_K C}{2^{\frac{3}{2}} h^{\frac{4}{3}}} \tag{4.47}$$

in the case where on one level, the resonance level, one uses a Krylov smoother with $O(k)$ steps, and otherwise two-step Jacobi. We see that, compared to the case of the Laplacian, each V-cycle has more than linear complexity with respect to the dimension of the finest system, $1/h$: with Jacobi on the resonance level, one pays $O(h^{-2})$

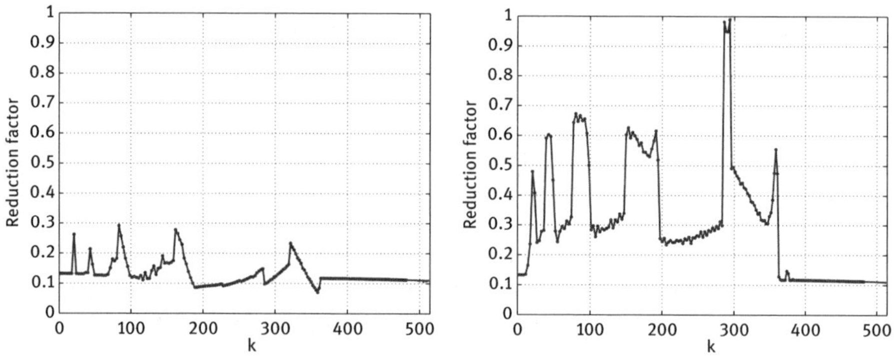

Figure 4.26: V-cycle with 6 levels using two-step Jacobi smoothing and Krylov smoothing on the resonance level and modified wave number for problem (4.43) (left) and problem (4.44) (right) with mesh width $h = 1/256$.

to have a fixed iteration number independent of the mesh size h, and hence k, and with Krylov smoothing on the resonance level one pays $O(h^{-4/3})$ per V-cycle to have a fixed iteration number independent of h and k. This latter result seems very much acceptable, however, for the difficult case of the Helmholtz equation: one pays $O(h^{-4/3})$ instead of $O(h^{-1})$ per V-cycle, for iteration numbers independent of h and k.

We now show numerical experiments for multigrid V-cycles over six levels applied to the Helmholtz equation using the modifications described in this chapter. We start by showing error reduction rates in Figure 4.26. On the left, we see that, for the problem (4.43) with pure Dirichlet boundary conditions, for which our analysis of the two-grid cycle was performed, the addition of further grid levels still gives uniform error reduction for all wave numbers k, by a factor bounded by about 0.3.

In Figure 4.26(right), we also show a numerical experiment for a situation not covered by our analysis, namely problem (4.44) which features a Sommerfeld radiation condition at one boundary. We have observed that the two-step Jacobi smoother, whose parameters were tuned to the Dirichlet case in Section 2.3, are not optimal for this case and that the modified wave number had to be restricted to grids for which $kh < 2.3$, a somewhat better resolution than $\lambda/h = 5$, to obtain a convergent scheme. Extending the analysis to this situation is more difficult as the discrete operator is no longer normal and, therefore, pure eigenvalue analysis of the smoother is not sufficient. Nevertheless, using the shift of the wave number and smoothing techniques developed for the Dirichlet problem, we nonetheless obtain a convergent multigrid V-cycle across the full range of wave numbers. Combining this type of solver with an (outer) Krylov acceleration could lead to an effective solver.

We finally show a sequence of realistic numerical experiments, starting on a fine grid with enough resolution to give an accurate discrete approximation, which means

Table 4.3: Measured reduction rates for two multigrid variants on a sequence of problems of type (4.43) with k^2h^3 held constant to bound pollution error (the smoothing on the resonance level employs $O(k^2)$ steps for two-step Jacobi and $O(k)$ for GMRES smoothing).

h	1/128	1/256	1/512	1/1024	1/2048
k	63.74	101.43	161.12	255.38	403.04
		Two-step Jacobi smoothing on the resonance level			
Two-grid	0.0891	0.1147	0.1111	0.1605	0.1448
Three-grid	0.1177	0.1216	0.1291	0.1686	0.1579
Four-grid	0.1091	0.1291	0.3157	0.2146	0.1604
Five-grid	0.1091	0.1291	0.3157	0.1649	0.1679
Six-grid	0.1091	0.1291	0.3157	0.1649	0.1679
		GMRES smoothing on resonance level			
Two-grid	0.0891	0.1147	0.1111	0.1605	0.1448
Three-grid	0.1177	0.1216	0.1291	0.1686	0.1580
Four-grid	0.1107	0.1187	0.3157	0.2146	0.1604
Five-grid	0.1107	0.1195	0.3157	0.2086	0.1686
Six-grid	0.1107	0.1195	0.3157	0.2086	0.1686

Table 4.4: Measured reduction rates for two multigrid variants on a sequence of problems of type (4.44) with k^2h^3 held constant to bound pollution error (the smoothing on the resonance level employs $O(k^2)$ steps for two-step Jacobi and $O(k)$ for GMRES smoothing).

h	1/128	1/256	1/512	1/1024	1/2048
k	63.74	101.43	161.12	255.38	403.04
		Two-step Jacobi smoothing on resonance level			
Two-grid	0.0936	0.1249	0.1442	0.1656	0.1618
Three-grid	0.2884	0.2187	0.1794	0.1899	0.1916
Four-grid	0.2864	0.3926	0.6222	0.2714	0.2370
Five-grid	0.2864	0.3963	0.6222	0.2761	0.4117
Six-grid	0.2864	0.3963	0.6222	0.2761	0.4117
		GMRES smoothing on resonance level			
Two-grid	0.0936	0.1249	0.1442	0.1656	0.1618
Three-grid	0.2884	0.2187	0.1794	0.1899	0.1916
Four-grid	0.3077	0.2837	0.6222	0.2714	0.2370
Five-grid	0.3077	0.2837	0.6222	0.2838	0.2893
Six-grid	0.3077	0.2838	0.6222	0.2838	0.2893

to use a scaling of k and h that bounds the pollution error of our centered finite-difference discretization scheme. As shown in [3, 35, 36], this requires that k^3h^2 remain bounded as the wave number is increased and the fine mesh refined.

We begin with a mesh width of $h = 1/64$ and $k = 40$, yielding a resolution of about ten points per wavelength. We then choose the wave numbers on the sequence of successively halved meshes with $h = 1/128, \dots, h = 1/2048$ so that k^3h^2 has about the same value for all problems. We show measured reduction rates for two variants of our multigrid method employing V-cycles containing two through six grid levels for the 1D model problem (4.43) in Table 4.3 and the Sommerfeld problem (4.44) in Table 4.4. The first variant employs $O(k^2)$ two-step Jacobi smoothing steps and the second $O(k)$ Krylov smoothing steps on the resonance level using GMRES. Spanning a range of wave numbers k from 63.74 to 403.04, which corresponds to roughly 12 and 64 full waves across the domain $\Omega = (0, 1)$, respectively, we obtain in the case of Dirichlet boundary conditions reduction rates around 0.1 for the two-grid method to at worst 0.32 for the six-grid method. For problem (4.44) with the Sommerfeld boundary condition, we obtain for the two-grid method again reduction rates of about 0.1, and at worst 0.62 for the six-grid method. In all our results, while the error reduction rates can deteriorate a bit when the resonance level becomes part of the multigrid hierarchy, they remain stable as one increases the number of levels further, which shows that the methods are now clearly independent of the number of levels used, crossing the resonance level in a stable fashion. We note that we have obtained even somewhat better rates when standard Jacobi smoothing is used in place of our two-step Jacobi smoother on sufficiently fine grids.

6 Conclusions

We have explained in this chapter in detail for a 1D model problem why one cannot simply use a standard multigrid method to solve discretized Helmholtz problems. Both the smoother and the coarse-grid correction will fail in that case. Standard smoothers amplify low-frequency error components, and are thus not stable, a fact that is well known. The coarse-grid correction fails because of the incorrect dispersion relation (phase error) on coarser and coarser grids, a fact that is less well known.

We then designed a two-step Jacobi smoother which is convergent for all levels, and this required scheduling specific numbers of smoothing steps depending on the wave number k and the mesh size h on the current level, in order to obtain uniform error reduction in the high-frequency modes. On one particular level, which we call the resonance level, an exceedingly high number of smoothing steps is needed, namely $O(k^2)$ steps, if one insists on uniform error reduction in the high-frequency components on that level. A better approach is to choose a more suitable higher-degree polynomial smoother on the resonance level, and we used GMRES in this case, and then "only" $O(k)$ smoothing steps are needed for a uniform error reduction in the high-frequency components.

In order to alleviate the dispersion relation problem on coarser meshes, we introduced a shift in the wave number, adapted to the mesh size h of the level. In one spatial dimension, we can obtain uniform error reduction with the two and multigrid cycle using this technique, and we showed that the overall complexity of our multigrid algorithm is $O(h^{-4/3})$ when we use on the resonance level a Krylov smoother, compared to the optimal complexity of multigrid for the Laplacian alone, which is $O(h^{-1})$.

We illustrated our results with numerical experiments in one spatial dimension, which illustrates well that a fixed number of iterations suffice to solve Helmholtz equations in one dimension with this adapted multigrid method, still only using standard, but adapted components, independently of the wave number k and the mesh size h.

In higher dimensions, our smoothing analysis still applies, and the two-step Jacobi smoother will work well, as will the Krylov smoother on the resonance level. A perfect dispersion relation correction, however, will not be possible any more with only a shift of the wave number on coarser and coarser levels, since the dispersion relation is now a curve in two dimensions, or a surface in three dimensions. We have, however, observed that higher-order discretization schemes lead to nearly spherical dispersion surfaces. We are currently analyzing how much a scalar dispersion correction can still help in higher dimension; these results will appear in a future publication.

References

[1] Mark Ainsworth, Discrete Dispersion Relation for hp-Version Finite Element Approximation at High Wave Number, *SIAM J. Numer. Anal.* 42 (2004), 553–575.

[2] Tuomas Airaksinen, Erkki Heikkola, Anssi Pennanen, and Jari Toivanen, An algebraic multigrid based shifted-Laplacian preconditioner for the Helmholtz equation, *Journal of Computational Physics* 226 (2007), 1196–1210.

[3] Ivo M. Babuska and Stefan A. Sauter, Is the pollution effect of the FEM avoidable for the Helmholtz equation considering high wave numbers?, *SIAM review* 42 (2000), 451–484.

[4] Randolph E. Bank, A comparison of two multilevel iterative methods for nonsymmetric and indefinite elliptic finite element equations, 18 (1981), 724–743.

[5] A. Bayliss, C. I. Goldstein, and E. Turkel, An iterative method for the Helmholtz equation, *J. Comput. Phys.* 49 (1983), 443–457.

[6] Alvin Bayliss, Charles I. Goldstein, and Eli Turkel, The numerical solution of the Helmholtz equation for wave propagation problems in underwater acoustics, 11 (1985), 655–665.

[7] Folkmar A. Bornemann and Rolf Krause, Classical and Cascadic Multigrid— A Methodological Comparison, in: *Ninth International Conference on Domain Decomposition Methods* (Petter E. Bjørstadt, Magne E. Espedal and David E. Keyes, eds.), pp. 64–71, 1998.

[8] James H. Bramble, Joseph E. Pasciak, and Jinchao Xu, The Analysis of Multigrid Algorithms for Nonsymmetric and Indefinite Elliptic Problems, 51 (1988), 389–414.

[9] A. Brandt and S. Ta'asan, Multigrid Methods for Nearly Singular and Slightly Indefinite Problems, in: *Multigrid Methods II*, pp. 99–121, Springer, 1986.

[10] Achi Brandt and Oren E. Livne, *Multigrid Techniques, 1984 Guide with Applications to Fluid Dynamics, Revised Edition*, Classics in Applied Mathematics 67, SIAM, 2011.

[11] Achi Brandt and Irene Livshits, Wave-ray multigrid method for standing wave equations, *Electron. Trans. Numer. Anal.* 6 (1997), 162–181.

[12] William L. Briggs, Van Emden Henson, and Steve F. McCormick, *A Multigrid Tutorial*, second ed, SIAM, 2000.

[13] Gary C. Cohen, *Higher-Order Numerical Methods for Transient Wave Equations*, Scientific Computation, Springer, Berlin, Heidelberg, 2002.

[14] David L. Colton and Rainer Kress, *Integral Equation Methods in Scattering Theory*, Wiley, New York, 1983.

[15] Siegfried Cools and Wim Vanroose, Local Fourier Analysis of the Complex Shifted Laplacian preconditioner for Helmholtz problems, *arXiv preprint arXiv:1112.5346* (2011).

[16] B. Després, *Méthodes de décomposition de demains pour les problèms de propagation d'ondes en régime harmonique*, Ph.D. thesis, Université Paris IX Dauphine, 1991.

[17] Howard C. Elman, Oliver G. Ernst, and Dianne P. O'Leary, A Multigrid Method Enhanced by Krylov Subspace Iteration for Discrete Helmholtz Equations, *SIAM J. Sci. Comp.* 23 (2001), 1290–1314.

[18] Björn Engquist and Lexing Ying, Sweeping preconditioner for the Helmholtz equation: hierarchical matrix representation, *Communications on pure and applied mathematics* 64 (2011), 697–735.

[19] Björn Engquist and Lexing Ying, Sweeping preconditioner for the Helmholtz equation: moving perfectly matched layers, *Multiscale Modeling & Simulation* 9 (2011), 686–710.

[20] Y. A. Erlangga, Advances in Iterative Methods and Preconditioners for the Helmholtz Equation, *Archives Comput. Methods in Engin.* 15 (2008), 37–66.

[21] Y. A. Erlangga, C. Vuik, and C. W. Oosterlee, On a class of preconditioners for solving the Helmholtz equation, *Applied Numerical Mathematics* 50 (2004), 409–425.

[22] O. G. Ernst and M. J. Gander, *Why it is Difficult to Solve Helmholtz Problems with Classical Iterative Methods*, Numerical Analysis of Multiscale Problems (I. Graham, T. Hou, O. Lakkis, and R. Scheichl, eds.), Springer Verlag, 2012, pp. 325–363.

[23] C. Farhat, P. Avery, R. Tezaur, and J. Li, FETI-DPH: a dual-primal domain decomposition method for accoustic scattering, *Journal of Computational Acoustics* 13 (2005), 499–524.

[24] Charbel Farhat, Philip Avery, Radek Tesaur, and Jing Li, FETI-DPH: a dual-primal domain decomposition method for acoustic scattering, *Journal of Computational Acoustics* 13 (2005), 499–524.

[25] Charbel Farhat, Antonini Macedo, and Radek Tezaur, FETI-H: a Scalable Domain Decomposition Method for High Frequency Exterior Helmholtz Problem, in: *Eleventh International Conference on Domain Decomposition Method* (Choi-Hong Lai, Petter Bjørstad, Mark Cross, and Olof Widlund, eds.), pp. 231–241, DDM.ORG, 1999.

[26] M. J. Gander and F. Nataf, AILU: A preconditioner based on the analytic factorization of the elliptic operator, *Num. Lin. Alg. and Appl.* 7 (2000), 543–567.

[27] M. J. Gander and H. Zhang, Domain decomposition methods for the Helmholtz equation: a numerical investigation., in: *Domain Decomposition Methods in Science and Engineering XX, Lecture Notes in Computational Science and Engineering* (R. Bank, M. Holst, and J. Xu, eds.), San Diego, p. *to appear*, Springer-Verlag, 2012.

[28] Martin J. Gander, Laurence Halpern, and Frédéric Magoulès, An Optimized Schwarz Method with Two-Sided Robin Transmission Conditions for the Helmholtz Equation, *Int. J. for Num. Meth. in Fluids* 55 (2007), 163–175.

[29] Martin J. Gander, Frédéric Magoulès, and Frédéric Nataf, Optimized Schwarz Methods without Overlap for the Helmholtz Equation, *SIAM J. Sci. Comput.* 24 (2002), 38–60.
[30] Martin J. Gander and Frédéric Nataf, An Incomplete LU Preconditioner for Problems in Acoustics, *Journal of Computational Acoustics* 13 (2005), 1–22.
[31] M. J. Gander, Ivan G. Graham, and Euan A. Spence, Analysis of the Shifted Laplace Preconditioner for the Helmholtz Equation, in preparation.
[32] M. B. Van Gijzen, Y. A. Erlangga, and C. Vuik, Spectral analysis of the discrete Helmholtz operator preconditioned with a shifted Laplacian, *SIAM J. Sci. Comput.* 29 (2007), 1942–1958.
[33] Wolfgang Hackbusch, *Multi-Grid Methods and Applications*, Springer-Verlag, 1985.
[34] G. Ronald Hadley, A complex Jacobi iterative method for the indefinite Helmholtz equation, *Journal of Computational Physics* 203 (2005), 358–370.
[35] Frank Ihlenburg and Ivo Babuška, Finite Element Solution to the Helmholtz Equation with High Wave Number. Part I: The h-version of the FEM, *Computer Methods in Applied Mechanics and Engineering* 39 (1995), 9–37.
[36] Frank Ihlenburg and Ivo Babuška, Finite Element Solution of the Helmholtz Equation with High Wave Number Part II: The h-p Version of The FEM, *SIAM J. Numer. Anal.* 34 (1997), 315–358.
[37] Seongjai Kim and Soohyun Kim, Multigrid simulation for high-frequency solutions of the Helmholtz problem in heterogeneous media, *SIAM Journal on Scientific Computing* 24 (2002), 684–701.
[38] B. Lee, T. Manteuffel, S. McCormick, and J. Ruge, First-order system least squares (FOSLS) for the Helmholtz equation, *SIAM J. Sci. Comp.* 21 (2000), 1927–1949.
[39] Irene Livshits, *Multigrid Solvers for Wave Equations*, Ph.D. thesis, Bar-Ilan University, Ramat-Gan, Israel, 1996.
[40] Irene Livshits and Achi Brandt, Accuracy properties of the wave-ray multigrid algorithm for Helmholtz equations, *SIAM Journal on Scientific Computing* 28 (2006), 1228–1251.
[41] Lewis Fry Richardson, On the Approximate Arithmetical Solution by Finite Differences of Physical Problems Involving Differential Equations, with an Application to the Stresses in a Masonry Dam, *Proceedings of the Royal Society of London. Series A* 83 (1910), 335–336.
[42] Yair Shapira, Multigrid Techniques for Highly Indefinite Equations, in: *Proceedings 1995 Copper Mountain Conference on Multigrid Methods*, 1995, available at http://www.c3.lanl.gov/ yairs/.
[43] Richard Vynne Southwell, *Relaxation methods in engineering science : a treatise on approximate computation*, Oxford University Press, 1940.
[44] John C. Strikwerda, *Finite Difference Schemes and Partial Differential Equations*, 2nd ed, SIAM, Philadelphia, 2004.
[45] Ulrich Trottenberg, C. Cornelis W. Oosterlee, and Anton Schüller, *Multigrid*, Academic Press, 2001.
[46] Petr Vanek, Jan Mandel, and Marian Brezina, Two-level algebraic multigrid for the Helmholtz problem, *Contemporary Mathematics* 218 (1998), 349–356.
[47] Wim Vanroose, Bram Reps, and Hisham bin Zubair, A polynomial multigrid smoother for the iterative solution of the heterogeneous Helmholtz problem, *arXiv preprint arXiv:1012.5379* (2010).
[48] Wim Vanroose, Bram Reps, Hisham Bin Zubair et al., Multigrid preconditioners based on polynomial smoothers for the Helmholtz equation with absorbing layers, in: *International Conference On Preconditioning Techniques For Scientific And Industrial Applications, Preconditioning 2011*, 2011.
[49] Richard S. Varga, *Matrix Iterative Analysis*, Prentice-Hall, Englewood-Cliffs, NJ, 1962.
[50] Harry Yserentant, On the Multilevel Spltting of Finite Element Spaces for Indefinite Elliptic Boundary Value Problems, 23 (1986), 581–595.

Marcus J. Grote and Teodora Mitkova

Explicit local time-stepping methods for time-dependent wave propagation

Abstract: Semidiscrete Galerkin formulations of transient wave equations, either with conforming or discontinuous Galerkin finite element discretizations, typically lead to large systems of ordinary differential equations. When explicit time integration is used, the time-step is constrained by the smallest elements in the mesh for numerical stability; possibly a high price to pay. To overcome that overly restrictive stability constraint on the time-step, yet without resorting to implicit methods, explicit local time-stepping schemes (LTS) are presented here for transient wave equations either with or without damping. In the undamped case, leap-frog-based LTS methods lead to high-order explicit LTS schemes, which conserve the energy. In the damped case, when energy is no longer conserved, Adams–Bashforth-based LTS methods also lead to explicit LTS schemes of arbitrarily high accuracy. When combined with a finite-element discretization in space with an essentially diagonal mass matrix, the resulting time-marching schemes are fully explicit and thus inherently parallel. Numerical experiments with continuous and discontinuous Galerkin finite-element discretizations validate the theory and illustrate the usefulness of these local time-stepping methods.

Keywords: time-dependent waves, damped waves, finite element methods, mass lumping, discontinuous Galerkin methods, explicit time integration, adaptive refinement, local time-stepping

2010 Mathematics Subject Classification: 65N30

Marcus J. Grote: Institute of Mathematics, University of Basel, Rheinsprung 21, 4051 Basel, Switzerland, marcus.grote@unibas.ch
Teodora Mitkova: Department of Mathematics, University of Fribourg, Chemin du Musée 23, 1700 Fribourg, Switzerland, teodora.mitkova@unifr.ch

1 Introduction

The efficient numerical simulation of transient wave phenomena is of fundamental importance in a wide range of applications from acoustics, electromagnetics or elasticity. Although classical finite difference methods remain a viable approach in rect-

This work was partly supported by the Swiss National Science Foundation. We thank Manuela Utzinger and Max Rietmann for their help with the numerical experiments.

angular geometry on Cartesian meshes, their usefulness is quite limited in the presence of complex geometry, such as cracks, sharp corners or irregular material interfaces. In contrast, finite element methods (FEMs) easily handle unstructured meshes and local refinement. Moreover, their extension to high order is straightforward, a key feature to keep numerical dispersion minimal.

Semidiscrete finite element Galerkin approximations typically lead to a system of ordinary differential equations. However, if explicit time-stepping is subsequently employed, the mass matrix arising from the spatial discretization by standard conforming finite elements must be inverted at each time-step: a major drawback in terms of efficiency. To overcome that difficulty, various "mass lumping" techniques have been proposed, which effectively replace the mass matrix by a diagonal approximation. While straightforward for piecewise linear elements [6, 32], mass lumping techniques require special quadrature rules and additional degrees of freedom at higher order to preserve the accuracy and guarantee numerical stability [10, 20].

Discontinuous Galerkin (DG) methods offer an attractive and increasingly popular alternative for the spatial discretization of time-dependent hyperbolic problems [1, 7, 21, 22, 30, 37]. Not only do they accommodate elements of various types and shapes, irregular nonmatching grids, and even locally varying polynomial order, and hence offer greater flexibility in the mesh design. They also lead to a block-diagonal mass matrix, with the block size equal to the number of degrees of freedom per element; in fact, for a judicious choice of (locally orthogonal) shape functions, the mass matrix is truly diagonal. Thus, when a spatial DG discretization is combined with explicit time integration, the resulting time-marching scheme will be truly explicit and inherently parallel.

In the presence of complex geometry, adaptivity and mesh refinement are certainly keys for the efficient numerical simulation of wave phenomena. However, locally refined meshes impose severe stability constraints on explicit time-marching schemes, where the maximal time-step allowed by the CFL condition is dictated by the smallest elements in the mesh. When mesh refinement is restricted to a small region, the use of implicit methods or a very small time-step in the entire computational domain is a very high price to pay. To overcome this overly restrictive stability constraint, various local time-stepping (LTS) schemes [12, 13, 17] were developed, which use either implicit time-stepping or explicit smaller time-steps, but only where the smallest elements in the mesh are located.

Since DG methods are inherently local, they are particularly well suited for the development of explicit local time-stepping schemes [30]. By combining the symplectic Störmer–Verlet method with a DG discretization, Piperno derived a symplectic LTS scheme for Maxwell's equations in a nonconducting medium [35], which is explicit and second-order accurate. In [34], Montseny et al. combined a similar recursive integrator with discontinuous hexahedral elements. Starting from the so-called arbitrary high-order derivatives (ADER) DG approach, alternative explicit LTS methods for

Maxwell's equations [39] and for elastic wave equations [18] were proposed. In [19], the LTS approach from Collino et al. [12, 13] was combined with a DG-FE discretization for the numerical solution of symmetric first-order hyperbolic systems. Based on energy conservation, that LTS approach is second-order and explicit inside the coarse and the fine mesh; at the interface, however, it nonetheless requires at every time-step the solution of a linear system. More recently, Constantinescu and Sandu devised multirate explicit methods for hyperbolic conservation laws, which are based on both Runge–Kutta and Adams–Bashforth schemes combined with a finite volume discretization [14, 15]. Again these multirate schemes are limited to second-order accuracy.

Starting from the standard leap-frog method, Diaz and Grote proposed energy conserving fully explicit LTS integrators of arbitrarily high accuracy for the classical wave equation [16]; that approach was extended to Maxwell's equations in [25] for nonconductive media. By blending the leap-frog and the Crank–Nicolson methods, a second-order LTS scheme was also derived there for (damped) electromagnetic waves in conducting media, yet this approach cannot be readily extended beyond order two. To achieve arbitrarily high accuracy in the presence of dissipation, while remaining fully explicit, explicit LTS methods for damped wave equations based on Adams–Bashforth (AB) multistep schemes were proposed in [26] – see also [27]. They can also be interpreted as particular approximations of exponential-Adams multistep methods [31].

The rest of the chapter is organized as follows. In Section 2, we first recall the standard continuous, the symmetric interior penalty (IP) DG and the nodal DG formulations. In Section 3, we consider leap-frog-based LTS methods, both for the undamped and the damped wave equation. In the undamped case, we show how to derive explicit LTS methods of the arbitrarily high order; these methods also conserve a discrete version of the energy. In the damped case, we present a second-order LTS method by blending the leap-frog and the Crank–Nicolson scheme; however, this approach does not easily extend to higher order. To achieve arbitrarily high accuracy even in the presence of dissipation, we then consider LTS methods based on Adams–Bashforth multistep schemes in Section 4. Finally in Section 5, we present numerical experiments in one and two space dimensions, which validate the theory and underpin both the stability properties and the usefulness of these high-order explicit LTS schemes.

2 Finite element discretizations for the wave equation

We consider the damped wave equation

$$\begin{aligned} u_{tt} + \sigma u_t - \nabla \cdot (c^2 \nabla u) &= f \quad \text{in } \Omega \times (0,T)\,, \\ u(\cdot,t) &= 0 \quad \text{on } \partial\Omega \times (0,T)\,, \\ u(\cdot,0) = u_0\,, \quad u_t(\cdot,0) &= v_0 \quad \text{in } \Omega\,, \end{aligned} \tag{5.1}$$

where Ω is a bounded Lipschitz domain in $\mathbb{R}^d$, $d = 1,2,3$. Here, $f \in L^2(0,T;L^2(\Omega))$ is a (known) source term, while $u_0 \in H_0^1(\Omega)$ and $v_0 \in L^2(\Omega)$ are prescribed initial conditions. At the boundary, $\partial\Omega$, we impose a homogeneous Dirichlet boundary condition, for simplicity. We assume that the damping coefficient, $\sigma = \sigma(x)$ and the speed of propagation $c = c(x)$ are piecewise smooth and satisfy the bounds

$$0 \le \sigma(x) \le \sigma^* < \infty\,, \quad 0 < c_* \le c(x) \le c^* < \infty\,, \quad x \in \overline{\Omega}\,.$$

We shall now discretize (5.1) in space by using any one of the following three distinct FE discretizations: continuous (H^1-conforming) finite elements with mass lumping, a symmetric IP-DG discretization, or a nodal DG method. Thus, we consider shape-regular meshes $\mathcal{T}_h$ that partition the domain Ω into disjoint elements K, such that $\overline{\Omega} = \cup_{K \in \mathcal{T}_h} \overline{K}$. The elements are triangles or quadrilaterals in two space dimensions, and tetrahedra or hexahedra in three dimensions, respectively. The diameter of element K is denoted by h_K and the mesh size, h, is given by $h = \max_{K \in \mathcal{T}_h} h_K$.

2.1 Continuous Galerkin formulation

The continuous (H^1-conforming) Galerkin formulation of (5.1) starts from its weak formulation: find $u \in [0,T] \to H_0^1(\Omega)$ such that

$$\begin{aligned} (u_{tt}, \varphi) + (\sigma u_t, \varphi) + (c \nabla u, c \nabla \varphi) &= (f, \varphi) \quad \forall\, \varphi \in H_0^1(\Omega)\,,\ t \in (0,T)\,, \\ u(\cdot,0) = u_0\,, \quad u_t(\cdot,0) &= v_0\,, \end{aligned} \tag{5.2}$$

where $(\cdot,\cdot)$ denotes the standard L^2 inner product over Ω. It is well-known that (5.2) is well-posed and has a unique solution [33].

For a given partition $\mathcal{T}_h$ of Ω, assumed polygonal (2D) or polyhedral (3D) for simplicity, and an approximation order $\ell \ge 1$, we shall approximate the solution $u(\cdot,t)$ of (5.2) in the finite element space

$$V^h := \left\{ \varphi \in H_0^1(\Omega) \ :\ \varphi|_K \in S^\ell(K)\ \forall\, K \in \mathcal{T}_h \right\}\,,$$

where $S^\ell(K)$ is the space $\mathcal{P}^\ell(K)$ of polynomials of total degree at most ℓ on K if K is a triangle or a tetrahedra, or the space $\mathcal{Q}^\ell(K)$ of polynomials of degree at most ℓ in each variable on K if K is a parallelogram or a parallelepiped. Here, we consider the following semidiscrete Galerkin approximation of (5.2): find $u^h : [0,T] \to V^h$ such that

$$\begin{aligned} (u^h_{tt}, \varphi) + (\sigma u^h_t, \varphi) + (c\nabla u^h, c\nabla \varphi) &= (f, \varphi) \quad \forall\, \varphi \in V^h,\ t \in (0,T), \\ u^h(\cdot, 0) = \Pi_h u_0, \quad u^h_t(\cdot, 0) &= \Pi_h v_0 . \end{aligned} \tag{5.3}$$

Here, Π_h denotes the L^2-projection onto V^h.

The semidiscrete formulation (5.3) is equivalent to the second-order system of ordinary differential equations

$$\begin{aligned} \mathbf{M}\frac{d^2\mathbf{U}}{dt^2}(t) + \mathbf{M}_\sigma \frac{d\mathbf{U}}{dt}(t) + \mathbf{K}\mathbf{U}(t) &= \mathbf{F}(t), \quad t \in (0,T), \\ \mathbf{M}\mathbf{U}(0) = u^h_0, \quad \mathbf{M}\frac{d\mathbf{U}}{dt}(0) &= v^h_0 . \end{aligned} \tag{5.4}$$

Here, $\mathbf{U}$ denotes the vector whose components are the time-dependent coefficients of the representation of u^h with respect to the finite element nodal basis of V_h, $\mathbf{M}$ the mass matrix, $\mathbf{K}$ the stiffness matrix, whereas $\mathbf{M}_\sigma$ denotes the mass matrix with weight σ. The matrix $\mathbf{M}$ is sparse, symmetric and positive definite, whereas the matrices $\mathbf{K}$ and $\mathbf{M}_\sigma$ are sparse, symmetric and, in general, only positive semidefinite. In fact, $\mathbf{K}$ is positive definite, unless Neumann boundary conditions would be imposed in (5.1) instead. Since we shall never need to invert $\mathbf{K}$, our derivation also applies to the semidefinite case with purely Neumann boundary conditions.

Usually, the mass matrix $\mathbf{M}$ is not diagonal, yet needs to be inverted at every time-step of any explicit time integration scheme. To overcome this difficulty, various mass lumping techniques have been developed [8–11], which essentially replace $\mathbf{M}$ with a diagonal approximation by computing the required integrals over each element K with judicious quadrature rules that do not affect the spatial accuracy [4].

2.2 Interior penalty discontinuous Galerkin formulation

Following [21] we briefly recall the symmetric IP DG formulation of (5.1). For simplicity, we assume in this section that the elements are triangles or parallelograms in two space dimensions and tetrahedra or parallelepipeds in three dimensions, respectively. Generally, we allow for irregular (k-irregular) meshes with hanging nodes [5]. We denote by $\mathcal{E}^{\mathcal{I}}_h$ the set of all interior edges of $\mathcal{T}_h$, by $\mathcal{E}^{\mathcal{B}}_h$ the set of all boundary edges of $\mathcal{T}_h$, and set $\mathcal{E}_h = \mathcal{E}^{\mathcal{I}}_h \cup \mathcal{E}^{\mathcal{B}}_h$. Here, we generically refer to any element of $\mathcal{E}_h$ as an "edge," that is a real edge in 2D and a face in 3D.

For a piecewise smooth function φ, we introduce the following trace operators. Let $e \in \mathcal{E}^{\mathcal{I}}_h$ be an interior edge shared by two elements K^+ and K^- with unit outward normal vectors $\mathbf{n}^\pm$, respectively. Denoting by $v^\pm$ the trace of v on $\partial K^\pm$ taken from within $K^{\pm}$, we define the jump and the average on e by

$$[\![\varphi]\!] := \varphi^+\mathbf{n}^+ + \varphi^-\mathbf{n}^-\,, \quad \{\!\{\varphi\}\!\} := (\varphi^+ + \varphi^-)/2\,.$$

On every boundary edge $e \in \mathcal{E}_h^{\mathcal{B}}$, we set $[\![\varphi]\!] := \varphi\mathbf{n}$ and $\{\!\{\varphi\}\!\} := \varphi$. Here, $\mathbf{n}$ is the outward unit normal to the domain boundary $\partial\Omega$.

For a piecewise smooth vector-valued function ψ, we analogously define the average across interior faces by $\{\!\{\psi\}\!\} := (\psi^+ + \psi^-)/2$, and on boundary faces we set $\{\!\{\psi\}\!\} := \psi$. The jump of a vector-valued function will not be used. For a vector-valued function ψ with continuous normal components across a face $e \in \mathcal{E}_h$, the trace identity

$$\varphi^+\,(\mathbf{n}^+ \cdot \psi^+) + \varphi^-\,(\mathbf{n}^- \cdot \psi^-) = [\![\varphi]\!] \cdot \{\!\{\psi\}\!\} \quad \text{on } e\,,$$

immediately follows from the above definitions.

For a given partition $\mathcal{T}_h$ of Ω and an approximation order $\ell \geq 1$, we wish to approximate the solution $u(t,\cdot)$ of (5.1) in the finite element space

$$V^h := \left\{\varphi \in L^2(\Omega) : \varphi|_K \in S^\ell(K)\ \ \forall K \in \mathcal{T}_h\right\},$$

where $S^\ell(K)$ is the space $\mathcal{P}^\ell(K)$ (for triangles or tetrahedra) or $\mathcal{Q}^\ell(K)$ (for quadrilaterals or hexahedra). Thus, we consider the following (semidiscrete) DG approximation of (5.1): find $u^h : [0,T] \to V^h$ such that

$$\begin{aligned} (u_{tt}^h,\varphi) + (\sigma u_t^h,\varphi) + a_h(u^h,\varphi) &= (f,\varphi) \quad \forall\,\varphi \in V^h, \quad t \in (0,T)\,, \\ u^h(\cdot,0) = \Pi_h u_0\,, \quad u_t^h(\cdot,0) &= \Pi_h v_0\,. \end{aligned} \tag{5.5}$$

Here, Π_h again denotes the L^2-projection onto V^h whereas the DG bilinear form $a_h(\cdot,\cdot)$, defined on $V^h \times V^h$, is given by

$$\begin{aligned} a_h(u,\varphi) := &\sum_{K\in\mathcal{T}_h}\int_K c^2\,\nabla u\cdot\nabla\varphi\,dx - \sum_{e\in\mathcal{E}_h}\int_e [\![u]\!]\cdot\{\!\{c^2\,\nabla\varphi\}\!\}\,dA \\ &- \sum_{e\in\mathcal{E}_h}\int_e [\![\varphi]\!]\cdot\{\!\{c^2\,\nabla u\}\!\}\,dA + \sum_{e\in\mathcal{E}_h}\int_e \mathsf{a}\,[\![u]\!]\cdot[\![\varphi]\!]\,dA\,. \end{aligned} \tag{5.6}$$

The last three terms in (5.6) correspond to jump and flux terms at element boundaries; they vanish when $u,\varphi,\in H_0^1(\Omega)\cap H^{1+m}(\Omega)$ for $m > \frac{1}{2}$. Hence, the above semidiscrete DG formulation (5.5) is consistent with the original continuous problem (5.2).

In (5.6) the function a penalizes the jumps of u and v over the faces of $\mathcal{T}_h$. To define it, we first introduce the functions h and c by

$$\mathsf{h}|_e = \begin{cases} \min\{h_{K^+},h_{K^-}\}, & e\in\mathcal{E}_h^{\mathcal{I}}, \\ h_K, & e\in\mathcal{E}_h^{\mathcal{B}}, \end{cases} \qquad \mathsf{c}|_e(x) = \begin{cases} \max\{c|_{K^+}(x),c|_{K^-}(x)\}, & e\in\mathcal{E}_h^{\mathcal{I}}, \\ c|_K(x), & e\in\mathcal{E}_h^{\mathcal{B}}. \end{cases}$$

Then, on each $e \in \mathcal{E}_h$, we set

$$\mathsf{a}|_e := \alpha\,\mathsf{c}^2\mathsf{h}^{-1}\,, \tag{5.7}$$

where α is a positive parameter independent of the local mesh sizes and the coefficient c. There exists a threshold value $\alpha_{\min} > 0$, which depends only on the shape regularity of the mesh and the approximation order ℓ such that for $\alpha \geq \alpha_{\min}$ the DG bilinear form a_h is coercive and, hence, the discretization is stable [2, 3]. Throughout the rest of the chapter, we shall assume that $\alpha \geq \alpha_{\min}$ so that the semidiscrete problem (5.5) has a unique solution that converges with the optimal order [21–24]. In [21, 24], a detailed convergence analysis and numerical study of the IP-DG method for (5.6) with $\sigma = 0$ was presented. In particular, optimal a-priori estimates in a DG-energy norm and the L^2-norm were derived. This theory immediately generalizes to the case $\sigma \geq 0$. For sufficiently smooth solutions, the IP-DG method (5.6) thus yields the optimal L^2-error estimate of order $\mathcal{O}(h^{\ell+1})$.

The semidiscrete IP-DG formulation (5.5) is equivalent to the second-order system of ordinary differential equations (5.4). Again, the mass matrix $\mathbf{M}$ is sparse, symmetric and positive definite. Yet because individual elements decouple, $\mathbf{M}$ (and $\mathbf{M}_\sigma$) is block-diagonal with block size equal to the number of degrees of freedom per element. Thus, $\mathbf{M}$ can be inverted at very low computational cost. In fact, for a judicious choice of (locally orthogonal) shape functions, $\mathbf{M}$ is truly diagonal.

2.3 Nodal discontinuous Galerkin formulation

Finally, we briefly recall the nodal discontinuous Galerkin formulation from [30] for the spatial discretization of (5.1) rewritten as a first-order system. To do so, we first let $v := u_t$, $\mathbf{w} := -\nabla u$, and thus we rewrite (5.1) as the first-order hyperbolic system

$$\begin{aligned} v_t + \sigma v + \nabla \cdot (c^2 \mathbf{w}) &= f && \text{in } \Omega \times (0,T)\,, \\ \mathbf{w}_t + \nabla v &= \mathbf{0} && \text{in } \Omega \times (0,T)\,, \\ v(\cdot,t) = 0\,,\ \mathbf{w}(\cdot,t) &= \mathbf{0} && \text{on } \partial\Omega \times (0,T)\,, \\ v(\cdot,0) = v_0\,,\ \mathbf{w}(\cdot,0) &= -\nabla u_0 && \text{in } \Omega\,, \end{aligned} \tag{5.8}$$

or in more compact notation as

$$\mathbf{q}_t + \Sigma\,\mathbf{q} + \nabla \cdot \mathcal{F}(\mathbf{q}) = \mathbf{S}\,, \tag{5.9}$$

with

$$\mathbf{q} = \begin{pmatrix} v \\ \mathbf{w} \end{pmatrix}, \quad \mathcal{F}(\mathbf{q}) = \begin{pmatrix} c^2\,\mathbf{w}^\top \\ v\,\mathbf{I}_{d\times d} \end{pmatrix}, \quad \Sigma = \begin{pmatrix} \sigma & \mathbf{0} \\ \mathbf{0} & \mathbf{0} \end{pmatrix}, \quad \mathbf{S} = \begin{pmatrix} f \\ \mathbf{0} \end{pmatrix}.$$

Following [30], we now consider the following nodal DG formulation of (5.9): find $\mathbf{q}^h : [0,T] \to \mathbf{V}^h$ such that

$$(\mathbf{q}_t^h, \psi) + (\Sigma\,\mathbf{q}^h, \psi) + \tilde{a}_h(\mathbf{q}^h, \psi) = (\mathbf{S}, \psi) \quad \forall\, \psi \in \mathbf{V}^h,\ t \in (0,T)\,. \tag{5.10}$$

Here $\mathbf{V}^h$ denotes the finite element space

$$\mathbf{V}^h := \left\{ \psi \in L^2(\Omega)^{d+1} : \psi|_K \in S^\ell(K)^{d+1} \ \ \forall K \in \mathcal{T}_h \right\}$$

for a given partition $\mathcal{T}_h$ of Ω and an approximation order $\ell \geq 1$. The nodal-DG bilinear form $\widetilde{a}_h(\cdot,\cdot)$ is defined on $\mathbf{V}^h \times \mathbf{V}^h$ as

$$\widetilde{a}_h(\mathbf{q},\psi) := \sum_{K\in\mathcal{T}_h} \int_K (\nabla \cdot \mathcal{F}(\mathbf{q})) \cdot \psi \, dx - \sum_{e\in\mathcal{E}_h} \int_e (\mathbf{n}\cdot\mathcal{F}(\mathbf{q}) - (\mathbf{n}\cdot\mathcal{F}(\mathbf{q}))^*) \cdot \psi \, dA \,,$$

where $(\mathbf{n}\cdot\mathcal{F}(\mathbf{q}))^*$ is a suitably chosen numerical flux in the unit normal direction $\mathbf{n}$. The semidiscrete problem (5.10) has a unique solution, which converges with optimal order in the L^2-norm [30].

The semidiscrete nodal DG formulation (5.10) is equivalent to the first-order system of ordinary differential equations

$$\mathbf{M}\frac{d\mathbf{Q}}{dt}(t) + \mathbf{M}_\sigma \mathbf{Q}(t) + \mathbf{C}\mathbf{Q}(t) = \mathbf{F}(t) \,, \quad t \in (0,T) \,. \tag{5.11}$$

Here $\mathbf{Q}$ denotes the vector whose components are the coefficients of $\mathbf{q}^h$ with respect to the finite element basis of $\mathbf{V}^h$ and $\mathbf{C}$ the DG stiffness matrix. Because the individual elements decouple, the mass matrices $\mathbf{M}$ and $\mathbf{M}_\sigma$ are sparse, symmetric, positive semidefinite and block-diagonal. Moreover, $\mathbf{M}$ is positive definite and can be inverted at very low computational cost.

3 Leap-frog-based LTS methods

Starting from the well-known second-order "leap-frog" scheme, we now derive an explicit second-order LTS scheme for undamped waves. By using the modified equation approach, we then derive an explicit fourth-order LTS method for undamped waves. Finally, by blending the leap-frog and the Crank–Nicolson methods, we also present a second-order LTS scheme for damped waves.

We consider the semidiscrete model equation

$$\mathbf{M}\frac{d^2\mathbf{U}}{dt^2}(t) + \mathbf{M}_\sigma \frac{d\mathbf{U}}{dt}(t) + \mathbf{K}\mathbf{U} = \mathbf{F}(t) \,, \tag{5.12}$$

where $\mathbf{M}$ and $\mathbf{M}_\sigma$ are symmetric positive definite matrices and $\mathbf{K}$ is a symmetric positive semidefinite matrix. Moreover, we assume that the mass matrix $\mathbf{M}$ is (block-) diagonal, as in (5.4). We remark, however, that the time integration techniques presented below are also applicable to other spatial discretizations of the damped wave equation that lead to the same semidiscrete form (5.12).

Because $\mathbf{M}$ is assumed essentially diagonal, $\mathbf{M}^{\frac{1}{2}}$ can be explicitly computed and inverted at low cost. Thus, we multiply (5.12) by $\mathbf{M}^{-\frac{1}{2}}$ to obtain

$$\frac{d^2\mathbf{z}}{dt^2}(t) + \mathbf{D}\frac{d\mathbf{z}}{dt}(t) + \mathbf{A}\mathbf{z}(t) = \mathbf{R}(t) \,, \tag{5.13}$$

with $\mathbf{z} = \mathbf{M}^{\frac{1}{2}}\mathbf{U}$, $\mathbf{D} = \mathbf{M}^{-\frac{1}{2}}\mathbf{M}_\sigma\mathbf{M}^{-\frac{1}{2}}$, $\mathbf{A} = \mathbf{M}^{-\frac{1}{2}}\mathbf{K}\mathbf{M}^{-\frac{1}{2}}$ and $\mathbf{R} = \mathbf{M}^{-\frac{1}{2}}\mathbf{F}$. Note that $\mathbf{A}$ is also sparse and symmetric positive semidefinite. For undamped waves, $\mathbf{D}$ vanishes and hence energy is conserved, whereas for damped waves $\mathbf{D}$ is nonzero and energy is dissipated. We shall distinguish these two situations in the derivation of local time-stepping schemes below.

3.1 Second-order method for undamped waves

For undamped waves, (5.13) reduces to

$$\frac{d^2\mathbf{z}}{dt^2} + \mathbf{A}\,\mathbf{z} = \mathbf{R}. \tag{5.14}$$

Since for any $f \in C^2$, we have

$$f(t+\Delta t) - 2\,f(t) + f(t-\Delta t) = \Delta t^2 \int_{-1}^{1} (1-|\theta|) f''(t+\theta\,\Delta t)\, d\theta\,, \tag{5.15}$$

the exact solution $\mathbf{z}(t)$ of (5.14) satisfies

$$\begin{aligned} &\mathbf{z}(t+\Delta t) - 2\,\mathbf{z}(t) + \mathbf{z}(t-\Delta t) \\ &= \Delta t^2 \int_{-1}^{1} (1-|\theta|)\,(\mathbf{R}(t+\theta\,\Delta t) - \mathbf{A}\mathbf{z}(t+\theta\,\Delta t))\, d\theta\,. \end{aligned} \tag{5.16}$$

The integral on the right-hand side of (5.16) represents a weighted average of $\mathbf{R}(s) - \mathbf{A}\,\mathbf{z}(s)$ over the interval $[t-\Delta t, t+\Delta t]$, which needs to be approximated in any numerical algorithm. If we approximate in (5.16) $\mathbf{A}\,\mathbf{z}(t+\theta\,\Delta t)$ and $\mathbf{R}(t+\theta\,\Delta t)$ by $\mathbf{A}\,\mathbf{z}(t)$ and $\mathbf{R}(t)$, respectively, and evaluate the remaining θ-dependent integral, we obtain the well-known second-order leap-frog scheme with time-step Δt,

$$\mathbf{z}_{n+1} - 2\,\mathbf{z}_n + \mathbf{z}_{n-1} = \Delta t^2\,(\mathbf{R}_n - \mathbf{A}\,\mathbf{z}_n)\,, \quad \mathbf{R}_n \simeq \mathbf{R}(t_n),\, \mathbf{z}_n \simeq \mathbf{z}(t_n)\,, \tag{5.17}$$

which, however, would require Δt to be comparable in size to the smallest elements in the mesh for numerical stability.

Following [16, 28], we instead split the vectors $\mathbf{z}(t)$ and $\mathbf{R}(t)$ as

$$\begin{aligned} \mathbf{z}(t) &= (\mathbf{I}-\mathbf{P})\mathbf{z}(t) + \mathbf{P}\mathbf{z}(t) = \mathbf{z}^{[\text{coarse}]}(t) + \mathbf{z}^{[\text{fine}]}(t)\,, \\ \mathbf{R}(t) &= (\mathbf{I}-\mathbf{P})\mathbf{R}(t) + \mathbf{P}\mathbf{R}(t) = \mathbf{R}^{[\text{coarse}]}(t) + \mathbf{R}^{[\text{fine}]}(t)\,, \end{aligned} \tag{5.18}$$

where the projection matrix $\mathbf{P}$ is diagonal. Its diagonal entries, equal to zero or one, identify the unknowns associated with the locally refined region, where smaller time-steps are needed. To circumvent the severe CFL restriction on Δt in the leap-frog scheme, we need to treat $\mathbf{z}^{[\text{fine}]}(t)$ and $\mathbf{R}^{[\text{fine}]}(t)$ differently from $\mathbf{z}^{[\text{coarse}]}(t)$ and $\mathbf{R}^{[\text{coarse}]}(t)$ in

$$\begin{aligned}
&\mathbf{z}(t+\Delta t) - 2\,\mathbf{z}(t) + \mathbf{z}(t-\Delta t) \\
&= \Delta t^2 \int_{-1}^{1} (1-|\theta|)\left\{\mathbf{R}^{[\text{coarse}]}(t+\theta\,\Delta t) + \mathbf{R}^{[\text{fine}]}(t+\theta\,\Delta t)\right. \\
&\quad \left. -\mathbf{A}\left(\mathbf{z}^{[\text{coarse}]}(t+\theta\,\Delta t) + \mathbf{z}^{[\text{fine}]}(t+\theta\,\Delta t)\right)\right\} d\theta\,.
\end{aligned} \tag{5.19}$$

Since we wish to use the standard leap-frog scheme in the coarse part of the mesh, we approximate the terms in (5.19) that involve $\mathbf{z}^{[\text{coarse}]}(t+\theta\,\Delta t)$ and $\mathbf{R}^{[\text{coarse}]}(t+\theta\,\Delta t)$ by their values at t, which yields

$$\begin{aligned}
\mathbf{z}(t+\Delta t) - 2\,\mathbf{z}(t) + \mathbf{z}(t-\Delta t) &\simeq \Delta t^2\,\{(\mathbf{I}-\mathbf{P})\mathbf{R}(t) - \mathbf{A}(\mathbf{I}-\mathbf{P})\mathbf{z}(t)\} \\
&+ \Delta t^2 \int_{-1}^{1} (1-|\theta|)\,\{\mathbf{PR}(t+\theta\Delta t) - \mathbf{APz}(t+\theta\Delta t)\}\,d\theta\,.
\end{aligned} \tag{5.20}$$

Note that $\mathbf{A}$ and $\mathbf{P}$ do not commute.

Next for fixed t, let $\widetilde{\mathbf{z}}(\tau)$ solve the differential equation

$$\begin{aligned}
\frac{d^2\widetilde{\mathbf{z}}}{d\tau^2}(\tau) &= (\mathbf{I}-\mathbf{P})\mathbf{R}(t) - \mathbf{A}(\mathbf{I}-\mathbf{P})\mathbf{z}(t) + \mathbf{PR}(t+\tau) - \mathbf{AP}\widetilde{\mathbf{z}}(\tau)\,, \\
\widetilde{\mathbf{z}}(0) &= \mathbf{z}(t)\,,\ \widetilde{\mathbf{z}}'(0) = \nu\,,
\end{aligned} \tag{5.21}$$

where ν will be specified below. Again from (5.15), we deduce that

$$\begin{aligned}
\widetilde{\mathbf{z}}(\Delta t) - 2\,\widetilde{\mathbf{z}}(0) + \widetilde{\mathbf{z}}(-\Delta t) &= \Delta t^2\,\{(\mathbf{I}-\mathbf{P})\mathbf{R}(t) - \mathbf{A}(\mathbf{I}-\mathbf{P})\mathbf{z}(t)\} \\
&+ \Delta t^2 \int_{-1}^{1} (1-|\theta|)\,\{\mathbf{PR}(t+\theta\Delta t) - \mathbf{AP}\widetilde{\mathbf{z}}(\theta\Delta t)\}\,d\theta\,.
\end{aligned} \tag{5.22}$$

From the comparison of (5.20) with (5.22), we infer that

$$\mathbf{z}(t+\Delta t) - 2\,\mathbf{z}(t) + \mathbf{z}(t-\Delta t) \simeq \widetilde{\mathbf{z}}(\Delta t) - 2\,\widetilde{\mathbf{z}}(0) + \widetilde{\mathbf{z}}(-\Delta t)\,,$$

or equivalently

$$\mathbf{z}(t+\Delta t) + \mathbf{z}(t-\Delta t) \simeq \widetilde{\mathbf{z}}(\Delta t) + \widetilde{\mathbf{z}}(-\Delta t)\,. \tag{5.23}$$

In fact from Taylor expansion and (5.14), we obtain

$$\begin{aligned}
\widetilde{\mathbf{z}}(\Delta t) + \widetilde{\mathbf{z}}(-\Delta t) &= 2\widetilde{\mathbf{z}}(0) + \widetilde{\mathbf{z}}''(0)\Delta t^2 + \mathcal{O}(\Delta t^4) \\
&= 2\mathbf{z}(t) + (\mathbf{R}(t) - \mathbf{Az}(t))\Delta t^2 + \mathcal{O}(\Delta t^4) \\
&= \mathbf{z}(t+\Delta t) + \mathbf{z}(t-\Delta t) + \mathcal{O}(\Delta t^4)\,.
\end{aligned}$$

Thus to advance $\mathbf{z}(t)$ from t to $t+\Delta t$, we shall evaluate $\widetilde{\mathbf{z}}(\Delta t) + \widetilde{\mathbf{z}}(-\Delta t)$ by solving (5.21) numerically.

To take advantage of the inherent symmetry in time and thereby reduce the computational effort even further, we now let

$$\mathbf{q}(\tau) = \tilde{\mathbf{z}}(\tau) + \tilde{\mathbf{z}}(-\tau)\,.$$

Then, $q(\tau)$ solves the differential equation

$$\frac{d^2\mathbf{q}}{d\tau^2}(\tau) = 2\left\{(\mathbf{I}-\mathbf{P})\mathbf{R}(t) - \mathbf{A}(\mathbf{I}-\mathbf{P})\mathbf{z}(t)\right\} + \mathbf{P}\left\{\mathbf{R}(t+\tau) + \mathbf{R}(t-\tau)\right\} - \mathbf{APq}(\tau)\,,$$
$$\mathbf{q}(0) = 2\mathbf{z}(t)\,,\ \mathbf{q}'(0) = 0\,, \tag{5.24}$$

with

$$\mathbf{z}(t+\Delta t) + \mathbf{z}(t-\Delta t) = \mathbf{q}(\Delta t) + \mathcal{O}(\Delta t^4)\,. \tag{5.25}$$

Note that $\mathbf{q}(\Delta t)$ does not depend on the value of ν. Now, we shall approximate the right-hand side of (5.16) by solving (5.24) on $[0, \Delta t]$, and then use (5.25) to compute $\mathbf{z}(t + \Delta t)$. Thus, we need the numerical value of $\mathbf{q}(\tau)$ only at Δt.

In summary, the second-order LTS algorithm for the solution of (5.14) computes $\mathbf{z}_{n+1} \simeq \mathbf{z}(t + \Delta t)$, given $\mathbf{z}_n$ and $\mathbf{z}_{n-1}$, as follows:

Algorithm 5.1 (LTS-LF2(p) Algorithm).
(1) *Set* $\mathbf{w} := (\mathbf{I}-\mathbf{P})\mathbf{R}_n - \mathbf{A}(\mathbf{I}-\mathbf{P})\mathbf{z}_n$ *and* $\mathbf{q}_0 := 2\mathbf{z}_n$.
(2) *Compute*

$$\mathbf{q}_{1/p} := \mathbf{q}_0 + \frac{1}{2}\left(\frac{\Delta t}{p}\right)^2 \left(2\mathbf{w} + 2\mathbf{PR}_{n,0} - \mathbf{APq}_0\right)\,.$$

(3) *For* $m = 1, \dots, p-1$, *compute*

$$\mathbf{q}_{(m+1)/p} := 2\mathbf{q}_{m/p} - \mathbf{q}_{(m-1)/p} + \left(\frac{\Delta t}{p}\right)^2 \left(2\mathbf{w} + \mathbf{P}(\mathbf{R}_{n,m} + \mathbf{R}_{n,-m}) - \mathbf{APq}_{m/p}\right)\,.$$

(4) *Compute* $\mathbf{z}_{n+1} := -\mathbf{z}_{n-1} + \mathbf{q}_1$.

Here, we have used the notations $\mathbf{R}_{n,m} \simeq \mathbf{R}(t_n + \tau_m)$ and $\mathbf{R}_{n,-m} \simeq \mathbf{R}(t_n - \tau_m)$, where $t_n = n\Delta t$ and $\tau_m = m\Delta\tau$; note that $\mathbf{R}_{n,0} \simeq \mathbf{R}(t_n + \tau_0) = \mathbf{R}(t_n) \simeq \mathbf{R}_n$. Steps 1–3 correspond to the numerical solution of (5.24) until $\tau = \Delta t$ with the leap-frog scheme, using the local time-step $\Delta\tau = \Delta t/p$. For $\mathbf{P} = \mathbf{0}$, that is without any local time-stepping, we recover the standard leap-frog scheme. If the fraction of nonzero entries in $\mathbf{P}$ is small, the overall cost is dominated by the computation of $\mathbf{w}$, which requires one multiplication by $\mathbf{A}(\mathbf{I}-\mathbf{P})$ per time-step Δt. All further matrix–vector multiplications by $\mathbf{AP}$ only affect those unknowns that lie inside the refined region, or immediately next to it.

Proposition 5.2. *For* $\mathbf{R}(t) \in C^2([0,T])$*, the local time-stepping method LTS-LF2(p) is second-order accurate.*

Proof. See [25]. □

To establish the stability of the LTS-LF2(p) scheme we consider the homogeneous case, $\mathbf{R}_n = \mathbf{0}$. Then, the standard leap-frog scheme (5.17) conserves the discrete energy

$$E^{n+\frac{1}{2}} = \frac{1}{2}\left[\left\langle\left(\mathbf{I} - \frac{\Delta t^2}{4}\mathbf{A}\right)\frac{\mathbf{z}_{n+1} - \mathbf{z}_n}{\Delta t}, \frac{\mathbf{z}_{n+1} - \mathbf{z}_n}{\Delta t}\right\rangle + \left\langle\mathbf{A}\frac{\mathbf{z}_{n+1} + \mathbf{z}_n}{2}, \frac{\mathbf{z}_{n+1} + \mathbf{z}_n}{2}\right\rangle\right]. \tag{5.26}$$

Here $E^{n+\frac{1}{2}} \simeq E(t_{n+\frac{1}{2}})$ and the angular brackets denote the standard Euclidean inner product. Since $\mathbf{A}$ is symmetric, the quadratic form in (5.26) is also symmetric. For sufficiently small Δt it is also positive semidefinite and hence yields a true energy.

To derive a necessary and sufficient condition for the numerical stability of the LTS-LF2(p) scheme, we exhibit a conserved discrete energy for the LTS-LF2(p) algorithm with $\mathbf{R}_n = \mathbf{0}$. Following [16], we first rewrite the LTS-LF2(p) scheme in "leap-frog manner."

Proposition 5.3. *The local time-stepping scheme LTS-LF2(p) with* $\mathbf{R}_{n,m} = \mathbf{0}$ *is equivalent to*

$$\mathbf{z}_{n+1} = 2\mathbf{z}_n - \mathbf{z}_{n-1} - \Delta t^2 \mathbf{A}_p \mathbf{z}_n\,,$$

where $\mathbf{A}_p$ *is defined by*

$$\mathbf{A}_p = \mathbf{A} - \frac{2}{p^2}\sum_{j=1}^{p-1}\left(\frac{\Delta t}{p}\right)^{2j} \alpha_j^p (\mathbf{AP})^j \mathbf{A} \tag{5.27}$$

and the constants α_j^p *are given by*

$$\begin{aligned}
\alpha_1^2 &= 1, \quad \alpha_1^3 = 6, \quad \alpha_2^3 = -1,\\
\alpha_1^{p+1} &= m^2 + 2\alpha_1^p - \alpha_1^{p-1},\\
\alpha_j^{p+1} &= 2\alpha_j^p - \alpha_j^{p-1} - \alpha_{j-1}^p, \quad j = 2,\dots,p-2,\\
\alpha_{p-1}^{p+1} &= 2\alpha_{p-1}^p - \alpha_{p-2}^p,\\
\alpha_p^{p+1} &= -\alpha_{p-1}^p\,.
\end{aligned}$$

Furthermore, the matrix $\mathbf{A}_p$ *is symmetric.*

Proof. See [16] and [25]. □

Proposition 5.4. *The local time-stepping scheme LTS-LF2(p) with* $\mathbf{R}_n = \mathbf{0}$ *conserves the energy*

$$E^{n+\frac{1}{2}} = \frac{1}{2}\left[\left\langle\left(\mathbf{I} - \frac{\Delta t^2}{4}\mathbf{A}_p\right)\frac{\mathbf{z}_{n+1}-\mathbf{z}_n}{\Delta t}, \frac{\mathbf{z}_{n+1}-\mathbf{z}_n}{\Delta t}\right\rangle \right. \\ \left. + \left\langle \mathbf{A}_p \frac{\mathbf{z}_{n+1}+\mathbf{z}_n}{2}, \frac{\mathbf{z}_{n+1}+\mathbf{z}_n}{2}\right\rangle\right]. \tag{5.28}$$

Proof. By symmetry of $\mathbf{A}_p$, this standard argument is similar the proof of (5.26); see also [16] for details. □

As a consequence, the LTS-LF2(p) is stable if $0 < (\Delta t^2/4)\lambda_{\max}(\mathbf{A}_p) < 1$; note that the matrix $\mathbf{A}_p$ itself also depends on Δt.

3.2 Fourth-order method for undamped waves

In the absence of damping, the wave equation corresponds to a separable Hamiltonian system. This fact explains the success of symplectic integrators, such as the Störmer–Verlet or the leap-frog method, when combined with a symmetric discretization in space. Indeed the fully discrete numerical scheme will then conserve (a discrete version of) the energy, too. Clearly, standard symplectic partitioned Runge–Kutta (Lobatto IIIA–IIIB pairs) or composition methods [28] can be used to achieve higher accuracy [36]. Because the Hamiltonian here is separable, those higher order versions will also remain explicit in time, like the Störmer–Verlet method. Since damped wave equations are linear, we instead opt for the even more efficient modified equation (ME) approach [38] in this section, which leads to explicit LTS of arbitrarily high (even) order.

Following the ME approach, we replace $\mathbf{A}\mathbf{z}(t + \theta\,\Delta t)$ in (5.16) by its Taylor expansion

$$\mathbf{A}\mathbf{z}(t+\theta\,\Delta t) = \mathbf{A}\left(\mathbf{z}(t) + \theta\,\Delta t\,\mathbf{z}'(t) + \frac{\theta^2\,\Delta t^2}{2}\mathbf{z}''(t) + \frac{\theta^3\,\Delta t^3}{6}\mathbf{z}'''(t)\right) + \mathcal{O}(\Delta t^4).$$

Then, the integrals involving odd powers of θ vanish. Next, by using that $\mathbf{z}''(t) = \mathbf{R}(t) - \mathbf{A}\mathbf{z}(t)$ and the Simpson quadrature rule for the term that involves $\mathbf{R}(t+\theta\,\Delta t)$, we obtain the fourth-order modified equation scheme

$$\frac{\mathbf{z}_{m+1} - 2\mathbf{z}_m + \mathbf{z}_{m-1}}{\Delta t^2} = \mathbf{R}_m - \mathbf{A}\mathbf{z}_m + \frac{\Delta t^2}{12}\mathbf{A}^2\mathbf{z}_m - \frac{\Delta t^2}{12}\mathbf{A}\mathbf{R}_m \\ + \frac{1}{3}\left(\mathbf{R}_{m-1/2} - 2\mathbf{R}_m + \mathbf{R}_{m+1/2}\right) + \mathcal{O}(\Delta t^4), \tag{5.29}$$

where $\mathbf{z}_m \simeq \mathbf{z}(t_m)$, $\mathbf{R}_m \simeq \mathbf{R}(t_m)$ and $\mathbf{R}_{m\pm 1/2} \simeq \mathbf{R}(t_m \pm \Delta t/2)$. Clearly, integration schemes of arbitrary (even) order can be obtained by using additional terms in the Taylor expansion. Since the maximal time-step allowed by the fourth-order ME method is about 70% times *larger* than that of the leap-frog scheme [10], the additional work needed for the improved accuracy is quite small; hence, the ME method is extremely efficient.

We now derive a fourth-order LTS method for (5.14). Similarly to the derivation in Section 3.1, we split the vectors $\mathbf{z}(t)$ and $\mathbf{R}(t)$ in (5.19) into a fine and a coarse part, and shall treat $\mathbf{z}^{[\text{fine}]}(t)$ and $\mathbf{R}^{[\text{fine}]}(t)$ differently from $\mathbf{z}^{[\text{coarse}]}(t)$ and $\mathbf{R}^{[\text{coarse}]}(t)$. We expand $\mathbf{z}^{[\text{coarse}]}(t+\theta\,\Delta t)$ in Taylor series as

$$\begin{aligned}\mathbf{z}^{[\text{coarse}]}(t+\theta\,\Delta t) = \mathbf{z}^{[\text{coarse}]}(t) + \theta\,\Delta t\,\frac{d\mathbf{z}^{[\text{coarse}]}}{dt}(t) + \frac{\theta^2\,\Delta t^2}{2}\,\frac{d^2\mathbf{z}^{[\text{coarse}]}}{dt^2}(t)\\ + \frac{\theta^3\,\Delta t^3}{6}\,\frac{d^3\mathbf{z}^{[\text{coarse}]}}{dt^3}(t) + \mathcal{O}(\Delta t^4)\end{aligned}$$

and insert it into (5.19). In (5.19), the integrals involving odd powers of θ vanish. By using

$$\frac{d^2\mathbf{z}^{[\text{coarse}]}}{dt^2}(t) = (\mathbf{I}-\mathbf{P})\frac{d^2\mathbf{z}}{dt^2}(t) = (\mathbf{I}-\mathbf{P})\mathbf{R}(t) - (\mathbf{I}-\mathbf{P})\mathbf{A}\mathbf{z}(t)$$

and the Simpson quadrature rule for the term in (5.19) that involves $\mathbf{R}^{[\text{coarse}]}(t+\theta\Delta t)$, we find that

$$\begin{aligned}&\mathbf{z}(t+\Delta t) - 2\,\mathbf{z}(t) + \mathbf{z}(t-\Delta t)\\ &= \Delta t^2\left\{(\mathbf{I}-\mathbf{P})\mathbf{R}(t) - \mathbf{A}(\mathbf{I}-\mathbf{P})\mathbf{z}(t)\right\} + \frac{\Delta t^4}{12}\mathbf{A}(\mathbf{I}-\mathbf{P})\mathbf{A}\mathbf{z}(t) \qquad (5.30)\\ &\quad - \frac{\Delta t^4}{12}\mathbf{A}(\mathbf{I}-\mathbf{P})\mathbf{R}(t) + \frac{\Delta t^2}{3}(\mathbf{I}-\mathbf{P})\left\{\mathbf{R}\left(t-\frac{\Delta t}{2}\right) - 2\mathbf{R}(t) + \mathbf{R}\left(t+\frac{\Delta t}{2}\right)\right\}\\ &\quad + \Delta t^2\int_{-1}^{1}(1-|\theta|)\left\{\mathbf{R}^{[\text{fine}]}(t+\theta\Delta t) - \mathbf{A}\mathbf{z}^{[\text{fine}]}(t+\theta\Delta t)\right\}d\theta\,.\end{aligned}$$

Hence, if $\mathbf{P}=\mathbf{0}$ we recover the standard ME scheme (5.29).

Similarly to Section 3.1, we now approximate the right-hand side of (5.30) by solving the following differential equation for $\tilde{\mathbf{z}}(\tau)$

$$\begin{aligned}\frac{d^2\tilde{\mathbf{z}}}{d\tau^2}(\tau) &= (\mathbf{I}-\mathbf{P})\mathbf{R}(t) - \mathbf{A}(\mathbf{I}-\mathbf{P})\mathbf{z}(t)\\ &\quad + \frac{1}{3}(\mathbf{I}-\mathbf{P})\left\{\mathbf{R}\left(t-\frac{\Delta t}{2}\right) - 2\mathbf{R}(t) + \mathbf{R}\left(t+\frac{\Delta t}{2}\right)\right\}\\ &\quad + \frac{\tau^2}{2}\mathbf{A}(\mathbf{I}-\mathbf{P})\mathbf{A}\mathbf{z}(t) - \frac{\tau^2}{2}\mathbf{A}(\mathbf{I}-\mathbf{P})\mathbf{R}(t) + \mathbf{P}\mathbf{R}(t+\tau) - \mathbf{A}\mathbf{P}\tilde{\mathbf{z}}(\tau)\,,\\ \tilde{\mathbf{z}}(0) &= \mathbf{z}(t)\,,\ \tilde{\mathbf{z}}'(0) = \nu\,,\end{aligned}$$

where ν will be specified below. Again, using Taylor expansions, we infer that

$$\mathbf{z}(t+\Delta t) + \mathbf{z}(t-\Delta t) = \tilde{\mathbf{z}}(\Delta t) + \tilde{\mathbf{z}}(-\Delta t) + \mathcal{O}(\Delta t^6)\,.$$

Again, the quantity $\widetilde{\mathbf{z}}(\Delta t) + \widetilde{\mathbf{z}}(-\Delta t)$ does not depend on the value of ν, which we set to zero. As in Section 3.1, we set $\mathbf{q}(\tau) := \widetilde{\mathbf{z}}(\tau) + \widetilde{\mathbf{z}}(-\tau)$, which solves the differential equation

$$\begin{aligned}\frac{d^2\mathbf{q}}{d\tau^2}(\tau) &= 2\left\{(\mathbf{I}-\mathbf{P})\mathbf{R}(t) - \mathbf{A}(\mathbf{I}-\mathbf{P})\mathbf{z}(t)\right\}\\ &\quad + \frac{2}{3}(\mathbf{I}-\mathbf{P})\left\{\mathbf{R}\left(t-\frac{\Delta t}{2}\right) - 2\mathbf{R}(t) + \mathbf{R}\left(t+\frac{\Delta t}{2}\right)\right\}\\ &\quad + \tau^2\mathbf{A}(\mathbf{I}-\mathbf{P})\mathbf{A}\mathbf{z}(t) - \tau^2\mathbf{A}(\mathbf{I}-\mathbf{P})\mathbf{R}(t)\\ &\quad + \mathbf{P}\left\{\mathbf{R}(t+\tau) + \mathbf{R}(t-\tau)\right\}\\ &\quad - \mathbf{A}\mathbf{P}\mathbf{q}(\tau),\\ \mathbf{q}(0) &= 2\mathbf{z}(t),\ \mathbf{q}'(0) = 0.\end{aligned} \tag{5.31}$$

Thus, we have

$$\mathbf{z}(t+\Delta t) + \mathbf{z}(t-\Delta t) = \mathbf{q}(\Delta t) + \mathcal{O}(\Delta t^6). \tag{5.32}$$

Now, we approximate the right-hand side of (5.16) by solving (5.31) with the fourth-order ME method on $[0, \Delta t]$ with a smaller time step $\Delta\tau = \Delta t/p$, and then use (5.32) to compute $\mathbf{z}(t+\Delta t)$.

In summary, the fourth-order LTS algorithm for the solution of (5.14) computes $\mathbf{z}_{n+1} \simeq \mathbf{z}(t+\Delta t)$, given $\mathbf{z}_n$ and $\mathbf{z}_{n-1}$, as follows:

Algorithm 5.5 (LTS-LFME4(p) Algorithm).

(1) *Set* $\mathbf{q}_0 := 2\mathbf{z}_n$, $w_1 := (\mathbf{I}-\mathbf{P})\mathbf{R}_n - \mathbf{A}(\mathbf{I}-\mathbf{P})\mathbf{z}_n$, $w_2 := \mathbf{A}(\mathbf{I}-\mathbf{P})\mathbf{A}\mathbf{z}_n - \mathbf{A}(\mathbf{I}-\mathbf{P})\mathbf{R}_n$ *and* $r_1 := \mathbf{R}_{n-1/2} - 2\mathbf{R}_n + \mathbf{R}_{n+1/2}$.

(2) *Compute*

$$u := 2w_1 + \frac{2}{3}(\mathbf{I}-\mathbf{P})r_1 + 2\mathbf{P}\mathbf{R}_{n,0} - \mathbf{A}\mathbf{P}\mathbf{q}_0$$

$$\mathbf{q}_{1/p} := \mathbf{q}_0 + \frac{1}{2}\left(\frac{\Delta t}{p}\right)^2 u + \frac{1}{24}\left(\frac{\Delta t}{p}\right)^4\left(2w_2 + 2\left(\frac{2}{\Delta t}\right)^2\mathbf{P}r_1 - \mathbf{A}\mathbf{P}u\right);$$

(3) *For* $m = 1, \dots, p-1$, *compute*

$$\begin{aligned}u_1 &:= 2w_1 + \frac{2}{3}(\mathbf{I}-\mathbf{P})r_1 + \left(\frac{m\Delta t}{p}\right)^2 w_2 + \mathbf{P}\left(\mathbf{R}_{n,m} + \mathbf{R}_{n,-m}\right) - \mathbf{A}\mathbf{P}\mathbf{q}_{m/p},\\ r &:= \mathbf{R}_{n,m-1/2} - 2\mathbf{R}_{n,m} + \mathbf{R}_{n,m+1/2} + \mathbf{R}_{n,-m-1/2}\\ &\quad - 2\mathbf{R}_{n,-m} + \mathbf{R}_{n,-m+1/2},\\ u_2 &:= 2w_2 + \left(\frac{2p}{m\Delta t}\right)^2 \mathbf{P}r - \mathbf{A}\mathbf{P}u_1,\\ \mathbf{q}_{(m+1)/p} &:= 2\mathbf{q}_{m/p} - \mathbf{q}_{(m-1)/p} + \left(\frac{\Delta t}{p}\right)^2 u_1 + \frac{1}{12}\left(\frac{\Delta t}{p}\right)^4 u_2.\end{aligned}$$

(4) *Compute* $\mathbf{z}_{n+1} := -\mathbf{z}_{n-1} + \mathbf{q}_1$.

Here, Steps 1–3 correspond to the numerical solution of (5.31) until $\tau = \Delta t$ with the ME approach using the local time-step $\Delta\tau = \Delta t/p$. The LTS-LFME4(p) algorithm requires three – two, without sources – multiplications by $\mathbf{A}(\mathbf{I} - \mathbf{P})$ and $2p$ further multiplications by $\mathbf{AP}$. For $\mathbf{P} = \mathbf{0}$, that is without any local time-stepping, the algorithm reduces to the modified equation scheme (5.29) above.

3.3 Second-order leap-frog/Crank–Nicolson-based method for damped waves

We shall now derive a second-order LTS method for (5.13) in a general form with $\mathbf{D} \neq \mathbf{0}$. In contrast to the time-stepping scheme presented in Section 3.1 for the case $\mathbf{D} = \mathbf{0}$, we are now faced with several difficulties due to the additional $\mathbf{Dz}'(t)$ term. First, we shall treat that term implicitly to avoid any additional CFL restriction; else, the stability condition will be more restrictive than that with the LTS-LF2(p) scheme, depending on the magnitude of σ. Note that very large values of σ will affect the CFL stability condition of any explicit method regardless of the use of local time-stepping. Nevertheless, the resulting scheme will be explicit, since $\mathbf{D}$ is essentially a diagonal matrix. Second, we can no longer take advantage of any inherent symmetry in time of the solution. Third, to avoid any loss of accuracy, we must carefully initialize the LTS scheme, which again is based on the highly efficient (two-step) leap-frog method.

The exact solution $\mathbf{z}(t)$ of (5.13) satisfies

$$\begin{aligned} &\mathbf{z}(t + \Delta t) - 2\,\mathbf{z}(t) + \mathbf{z}(t - \Delta t) + \frac{\Delta t}{2}\mathbf{D}\,(\mathbf{z}(t + \Delta t) - \mathbf{z}(t - \Delta t)) \\ &= \Delta t^2 \int_{-1}^{1} (1 - |\theta|)\,(\mathbf{R}(t + \theta\,\Delta t) - \mathbf{A}\,\mathbf{z}(t + \theta\,\Delta t))\,d\theta + \mathcal{O}(\Delta t^4)\,. \end{aligned} \tag{5.33}$$

To derive a second-order LTS method for (5.13), we now split the vectors $\mathbf{z}(t)$ and $\mathbf{R}(t)$ as in (5.18) and approximate the integrands in (5.33) as follows:

$$\begin{aligned} \mathbf{R}^{[\mathrm{coarse}]}(t + \theta\,\Delta t) + \mathbf{R}^{[\mathrm{fine}]}(t + \theta\,\Delta t) &\simeq \mathbf{R}^{[\mathrm{coarse}]}(t) + \mathbf{PR}(t + \theta\Delta t)\,, \\ \mathbf{A}\left(\mathbf{z}^{[\mathrm{coarse}]}(t + \theta\,\Delta t) + \mathbf{z}^{[\mathrm{fine}]}(t + \theta\,\Delta t)\right) &\simeq \mathbf{Az}^{[\mathrm{coarse}]}(t) + \mathbf{APz}(\theta\Delta t)\,. \end{aligned}$$

We thus have

$$\begin{aligned} &\mathbf{z}(t + \Delta t) - 2\,\mathbf{z}(t) + \mathbf{z}(t - \Delta t) + \frac{\Delta t}{2}\mathbf{D}\,(\mathbf{z}(t + \Delta t) - \mathbf{z}(t - \Delta t)) \\ &\simeq \Delta t^2\,\{(\mathbf{I} - \mathbf{P})\mathbf{R}(t) - \mathbf{A}(\mathbf{I} - \mathbf{P})\mathbf{z}(t)\} \\ &\quad + \Delta t^2 \int_{-1}^{1} (1 - |\theta|)\,\{\mathbf{PR}(t + \theta\Delta t) - \mathbf{APz}(\theta\Delta t)\}\,d\theta\,. \end{aligned} \tag{5.34}$$

Next for fixed t, let $\widetilde{\mathbf{z}}(\tau)$ solve the differential equation

$$\frac{d^2\widetilde{\mathbf{z}}}{d\tau^2}(\tau) + \mathbf{D}\frac{d\widetilde{\mathbf{z}}}{d\tau}(\tau) = (\mathbf{I}-\mathbf{P})\mathbf{R}(t) - \mathbf{A}(\mathbf{I}-\mathbf{P})\mathbf{z}(t) + \mathbf{P}\mathbf{R}(t+\tau) - \mathbf{A}\mathbf{P}\widetilde{\mathbf{z}}(\tau)\,,$$
$$\widetilde{\mathbf{z}}(0) = \mathbf{z}(t)\,,\ \widetilde{\mathbf{z}}'(0) = \nu\,, \tag{5.35}$$

where ν will be specified below. Since the exact solution $\widetilde{\mathbf{z}}(t)$ of (5.35) satisfies

$$\begin{aligned}&\widetilde{\mathbf{z}}(\Delta t) - 2\,\widetilde{\mathbf{z}}(0) + \widetilde{\mathbf{z}}(-\Delta t) + \frac{\Delta t}{2}\mathbf{D}\left(\widetilde{\mathbf{z}}(\Delta t) - \widetilde{\mathbf{z}}(-\Delta t)\right)\\ &= \Delta t^2\left\{(\mathbf{I}-\mathbf{P})\mathbf{R}(t) - \mathbf{A}(\mathbf{I}-\mathbf{P})\mathbf{z}(t)\right\}\\ &+ \Delta t^2\int_{-1}^{1}(1-|\theta|)\left\{\mathbf{P}\mathbf{R}(t+\theta\Delta t) - \mathbf{A}\mathbf{P}\widetilde{\mathbf{z}}(\theta\Delta t)\right\}\,d\theta\,,\end{aligned} \tag{5.36}$$

from the comparison of (5.34) and (5.36), we infer that

$$\begin{aligned}&\mathbf{z}(t+\Delta t) + \mathbf{z}(t-\Delta t) + \frac{\Delta t}{2}\mathbf{D}\left(\mathbf{z}(t+\Delta t) - \mathbf{z}(t-\Delta t)\right)\\ &\simeq \widetilde{\mathbf{z}}(\Delta t) + \widetilde{\mathbf{z}}(-\Delta t) + \frac{\Delta t}{2}\mathbf{D}\left(\widetilde{\mathbf{z}}(\Delta t) - \widetilde{\mathbf{z}}(-\Delta t)\right)\,.\end{aligned} \tag{5.37}$$

In our local time-stepping scheme, we shall use the right-hand side of (5.37) to approximate the left-hand side. In doing so, we must carefully choose ν in (5.35) to minimize that approximation error. By using Taylor expansions and the fact that $\mathbf{z}$ and $\widetilde{\mathbf{z}}$ solve (5.13) and (5.35), respectively, we obtain

$$\begin{aligned}\mathbf{z}(t+\Delta t) + \mathbf{z}(t-\Delta t) &= 2\mathbf{z}(t) + \mathbf{z}''(t)\Delta t^2 + \mathcal{O}(\Delta t^4)\\ &= 2\mathbf{z}(t) + (\mathbf{R}(t) - \mathbf{A}\mathbf{z}(t) - \mathbf{D}\,\mathbf{z}'(t))\Delta t^2 + \mathcal{O}(\Delta t^4)\,,\\ \widetilde{\mathbf{z}}(\Delta t) + \widetilde{\mathbf{z}}(-\Delta t) &= 2\widetilde{\mathbf{z}}(0) + \widetilde{\mathbf{z}}''(0)\Delta t^2 + \mathcal{O}(\Delta t^4)\\ &= 2\mathbf{z}(t) + (\mathbf{R}(t) - \mathbf{A}\mathbf{z}(t) - \mathbf{D}\,\nu)\Delta t^2 + \mathcal{O}(\Delta t^4)\,,\end{aligned}$$

together with

$$\begin{aligned}\mathbf{z}(t+\Delta t) - \mathbf{z}(t-\Delta t) &= 2\mathbf{z}'(t)\,\Delta t + \mathcal{O}(\Delta t^3)\,,\\ \widetilde{\mathbf{z}}(\Delta t) - \widetilde{\mathbf{z}}(-\Delta t) &= 2\nu\,\Delta t + \mathcal{O}(\Delta t^3)\,.\end{aligned}$$

Hence for arbitrary ν, the right-hand side of (5.37) is not sufficiently accurate to approximate the left-hand side while preserving overall second-order accuracy. However, if we choose

$$\nu = \mathbf{z}'(t)$$

in (5.35), the $\mathcal{O}(\Delta t^2)$ terms in (5.37) cancel each other and overall second-order accuracy of the scheme can be achieved. Since the term on the right-hand side of (5.37) is not symmetric in time, unlike in the previous section (see (5.23) and (5.25)), we need to compute the value of $\tilde{\mathbf{z}}(\tau)$ both at $\tau = \Delta t$ and at $\tau = -\Delta t$.

For the numerical solution of (5.35), we shall use the leap-frog scheme with the local time-step $\Delta\tau = \Delta t/p$. Since the leap-frog scheme is a two-step method, we need a second-order approximation of $\tilde{\mathbf{z}}'(0) = \mathbf{z}'(t)$ during every initial local time-step. Since the value of $\mathbf{z}_{n+1}$ is still unknown at time $t = t_n$, we now derive a second-order approximation $\mathbf{z}'_n \simeq \mathbf{z}'(t)$ that uses only $\mathbf{z}_n$ and $\mathbf{z}_{n-1}$. First, we approximate

$$\mathbf{z}'_n \simeq \frac{\mathbf{z}'_{n-1/2} + \mathbf{z}'_{n+1/2}}{2}, \tag{5.38}$$

where both $\mathbf{z}'_{n-1/2} \simeq \mathbf{z}'(t - \Delta t/2)$ and $\mathbf{z}'_{n+1/2} \simeq \mathbf{z}'(t + \Delta t/2)$ are second-order approximations. By using second-order central differences for $\mathbf{z}'_{n-1/2}$,

$$\mathbf{z}'_{n-1/2} = \frac{\mathbf{z}_n - \mathbf{z}_{n-1}}{\Delta t} + \mathcal{O}(\Delta t^2), \tag{5.39}$$

and the differential equation (5.13) for $\mathbf{z}'_{n+1/2}$,

$$\frac{\mathbf{z}'_{n+1/2} - \mathbf{z}'_{n-1/2}}{\Delta t} + \mathbf{D}\mathbf{z}'_n = \mathbf{R}_n - \mathbf{A}\mathbf{z}_n + \mathcal{O}(\Delta t^2),$$

we obtain

$$\mathbf{z}'_{n+1/2} = \left(\mathbf{I} + \frac{\Delta t}{2}\mathbf{D}\right)^{-1} \left\{ \left(\mathbf{I} - \frac{\Delta t}{2}\mathbf{D}\right) \frac{\mathbf{z}_n - \mathbf{z}_{n-1}}{\Delta t} + \Delta t \mathbf{R}_n - \Delta t \mathbf{A}\mathbf{z}_n \right\} + \mathcal{O}(\Delta t^2). \tag{5.40}$$

Then, we insert (5.39), (5.40) into (5.38), which yields a second-order approximation of $\mathbf{z}'(t)$.

In summary, the second-order LTS algorithm for the solution of (5.13) computes $\mathbf{z}_{n+1} \simeq \mathbf{z}(t + \Delta t)$, for given $\mathbf{z}_n$ and $\mathbf{z}_{n-1}$, as follows:

Algorithm 5.6 (LTS-LFCN2(p) Algorithm).

(1) *Set* $\mathbf{w} := (\mathbf{I} - \mathbf{P})\mathbf{R}_n - \mathbf{A}(\mathbf{I} - \mathbf{P})\mathbf{z}_n$, $\tilde{\mathbf{z}}_0 := \mathbf{z}_n$ *and*

$$\mathbf{z}'_n := \frac{1}{2}\left[\frac{\mathbf{z}_n - \mathbf{z}_{n-1}}{\Delta t} + \left(\mathbf{I} + \frac{\Delta t}{2}\mathbf{D}\right)^{-1} \left\{ \left(\mathbf{I} - \frac{\Delta t}{2}\mathbf{D}\right) \frac{\mathbf{z}_n - \mathbf{z}_{n-1}}{\Delta t} + \Delta t \mathbf{R}_n - \Delta t \mathbf{A}\mathbf{z}_n \right\}\right].$$

(2) *Compute*

$$\tilde{\mathbf{z}}_{1/p} := \tilde{\mathbf{z}}_0 + \frac{\Delta t}{p}\mathbf{z}'_n + \frac{1}{2}\left(\frac{\Delta t}{p}\right)^2 (\mathbf{w} + \mathbf{P}\mathbf{R}_{n,0} - \mathbf{A}\mathbf{P}\tilde{\mathbf{z}}_0 - \mathbf{D}\mathbf{z}'_n) \quad \textit{and}$$

$$\tilde{\mathbf{z}}_{-1/p} := \tilde{\mathbf{z}}_0 - \frac{\Delta t}{p}\mathbf{z}'_n + \frac{1}{2}\left(\frac{\Delta t}{p}\right)^2 (\mathbf{w} + \mathbf{P}\mathbf{R}_{n,0} - \mathbf{A}\mathbf{P}\tilde{\mathbf{z}}_0 - \mathbf{D}\mathbf{z}'_n).$$

(3) *For $m = 1, \dots, p-1$, compute*

$$\widetilde{\mathbf{z}}_{(m+1)/p} := \left(\mathbf{I} + \frac{\Delta t}{2p}\mathbf{D}\right)^{-1} \left\{ 2\widetilde{\mathbf{z}}_{m/p} - \left(\mathbf{I} - \frac{\Delta t}{2p}\mathbf{D}\right)\widetilde{\mathbf{z}}_{(m-1)/p} + \left(\frac{\Delta t}{p}\right)^2 \left(\mathbf{w} + \mathbf{P}\mathbf{R}_{n,m} - \mathbf{A}\mathbf{P}\widetilde{\mathbf{z}}_{m/p}\right) \right\}$$

and

$$\widetilde{\mathbf{z}}_{-(m+1)/p} := \left(\mathbf{I} - \frac{\Delta t}{2p}\mathbf{D}\right)^{-1} \left\{ 2\widetilde{\mathbf{z}}_{-m/p} - \left(\mathbf{I} + \frac{\Delta t}{2p}\mathbf{D}\right)\widetilde{\mathbf{z}}_{-(m-1)/p} + \left(\frac{\Delta t}{p}\right)^2 \left(\mathbf{w} + \mathbf{P}\mathbf{R}_{n,-m} - \mathbf{A}\mathbf{P}\widetilde{\mathbf{z}}_{-m/p}\right) \right\}.$$

(4) *Compute*

$$\mathbf{z}_{n+1} := \widetilde{\mathbf{z}}_1 + \left(\mathbf{I} + \frac{\Delta t}{2}\mathbf{D}\right)^{-1} \left(\mathbf{I} - \frac{\Delta t}{2}\mathbf{D}\right) \left(-\mathbf{z}_{n-1} + \widetilde{\mathbf{z}}_{-1}\right).$$

If σ is piecewise constant in each element, $\mathbf{M}$ and $\mathbf{M}_\sigma$ can be diagonalized simultaneously and hence the matrix $\mathbf{D}$ is diagonal. If σ varies in elements, $\mathbf{D}$ is a block-diagonal matrix and both $(\mathbf{I} \pm (\Delta t/2p)\mathbf{D})$ and $(\mathbf{I} \pm (\Delta t/2)\mathbf{D})$ can be explicitly inverted at low cost. In that sense, the LTS-LFCN2(p) scheme is truly explicit. Again, if the fraction of nonzero entries in $\mathbf{P}$ is small, the overall cost is dominated by the computation of $\mathbf{w}$ in step 1.

Proposition 5.7. *For $\mathbf{R}(t) \in C^2([0,T])$, the local time-stepping method LTS-LFCN2 is second-order accurate.*

Proof. See [25]. □

Remark 5.8. For $\sigma = 0$ ($\mathbf{D} = \mathbf{0}$), the LTS-LFCN2(p) algorithm coincides with the LTS-LF2(p) algorithm and thus also conserves the discrete energy (5.28). For $\sigma \neq 0$ and $p = 1$, i.e. no local mesh refinement, one can easily show that the energy is no longer conserved but decays with time (independently of σ) under the same CFL condition as in the case with $\sigma = 0$.

4 Adams–Bashforth-based LTS methods for damped waves

Starting from the standard leap-frog method, we proposed in Section 3 energy conserving fully explicit LTS integrators of arbitrarily high accuracy for undamped waves.

By blending the leap-frog and the Crank–Nicolson methods, a second-order LTS scheme was also derived there for damped waves, yet this approach cannot be readily extended beyond order two. To achieve arbitrarily high accuracy in the presence of damping, while remaining fully explicit, we shall derive here explicit LTS methods for damped wave equations based on Adams–Bashforth (AB) multistep schemes.

The H^1-conforming and the IP-DG finite-element discretizations of (5.1) presented in Section 2 lead to the second-order system of differential equations (5.4), whereas the nodal DG discretization leads to the first-order system of differential equations (5.11). In both (5.4) and (5.11) the mass matrix $\mathbf{M}$ is symmetric, positive definite and essentially diagonal; thus, $\mathbf{M}^{-1}$ or $\mathbf{M}^{-\frac{1}{2}}$ can be computed explicitly at a negligible cost. For simplicity, we restrict ourselves here to the homogeneous case, i.e. $\mathbf{F}(t) = \mathbf{0}$.

If we multiply (5.4) by $\mathbf{M}^{-\frac{1}{2}}$, we obtain (5.13). Thus, we can rewrite (5.13) as a first-order problem of the form

$$\frac{d\mathbf{y}}{dt}(t) = \mathbf{B}\mathbf{y}(t)\,, \tag{5.41}$$

with

$$\mathbf{y}(t) = \left(\mathbf{z}(t), \frac{d\mathbf{z}}{dt}(t)\right)^T, \quad \mathbf{B} = \begin{pmatrix} \mathbf{0} & \mathbf{I} \\ -\mathbf{A} & -\mathbf{D} \end{pmatrix}.$$

Similarly, we can also rewrite (5.11) in the form (5.41) with $\mathbf{y}(t) = \mathbf{Q}(t)$ and $\mathbf{B} = \mathbf{M}^{-1}(-\mathbf{M}_\sigma - \mathbf{C})$. Hence all three distinct finite-element discretizations from Section 2 lead to a semidiscrete system as in (5.41). Starting from explicit multistep AB methods, we shall now derive explicit LTS schemes of arbitrarily high accuracy for a general problem of the form (5.41).

First, we briefly recall the construction of the classical k-step (kth-order) Adams–Bashforth method for the numerical solution of (5.41) [29]. Let $t_i = i\Delta t$ and $\mathbf{y}_n, \mathbf{y}_{n-1}, \ldots, \mathbf{y}_{n-k+1}$ the numerical approximations to the exact solution $\mathbf{y}(t_n), \ldots, \mathbf{y}(t_{n-k+1})$. The solution of (5.41) satisfies

$$\mathbf{y}(t_n + \xi\Delta t) = \mathbf{y}(t_n) + \int_{t_n}^{t_n+\xi\Delta t} \mathbf{B}\mathbf{y}(t)\,dt\,, \quad 0 < \xi \le 1\,. \tag{5.42}$$

We now replace the unknown solution $\mathbf{y}(t)$ under the integral in (5.42) by the interpolation polynomial $p(t)$ through the points $(t_i, \mathbf{y}_i)$, $i = n-k+1, \ldots, n$. It is explicitly given in terms of backward differences

$$\nabla^0\mathbf{y}_n = \mathbf{y}_n\,, \quad \nabla^{j+1}\mathbf{y}_n = \nabla^j\mathbf{y}_n - \nabla^j\mathbf{y}_{n-1}$$

by

$$p(t) = p(t_n + s\Delta t) = \sum_{j=0}^{k-1} (-1)^j \begin{pmatrix} -s \\ j \end{pmatrix} \nabla^j\mathbf{y}_n\,.$$

Table 5.1: Coefficients $\gamma_j(\xi)$ for the explicit Adams–Bashforth methods.

j	0	1	2	3
$\gamma_j(\xi)$	ξ	$\frac{1}{2}\xi^2$	$\frac{1}{6}\xi^3 + \frac{1}{4}\xi^2$	$\frac{1}{24}\xi^4 + \frac{1}{6}\xi^3 + \frac{1}{6}\xi^2$

Table 5.2: Coefficients for the kth order Adams–Bashforth methods.

	α_0	α_1	α_2	α_3
$k = 2$	$\frac{3}{2}$	$-\frac{1}{2}$	0	0
$k = 3$	$\frac{23}{12}$	$-\frac{16}{12}$	$\frac{5}{12}$	0
$k = 4$	$\frac{55}{24}$	$-\frac{59}{24}$	$\frac{37}{24}$	$-\frac{9}{24}$

Integration of (5.42) with $\mathbf{y}(t)$ replaced by $p(t)$ then yields the approximation $\mathbf{y}_{n+\xi}$ of $\mathbf{y}(t_n + \xi\Delta t)$, $0 < \xi \le 1$,

$$\mathbf{y}_{n+\xi} = \mathbf{y}_n + \Delta t \mathbf{B} \sum_{j=0}^{k-1} \gamma_j(\xi) \nabla^j \mathbf{y}_n \,, \tag{5.43}$$

where the polynomials $\gamma_j(\xi)$ are defined as

$$\gamma_j(\xi) = (-1)^j \int_0^{\xi} \begin{pmatrix} -s \\ j \end{pmatrix} ds \,.$$

They are given in Table 5.1 for $j \le 3$. After expressing the backward differences in terms of $\mathbf{y}_{n-j}$ and setting $\xi = 1$ in (5.43), we recover the common form of the k-step Adams–Bashforth scheme [29]

$$\mathbf{y}_{n+1} = \mathbf{y}_n + \Delta t \mathbf{B} \sum_{j=0}^{k-1} \alpha_j \mathbf{y}_{n-j} \,, \tag{5.44}$$

where the coefficients α_j, $j = 0, \dots, k-1$ for the second-, third- and fourth-order ($k = 2, 3, 4$) Adams–Bashforth schemes are given in Table 5.2. For higher values of k we refer to [29].

Starting from the classical AB methods, we shall now derive LTS schemes of arbitrarily high accuracy for (5.41), which allow arbitrarily small time-steps precisely where small elements in the spatial mesh are located. To do so, we first split the unknown vector $\mathbf{y}(t)$ in two parts

$$\mathbf{y}(t) = (\mathbf{I} - \mathbf{P})\mathbf{y}(t) + \mathbf{P}\mathbf{y}(t) = \mathbf{y}^{[\text{coarse}]}(t) + \mathbf{y}^{[\text{fine}]}(t) \,,$$

where the matrix **P** is diagonal. Its diagonal entries, equal to zero or one, identify the unknowns associated with the locally refined region, where smaller time-steps are needed.

The exact solution of (5.41) again satisfies

$$\mathbf{y}(t_n + \xi\Delta t) = \mathbf{y}(t_n) + \int_{t_n}^{t_n+\xi\Delta t} \mathbf{B}\left(\mathbf{y}^{[\text{coarse}]}(t) + \mathbf{y}^{[\text{fine}]}(t)\right) dt\,, \quad 0 < \xi \le 1\,. \tag{5.45}$$

Since we wish to use the standard k-step Adams–Bashforth method in the coarse region, we approximate the term in (5.45) that involve $\mathbf{y}^{[\text{coarse}]}(t)$ as in (5.42), which yields

$$\mathbf{y}(t_n + \xi\Delta t) \approx \mathbf{y}_n + \Delta t\,\mathbf{B}(\mathbf{I} - \mathbf{P}) \sum_{j=0}^{k-1} \gamma_j(\xi)\nabla^j \mathbf{y}_n + \int_{t_n}^{t_n+\xi\Delta t} \mathbf{BPy}(t)\,dt\,. \tag{5.46}$$

To circumvent the severe stability constraint due to the smallest elements associated with $\mathbf{y}^{[\text{fine}]}(t)$, we shall now treat $\mathbf{y}^{[\text{fine}]}(t)$ differently from $\mathbf{y}^{[\text{coarse}]}(t)$. Hence, we instead approximate the integrand in (5.46) as

$$\int_{t_n}^{t_n+\xi\Delta t} \mathbf{BPy}(t)\,dt \approx \int_0^{\xi\Delta t} \mathbf{BP}\widetilde{\mathbf{y}}(\tau)\,d\tau\,,$$

where $\widetilde{\mathbf{y}}(\tau)$ solves the differential equation

$$\begin{aligned} \frac{d\widetilde{\mathbf{y}}}{d\tau}(\tau) &= \mathbf{B}(\mathbf{I} - \mathbf{P}) \sum_{j=0}^{k-1} \widetilde{\gamma}_j\left(\frac{\tau}{\Delta t}\right) \nabla^j \mathbf{y}_n + \mathbf{BP}\,\widetilde{\mathbf{y}}(\tau)\,, \\ \widetilde{\mathbf{y}}(0) &= \mathbf{y}_n\,, \end{aligned} \tag{5.47}$$

with coefficients

$$\widetilde{\gamma}_j(\xi) = \frac{d}{d\xi}\gamma_j(\xi) = \frac{d}{d\xi}\left((-1)^j \int_0^{\xi} \binom{-s}{j} ds\right) = (-1)^j \binom{-\xi}{j}\,. \tag{5.48}$$

The polynomials $\widetilde{\gamma}_j(\xi)$ are given in Table 5.3 for $j \le 3$. Replacing $\mathbf{y}(t)$ by $\widetilde{\mathbf{y}}(t)$ in

Table 5.3: The polynomial coefficients $\widetilde{\gamma}_j(\xi)$.

j	0	1	2	3
$\widetilde{\gamma}_j(\xi)$	1	ξ	$\frac{1}{2}\xi^2 + \frac{1}{2}\xi$	$\frac{1}{6}\xi^3 + \frac{1}{2}\xi^2 + \frac{1}{3}\xi$

(5.46), we obtain

$$\mathbf{y}(t_n + \xi\Delta t) \approx \mathbf{y}_n + \Delta t\, \mathbf{B}(\mathbf{I} - \mathbf{P}) \sum_{j=0}^{k-1} \gamma_j(\xi) \nabla^j \mathbf{y}_n + \int_0^{\xi\Delta t} \mathbf{BP}\tilde{\mathbf{y}}(\tau)\, d\tau\,. \tag{5.49}$$

By considering (5.47) in integrated form, we find that

$$\begin{aligned}\tilde{\mathbf{y}}(\xi\Delta t) &= \tilde{\mathbf{y}}(0) + \mathbf{B}(\mathbf{I} - \mathbf{P}) \sum_{j=0}^{k-1} \left(\int_0^{\xi\Delta t} \tilde{\gamma}_j \left(\frac{\tau}{\Delta t} \right) d\tau \right) \nabla^j \mathbf{y}_n + \int_0^{\xi\Delta t} \mathbf{BP}\, \tilde{\mathbf{y}}(\tau)\, d\tau \\ &= \mathbf{y}_n + \Delta t\, \mathbf{B}(\mathbf{I} - \mathbf{P}) \sum_{j=0}^{k-1} \gamma_j(\xi) \nabla^j \mathbf{y}_n + \int_0^{\xi\Delta t} \mathbf{BP}\, \tilde{\mathbf{y}}(\tau)\, d\tau\,.\end{aligned} \tag{5.50}$$

From the comparison of (5.49) and (5.50) we infer that

$$\mathbf{y}(t_n + \xi\Delta t) \approx \tilde{\mathbf{y}}(\xi\Delta t)\,.$$

Thus to advance $\mathbf{y}(t_n)$ from t_n to $t_n + \Delta t$, we shall evaluate $\tilde{\mathbf{y}}(\Delta t)$ by solving (5.47) on $[0, \Delta t]$ numerically. We solve (5.47) until $\tau = \Delta t$ again with a k-step Adams–Bashforth scheme, using a smaller time-step $\Delta\tau = \Delta t / p$, where p denotes the ratio of local refinement. For $m = 0, \ldots, p-1$ we then have

$$\begin{aligned}\tilde{\mathbf{y}}_{(m+1)/p} &= \tilde{\mathbf{y}}_{m/p} + \Delta\tau\, \mathbf{B}(\mathbf{I} - \mathbf{P}) \sum_{\ell=0}^{k-1} \alpha_\ell \sum_{j=0}^{k-1} \tilde{\gamma}_j \left(\frac{m - \ell}{p} \right) \nabla^j \mathbf{y}_n \\ &\quad + \Delta\tau\, \mathbf{BP} \sum_{\ell=0}^{k-1} \alpha_\ell \tilde{\mathbf{y}}_{(m-l)/p}\,,\end{aligned} \tag{5.51}$$

where α_ℓ, $\ell = 0, \ldots, k-1$ denote the coefficients of the classical k-step AB scheme (see Table 5.2). Finally, after expressing the backward differences in terms of $\mathbf{y}_{n-\ell}$, we find

$$\tilde{\mathbf{y}}_{(m+1)/p} = \tilde{\mathbf{y}}_{m/p} + \Delta\tau\, \mathbf{B}(\mathbf{I} - \mathbf{P}) \sum_{\ell=0}^{k-1} \beta_{m,\ell}\, \mathbf{y}_{n-\ell} + \Delta\tau\, \mathbf{BP} \sum_{\ell=0}^{k-1} \alpha_\ell\, \tilde{\mathbf{y}}_{(m-l)/p}\,, \tag{5.52}$$

where the constant coefficients $\beta_{m,\ell}$, $m = 0, \ldots, p-1$, $\ell = 0, \ldots, k-1$, satisfy

$$\beta_{m,\ell} = \sum_{i=0}^{k-1} \alpha_i \sum_{j=\ell}^{k-1} (-1)^\ell \binom{j}{\ell} \tilde{\gamma}_j \left(\frac{m - i}{p} \right), \tag{5.53}$$

with $\tilde{\gamma}_j$ defined in (5.48).

In summary, the LTS-AB$k(p)$ algorithm computes $\mathbf{y}_{n+1} \simeq \mathbf{y}(t_n + \Delta t)$, given $\mathbf{y}_n, \mathbf{y}_{n-1}, \ldots, \mathbf{y}_{n-k+1}$, $\mathbf{B}(\mathbf{I} - \mathbf{P})\mathbf{y}_{n-1}, \ldots, \mathbf{B}(\mathbf{I} - \mathbf{P})\mathbf{y}_{n-k+1}$ and $\mathbf{P}\mathbf{y}_{n-1/p}, \mathbf{P}\mathbf{y}_{n-2/p}, \ldots,$ $\mathbf{P}\mathbf{y}_{n-(k-1)/p}$ as follows:

Algorithm 5.9 (LTS-AB$k(p)$ Algorithm).
(1) *Set* $\tilde{\mathbf{y}}_0 := \mathbf{y}_n$, $\tilde{\mathbf{y}}_{-\ell/p} := \mathbf{P}\mathbf{y}_{n-\ell/p}$, $\ell = 1, \dots, k-1$.
(2) *Set* $\mathbf{w}_{n-\ell} := \mathbf{B}(\mathbf{I}-\mathbf{P})\mathbf{y}_{n-\ell}$, $\ell = 1, \dots, k-1$.
(3) *Compute* $\mathbf{w}_n := \mathbf{B}(\mathbf{I}-\mathbf{P})\mathbf{y}_n$.
(4) *For* $m = 0, \dots, p-1$, *compute*

$$\tilde{\mathbf{y}}_{(m+1)/p} := \tilde{\mathbf{y}}_{m/p} + \frac{\Delta t}{p} \sum_{\ell=0}^{k-1} \beta_{m,\ell} \mathbf{w}_{n-\ell} + \frac{\Delta t}{p} \mathbf{BP} \sum_{\ell=0}^{k-1} \alpha_\ell \tilde{\mathbf{y}}_{(m-l)/p} .$$

(5) *Set* $\mathbf{y}_{n+1} := \tilde{\mathbf{y}}_1$.

Steps 1–4 correspond to the numerical solution of (5.47) until $\tau = \Delta t$ with the k-step AB scheme, using the local time-step $\Delta\tau = \Delta t/p$. For $\mathbf{P} = \mathbf{0}$ or $p = 1$, that is without any local time-stepping, we thus recover the standard k-step Adams–Bashforth scheme. If the fraction of nonzero entries in $\mathbf{P}$ is small, the overall cost is dominated by the computation of $\mathbf{w}_n$ in Step 3, which requires one multiplications by $\mathbf{B}(\mathbf{I}-\mathbf{P})$ per time-step Δt. All further matrix–vector multiplications by $\mathbf{BP}$ only affect those unknowns that lie inside the refined region, or immediately next to it; hence, their computational cost remains negligible as long as the locally refined region contains a small part of Ω.

We have shown above how to derive LTS-AB$k(p)$ schemes of arbitrarily high accuracy. Since the third- and fourth-order LTS-AB$k(p)$ schemes are probably the most relevant for applications, we now describe the LTS-AB$k(p)$ schemes for $k = 3, 4$ and $p = 2$. Other examples of LTS Adams–Bashforth schemes are listed in [26].

For $k = 3$ and $p = 2$, the LTS-AB3(2) method reads:

$$\begin{aligned}
\tilde{\mathbf{y}}_{1/2} &= \mathbf{y}_n + \frac{\Delta t}{2}\mathbf{B}(\mathbf{I}-\mathbf{P})\left[\frac{17}{12}\mathbf{y}_n - \frac{7}{12}\mathbf{y}_{n-1} + \frac{2}{12}\mathbf{y}_{n-2}\right] \\
&\quad + \frac{\Delta t}{2}\mathbf{BP}\left[\frac{23}{12}\mathbf{y}_n - \frac{16}{12}\tilde{\mathbf{y}}_{-1/2} + \frac{5}{12}\mathbf{y}_{n-1}\right], \\
\mathbf{y}_{n+1} = \tilde{\mathbf{y}}_1 &= \tilde{\mathbf{y}}_{1/2} + \frac{\Delta t}{2}\mathbf{B}(\mathbf{I}-\mathbf{P})\left[\frac{29}{12}\mathbf{y}_n - \frac{25}{12}\mathbf{y}_{n-1} + \frac{8}{12}\mathbf{y}_{n-2}\right] \\
&\quad + \frac{\Delta t}{2}\mathbf{BP}\left[\frac{23}{12}\tilde{\mathbf{y}}_{1/2} - \frac{16}{12}\mathbf{y}_n + \frac{5}{12}\tilde{\mathbf{y}}_{-1/2}\right].
\end{aligned}$$

For the case with $k = 4$ and $p = 2$, we find the LTS-AB4(2) scheme:

$$\begin{aligned}
\tilde{\mathbf{y}}_{1/2} &= \mathbf{y}_n + \frac{\Delta t}{2}\mathbf{B}(\mathbf{I}-\mathbf{P})\left[\frac{297}{192}\mathbf{y}_n - \frac{187}{192}\mathbf{y}_{n-1} + \frac{107}{192}\mathbf{y}_{n-2} - \frac{25}{192}\mathbf{y}_{n-3}\right] \\
&\quad + \frac{\Delta t}{2}\mathbf{BP}\left[\frac{55}{24}\mathbf{y}_n - \frac{59}{24}\tilde{\mathbf{y}}_{-1/2} + \frac{37}{24}\mathbf{y}_{n-1} - \frac{9}{24}\tilde{\mathbf{y}}_{-3/2}\right], \\
\mathbf{y}_{n+1} = \tilde{\mathbf{y}}_1 &= \tilde{\mathbf{y}}_{1/2} + \frac{\Delta t}{2}\mathbf{B}(\mathbf{I}-\mathbf{P})\left[\frac{583}{192}\mathbf{y}_n - \frac{757}{192}\mathbf{y}_{n-1} + \frac{485}{192}\mathbf{y}_{n-2} - \frac{119}{192}\mathbf{y}_{n-3}\right] \\
&\quad + \frac{\Delta t}{2}\mathbf{BP}\left[\frac{55}{24}\tilde{\mathbf{y}}_{1/2} - \frac{59}{24}\mathbf{y}_n + \frac{37}{24}\tilde{\mathbf{y}}_{-1/2} - \frac{9}{24}\mathbf{y}_{n-1}\right].
\end{aligned}$$

Proposition 5.10. *The local time-stepping method LTS-ABk(p) is consistent of order k.*

Proof. See [26]. □

5 Numerical results

Here we present numerical experiments that validate the expected order of convergence of the above LTS methods and demonstrate their usefulness in the presence of complex geometry. First, we consider a simple one-dimensional test problem illustrate the stability properties of the different LTS schemes presented above and to show that they yield the expected overall rate of convergence when combined with a spatial finite-element discretization of comparable accuracy, independently of the number of local time-steps p used in the fine region. Then, we illustrate the versatility of our LTS schemes by simulating the propagation of a circular wave in a square cavity with a small sigma-shaped hole.

5.1 Stability

We consider the one-dimensional homogeneous damped wave equation (5.1) with constant wave speed $c = 1$ and damping coefficient $\sigma = 0$ on the interval $\Omega = [0, 6]$. Next, we divide Ω into three equal parts. The left and right intervals, $[0, 2]$ and $[4, 6]$, respectively, are discretized with an equidistant mesh of size h^{coarse}, whereas the interval $\Omega_f = [2, 4]$ is discretized with an equidistant mesh of size $h^{\text{fine}} = h^{\text{coarse}}/p$. Hence, the two outer intervals correspond to the coarse region whereas the inner interval $[2, 4]$ to the refined region. In [16], we have studied numerically the stability of the LTS-LF2(p) and the LTS-LFME4(p) methods. To determine the range of values Δt for which the LTS-LF2(p) scheme is stable, the eigenvalues of $(\Delta t^2/4)\mathbf{A}_p$ ($\mathbf{A}_p$ is defined by (5.27)) for varying $\Delta t/\Delta t_{\text{LF}}$ are computed, where Δt_{LF} denotes the largest time-step allowed by the standard leap-frog method. The LTS-LF2(p) scheme is stable for any particular Δt if all corresponding eigenvalues lie between zero and one; otherwise, it is unstable. We have observed that the largest time step allowed by the LTS-LF2(p) scheme is only about 60% of Δt_{LF}. A slight extension (overlap) of the region where local time steps are used into that part of the mesh immediately adjacent to the refined region typically improves the stability of the LTS-LF2(p) scheme. Moreover, the numerical results suggested that an overlap by one element when combined with a $\mathcal{P}^1$ continuous FE discretization (with mass lumping), or by two elements when combined with a IP-DG discretization, permits the use of the maximal (optimal) time step Δt_{LF}. The numerical results also suggested that an overlap by one element for the IP-DG discretization is needed for the optimal CFL stability condition of the LTS-LFME4(p) independently of p. Remarkably, no overlap is needed for the LTS-LFME4(p) scheme to remain stable with the optimal time-step when combined with the continuous $\mathcal{P}^3$ elements.

In [26], we have considered the above one-dimensional problem with $\sigma = 0.1$. We have written the LTS-ABk(p) scheme as a one-step method and then studied numerically it stability when combined with a spatial finite-element discretization of comparable accuracy. For a spatial discretization with standard continuous, IP-DG or nodal DG finite elements, we have obtained that the maximal time-step Δt_p allowed by the LTS-AB 2(p) scheme is about 80 % of the optimal time-step Δt_{AB2} (the largest time-step allowed by the standard two-step AB method) independently of h, p and σ; moreover, the CFL stability condition of the LTS-AB 3(p) and LTS-AB 4(p) schemes is optimal for all h, p and σ.

5.2 Convergence

We consider the one-dimensional homogeneous model problems (5.1) and (5.9) with constant wave speed $c = 1$ and damping coefficient $\sigma = 0.1$ on the interval $\Omega = (0, 6)$. The initial conditions are chosen to yield the exact solution

$$\begin{aligned} u(x,t) &= \frac{2e^{-\frac{\sigma t}{2}}}{\sqrt{4\pi^2 - \sigma^2}} \sin(\pi x) \sin\left(\frac{t}{2}\sqrt{4\pi^2 - \sigma^2}\right), \\ v(x,t) &= \frac{\partial u}{\partial t}(x,t), \quad \mathbf{w}(x,t) = -\nabla u(x,t). \end{aligned}$$

Again, we divide Ω into three equal parts. The left and right intervals, $[0, 2]$ and $[4, 6]$, respectively, are discretized with an equidistant mesh of size h^{coarse}, whereas the interval $[2, 4]$ is discretized with an equidistant mesh of size $h^{\text{fine}} = h^{\text{coarse}}/p$. Hence, the two outer intervals correspond to the coarse region and the inner interval $[2, 4]$ to the refined region.

First, we consider a $\mathcal{P}^3$ continuous FE discretization with mass lumping and a sequence of increasingly finer meshes. For every time-step Δt, we shall take $p \geq 2$ local steps of size $\Delta\tau = \Delta t/p$ in the refined region, with the fourth-order time-stepping scheme LTS-AB4(p). The first three time-steps of each LTS-AB4(p) scheme are initialized by using the exact solution. According to our results on stability, we set $\Delta t = \Delta t_{AB4}$, the corresponding largest possible time-step allowed by the AB approach of order four on an equidistant mesh with $h = h^{\text{coarse}}$. As we systematically reduce the global mesh size h^{coarse}, while simultaneously reducing Δt, we monitor the L^2 space-time error in the numerical solution $\|u(\cdot, T) - u^h(\cdot, T)\|_{L^2(\Omega)}$ at the final time $T = 10$. In frame (a) of Figure 5.1, the numerical error is shown vs. the mesh size $h = h^{\text{coarse}}$: regardless of the number of local time-steps $p = 2, 5$ or 7, the numerical method converges with order four.

We now repeat the same experiment with the IP-DG ($\alpha = 20$ in (5.7)) and the nodal DG discretizations with $\mathcal{P}^3$-elements. As shown in frames (b) and (c) of Fig-

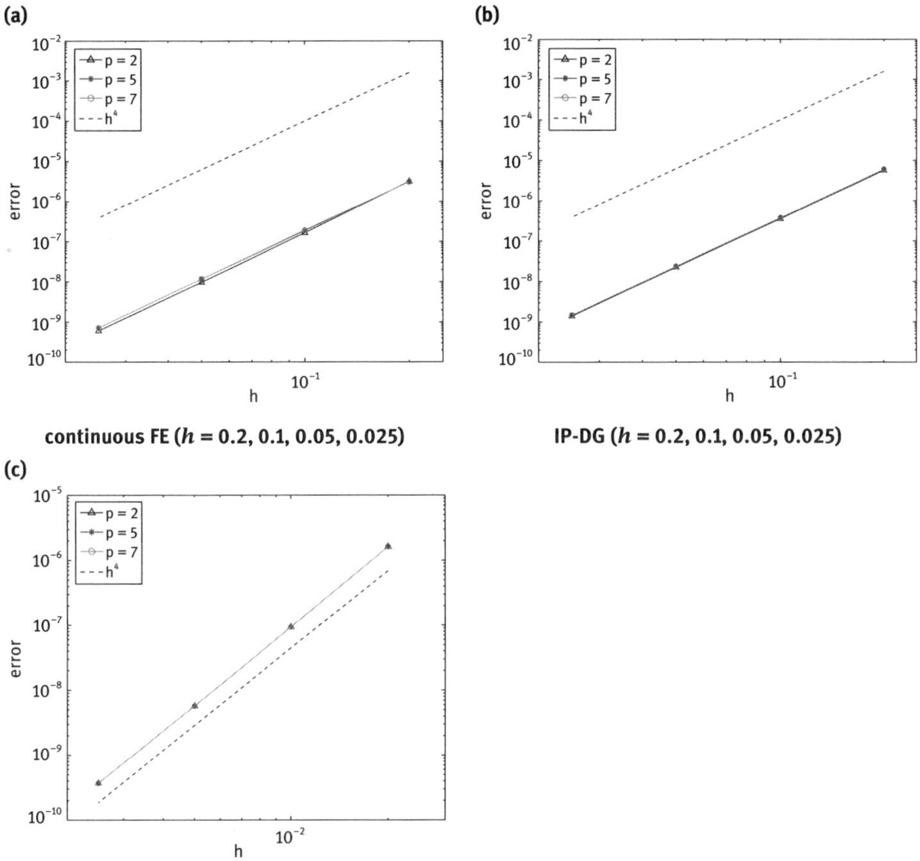

continuous FE ($h = 0.2, 0.1, 0.05, 0.025$)

IP-DG ($h = 0.2, 0.1, 0.05, 0.025$)

nodal DG ($h = 0.02, 0.01, 0.005, 0.0025$)

Figure 5.1: LTS-AB4(p) error vs. $h = h^{\text{coarse}}$ for $\mathcal{P}^3$ finite elements with $p = 2, 5, 7$.

ure 5.1, the LTS-AB4(p) method again yields overall fourth-order convergence independently of p.

Remark 5.11. We have obtained similar convergence results for other values of p and σ. In summary, we observe the optimal rates convergence of order k for the LTS-ABk(p) schemes as well as for the LTS-LF2(p), LTS-LSME4(p) and LTS-LFCN2(p) schemes, regardless of the spatial FE discretization and independently of the number of local time-steps p and the damping coefficient σ. For more details, we refer to [16, 25, 26].

5.3 Two-dimensional example

To illustrate the usefulness of the LTS method presented above, we consider (5.1) in a square cavity $\Omega = (0,1)^2$, with a small sigma-shaped hole – see Figure 5.2. We set the constant wave speed $c = 1$ and the damping coefficient

$$\sigma(\mathbf{x}) = \begin{cases} 10, & x_2 < 0.5 \\ 0.1, & \text{otherwise}. \end{cases}$$

We impose homogeneous Neumann conditions on the boundary of Ω and choose as initial conditions

$$u_0(\mathbf{x}) = \exp\left(-\|\mathbf{x} - \mathbf{x}_0\|^2/r^2\right), \quad v_0(\mathbf{x}) = 0,$$

where $\mathbf{x}_0 = (0.45, 0.55)$ and $r = 0.012$.

For the spatial discretization we opt for the $\mathcal{P}^2$ continuous finite elements with mass lumping. First, Ω is discretized with triangles of minimal size $h^{\text{coarse}} = 0.03$. However, such triangles do not resolve the small geometric features of the sigma-shaped hole, which require $h^{\text{fine}} \approx h^{\text{coarse}}/7$, as shown in Figure 5.3. Then, we successively refine the entire mesh three times, each time splitting every triangle into four. Since the initial mesh in Ω is unstructured, the boundary between the fine and

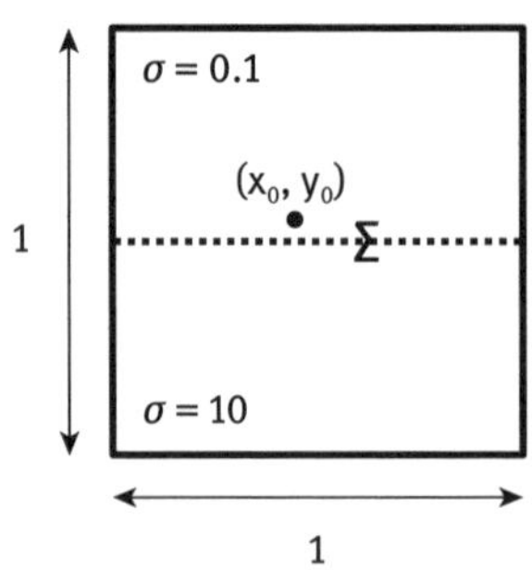

Figure 5.2: Two-dimensional example: the computational domain Ω.

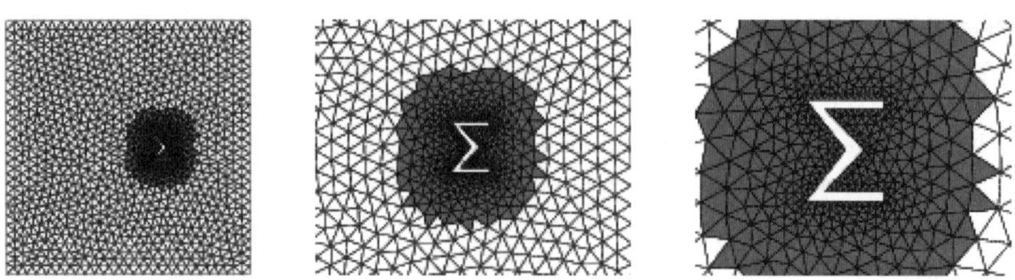

Figure 5.3: The triangular initial mesh at various magnification rates: the darker triangles belong to the "fine" mesh.

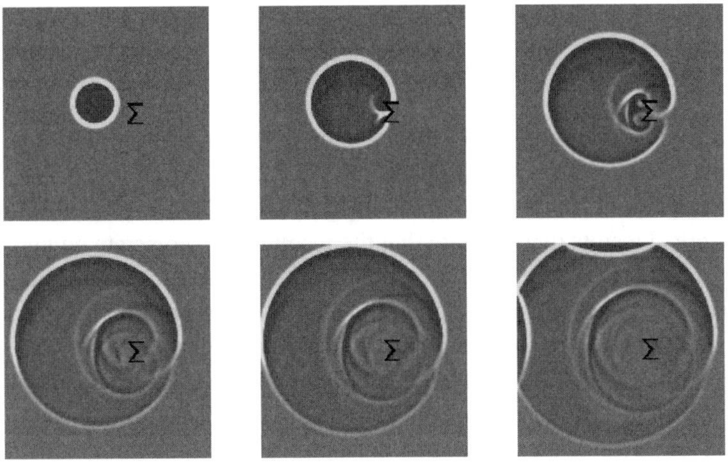

Figure 5.4: (See Color Plates for a color version.) Gaussian pulse penetrating a cavity with a small sigma-shaped hole. The solution is shown at times $t = 0.1, 0.2, 0.3, 0.4, 0.44$ and 0.5.

coarse mesh is not well-defined. Given h^{coarse}, here the fine mesh corresponds to all triangles with $h < 0.75h^{\text{coarse}}$ in size, that is the darker triangles in Figure 5.3. The corresponding degrees of freedom in the finite element solution are then selected merely by setting to one the corresponding diagonal entries of the matrix $\mathbf{P}$.

For the time discretization, we choose the third-order LTS-AB3(7) time-stepping scheme with $p = 7$, which for every time-step Δt takes seven local time-steps inside the refined region. Thus, the numerical method is third-order accurate in both space and time under the CFL condition $\Delta t = 0.07\,h^{\text{coarse}}$, determined experimentally. If instead the same (global) time-step Δt was used everywhere inside Ω, it would have to be about seven times smaller than necessary in most of Ω. As a starting procedure, we employ a standard fourth-order Runge–Kutta scheme.

In Figure 5.4, snapshots of the numerical solution are shown at different times. The circular wave, initiated by the Gaussian pulse, propagates outward until it impinges first on the sigma-shaped hole and later on the upper and left boundaries of Ω. The reflected waves move back into Ω while multiple reflections occur both at the obstacle and along the interface at $x_2 = 0.5$. As the waves cross that interface and penetrate the lower part of Ω, they are strongly damped.

6 Concluding remarks

Starting from the classical leap-frog (LF) or Adams–Bashforth (AB) methods, we have presented explicit local time-stepping (LTS) schemes for wave equations, either with or without damping. By allowing arbitrarily small time-steps precisely where the smallest elements in the mesh are located, these LTS schemes circumvent the crippling effect of locally refined meshes on explicit time integration.

When combined with a spatial finite-element discretization with an essentially diagonal mass matrix, the resulting LTS schemes remain fully explicit. Here three such finite-element discretizations were considered: standard H^1-conforming finite elements with mass-lumping, an IP-DG formulation, and nodal DG finite elements. In all cases, our numerical results demonstrate that the resulting fully discrete numerical schemes yield the expected space-time optimal convergence rates. Moreover, the LTS-AB(k) schemes of order $k \geq 3$ have optimal CFL stability properties regardless of the mesh size h, the global to local step-size ratio p, or the dissipation σ. Otherwise, the CFL condition of the LTS scheme may be suboptimal; then, by including a small overlap of the fine and the coarse region, the CFL condition of the resulting LTS scheme can be significantly enhanced.

Since the LTS methods presented here are truly explicit, their parallel implementation is straightforward. Let Δt denote the time-step imposed by the CFL condition in the coarser part of the mesh. Then, during every (global) time-step Δt, each local time-step of size $\Delta t / p$ inside the fine region of the mesh simply corresponds to sparse matrix–vector multiplications that only involve degrees of freedom associated with the fine region of the mesh. Those "fine" degrees of freedom can be selected individually and without any restriction by setting the corresponding entries in the diagonal projection matrix P to one; in particular, no adjacency or coherence in the numbering of the degrees of freedom is assumed. Hence the implementation is straightforward and requires no special data structures.

In the presence of multilevel mesh refinement, each local time-step in the fine region can itself include further local time-steps inside a smaller subregion with an even higher degree of local mesh refinement. The explicit local time-stepping schemes developed here for the scalar damped wave equation immediately apply to other damped wave equations, such as in electromagnetics or elasticity; in fact, they can be used for general linear first-order hyperbolic systems.

References

[1] M. Ainsworth, P. Monk, W. Muniz, Dispersive and dissipative properties of discontinuous Galerkin finite element methods for the second-order wave equation, *J. Sci. Comput.* **27** (2006), 5–40.

[2] D. N. Arnold, F. Brezzi, B. Cockburn, and L. D. Marini, Unified analysis of discontinuous Galerkin methods for elliptic problems, *SIAM J. Numer. Anal.* **39** (2002), 1749–1779.

[3] C. Agut and J. Diaz, Stability analysis of the Interior Penalty Discontinuous Galerkin method for the wave equation, *INRIA Research Report* **7494** (2010).

[4] G. Baker and V. Douglas, The effect of quadrature errors on finite element approximations for second order hyperbolic equations, *SIAM J. Numer. Anal.* **13** (1976), 577–598.

[5] A. Buffa, I. Perugia, and T. Warburton, The Mortar-Discontinuous Galerkin method for the 2D Maxwell eigenproblem, *J. Sci. Comp.* **40** (2009), 86–114.
[6] P. Ciarlet, The Finite Element Method for Elliptic Problems, North-Holland, Amsterdam, 1978.
[7] B. Cockburn, C.-W. Shu, TVB Runge–Kutta local projection discontinuous Galerkin method for conservation laws II: General framework, *Math. Comp.* **52** (1989), 411–435.
[8] G. Cohen, P. Joly, and N. Tordjman, Construction and analysis of higher order finite elements with mass lumping for wave equation, *in Proceedings of Second International Conference on Mathematical and Numerical Aspects of Wave Propagation Phenomena, SIAM,* (1993), 152–160.
[9] G. Cohen, P. Joly, and N. Tordjman, Higher-order finite elements with mass-lumping for the 1D wave equation, *Finite Elem. Anal. Des.* **16** (1994), 329–336.
[10] G. Cohen, P. Joly, J. Roberts, and N. Tordjman, Higher order triangular finite elements with mass lumping for the wave equation, *SIAM J. Numer. Anal.* **38** (2001), 2047–2078.
[11] G. Cohen, High-Order Numerical Methods for Transient Wave Equations, Springer-Verlag, 2002.
[12] F. Collino, T. Fouquet, P. Joly, A conservative space-time mesh refinement method for the 1-D wave equation. I. Construction, *Numer. Math.* **95** (2003), 197–221.
[13] F. Collino, T. Fouquet, P. Joly, Conservative space-time mesh refinement method for the FDTD solution of Maxwell's equations, *J. Comput. Phys.* **211** (2006), 9–35.
[14] E. Constantinescu, A. Sandu, Multirate timestepping methods for hyperbolic conservation laws, *J. Sci. Comput.* **33** (2007), 239–278.
[15] E. Constantinescu, A. Sandu, Multirate explicit Adams methods for time integration of conservation laws, *J. Sci. Comput.* **38** (2009), 229–249.
[16] J. Diaz and M. J. Grote, Energy conserving explicit local time-stepping for second-order wave equations, *SIAM J. Sci. Comput.* **31** (2009), 1985–2014.
[17] V. Dolean, H. Fahs, L. Fezoui, S. Lanteri, Locally implicit discontinuous Galerkin method for time domain electromagnetics, *J. Comput. Phys.* **229** (2010), 512–526.
[18] M. Dumbser, M. Käser, E. Toro, An arbitrary high-order Discontinuous Galerkin method for elastic waves on unstructured meshes - V. Local time stepping and p-adaptivity, *Geophys. J. Int.* **171** (2007), 695–717.
[19] A. Ezziani, P. Joly, Local time stepping and discontinuous Galerkin methods for symmetric first order hyperbolic systems, *J. Comput. Appl. Math.* **234** (2010), 1886–1895.
[20] F. X. Giraldo, M. A. Taylor, A diagonal-mass-matrix triangular-spectral-element method based on cubature points, *J. Engrg. Math.* **56** (2006), 307–322.
[21] M. J. Grote, A. Schneebeli, and D. Schötzau, Discontinuous Galerkin finite element method for the wave equation, *SIAM J. Numer. Anal.* **44** (2006), 2408–2431.
[22] M. J. Grote, A. Schneebeli, and D. Schötzau, Interior penalty discontinuous Galerkin method for Maxwell's equations: Energy norm error estimates, *J. Comput. Appl. Math.* **204** (2007), 375–386.
[23] M. J. Grote, A. Schneebeli, and D. Schötzau, Interior penalty discontinuous Galerkin method for Maxwell's equations: Optimal L^2-norm error estimates, *IMA J. Numer. Anal.* **28** (2008), 440–468.
[24] M. J. Grote and D. Schötzau, Optimal Error Estimates for the Fully Discrete Interior Penalty DG Method for the Wave Equation, *J. Sci. Comput.* **40** (2009), 257–272.
[25] M. J. Grote and T. Mitkova, Explicit local time-stepping for Maxwell's equations, *J. Comput. App. Math.* **234** (2010), 3283–3302.
[26] M. J. Grote and T. Mitkova, High-order explicit local time-stepping methods for damped wave equations, *J. Comput. App. Math.*, **239** (2013), 270–289.

[27] T. Hagstrom, T. Warburton, and L. C. Wilcox, Accelerating the discontinuous Galerkin time-domain method, *Oberwolfach Report* **36/2008** (2008), 2084–2086.

[28] E. Hairer, C. Lubich, and G. Wanner, Geometric Numerical Integration, Structure-Preserving Algorithums for Ordinary Differential Equations, Springer, 2002.

[29] E. Hairer, S. Nørsett, and G. Wanner, Solving Ordinary Differential Equations I: Nonstiff Problems, Springer, 2000.

[30] J. S. Hesthaven and T. Warburton, Nodal Discontinuous Galerkin Methods, Springer, 2008.

[31] M. Hochbruck and A. Ostermann, Exponential multistep methods of Adams-type, *BIT Numer. Math.* **51** (2011), 889–908.

[32] T. Hughes, The finite element method : linear static and dynamic finite element analysis, Dover Publications, 2000.

[33] J. L. Lions and E. Magenes, Non-homogeneous Boundary Value Problems and Applications, Vol. I, Springer-Verlag, 1972.

[34] E. Montseny, S. Pernet, X. Ferriéres, G. Cohen, Dissipative terms and local time-stepping improvements in a spatial high order Discontinuous Galerkin scheme for the time-domain Maxwell's equations, *J. Comp. Physics* **227** (2008), 6795–6820.

[35] S. Piperno, Symplectic local time-stepping in non-dissipative DGTD methods applied to wave propagation problems, *M2AN Math. Model. Numer. Anal.* **40** (2006), 815–841.

[36] R. Rieben, D. White, and G. Rodrigue, High-order symplectic integration methods for finite element solutions to time dependent Maxwell equations, *IEEE Trans. Antennas Propagat.* **52** (2004), 2190–2195.

[37] B. Rivière, M. F. Wheeler, Discontinuous Finite Element Methods for Acoustic and Elastic Wave Problems, *Contemporary Mathematics* **329** (2003), 271–282.

[38] G. R. Shubin and J. B. Bell, A modified equation approach to constructing fourth order methods for acoustic wave propagation, *SIAM J. Sci. Stat. Comput.* **8** (1987), 135–151.

[39] A. Taube, M. Dumbser, C.-D. Munz, and R. Schneider, A high-order discontinuous Galerkin method with time-accurate local time stepping for the Maxwell equations, *Int. J. Numer. Model.* **22** (2009), 77–103.

Frédéric Nataf

Absorbing boundary conditions and perfectly matched layers in wave propagation problems

Abstract: In this chapter, we discuss different techniques to solve numerically wave propagation phenomena in unbounded domains. We present in two ways a unified and simple way to restrict the computation to a finite domain: absorbing (or artificial) boundary conditions and perfectly matched layers. The intent is to give the possibility to the reader to grasp easily similarities and differences between these two truncation techniques. It should also allow the reader to adapt a truncation technique to the peculiarities of his physical modeling.

Keywords: artificial boundary condition, perfectly matched layer

2010 Mathematics Subject Classification: 35L05,65Mxx

Frédéric Nataf: Laboratory J.L. Lions, UPMC and CNRS, Paris, France, nataf@ann.jussieu.fr

1 Introduction

In order to model numerically a wave propagation phenomenon in a domain which extends to infinity, it is common practice to truncate the computational domain to a finite domain. The newly formed external boundary is somewhat artificial. To reduce undesired numerical reflections, there is a need for a special treatment at these boundaries. The solution must be as close as possible to the restriction of the solution that would have been computed in the original physical infinite domain. The most commonly used grid truncation techniques are so-called artificial (or absorbing) boundary conditions (ABC), and perfectly matched layer (PML), formulations. The ABC technique was first formalized by Engquist and Majda in a seminal 1977 paper in [7]. The PML concept was introduced by Berenger in 1994 [5]. ABC techniques are more general than PML. However, PML (which is an absorbing region rather than a boundary condition per se) can provide orders-of-magnitude smaller reflections. Both methods are widely used in acoustic, electromagnetic or elastodynamic simulation software codes. Their precise formulations and efficiency depend strongly on the underlying partial differential equations that are discretized. For example, a stable and quite efficient PML for the nondissipative wave equation may surprisingly lead to unstable computations if used in conjunction with a damped wave equation. We present in this chapter a unified and simple approach to the design and study of both ABC and PML formulations. The intent is to give the reader the possibility to grasp easily similarities and differences between these two techniques. It should also allow the reader to adapt a truncation technique to the peculiarities of his physical modeling.

2 ABC

Homogeneous Dirichlet or Neumann boundary conditions for the wave equation lead to a total reflection of an impinging wave on the artificial boundary of a computational domain. We present here the method introduced by Engquist and Majda [7] to design absorbing conditions that are easy to implement and yield a small reflection coefficient on the truncation boundary. The wave equation is solved in a domain exterior to an open bounded subset Ω of $\mathbb{R}^2$ denoted by Ω^c:

$$\begin{aligned}
&\frac{\partial^2 u}{\partial t^2} - c^2\left(\frac{\partial^2 u}{\partial x^2} + \frac{\partial^2 u}{\partial y^2}\right) = f(t,x,y), \quad (x,y) \in \Omega^c,\ t > 0 \\
&u(0,x,y) = u_0(x,y) \quad \text{in } \Omega^c \\
&\frac{\partial u}{\partial t}(0,x,y) = u_1(x,y) \quad \text{in } \Omega^c \\
&u = 0 \quad \text{on } \partial\Omega \\
&u \quad \text{satisfies a radiation boundary condition at infinity.}
\end{aligned} \tag{6.1}$$

The origin of the coordinate system is such that Ω is a subset of the half-space of negative x. We assume that the initial conditions u_0 and u_1 and the right-hand side f have a compact support in this half-space. The computational domain is truncated on its right by the artificial boundary $x = 0$, denoted by Γ (Figure 6.1). It is not possible to use as an ABC a Dirichlet boundary condition on Γ since the solution is not known on Γ. The idea is to design the ABC by finding a relation satisfied by the solution on Γ that does not depend on the values of the right-hand sides. This integrodifferential relation defines an exact ABC which will be approximated by a partial differential equation.

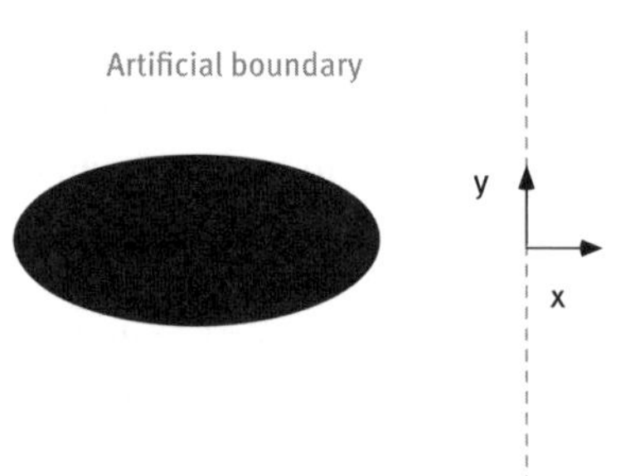

Figure 6.1: Truncation of the computational domain to a half-plane.

2.1 Exact ABC

We first study the solution u in the right half-space $x \geq 0$ independently of the boundary condition satisfied at $x = 0$. We have

$$\frac{\partial^2 u}{\partial t^2} - c^2 \left(\frac{\partial^2 u}{\partial x^2} + \frac{\partial^2 u}{\partial y^2} \right) = 0, \quad x \geq 0, \quad y \in \mathbb{R}, \quad t > 0$$
$$u(0, x, y) = 0, \quad x \geq 0, \quad y \in \mathbb{R}$$
$$\frac{\partial u}{\partial t}(0, x, y) = 0, \quad x \geq 0, \quad y \in \mathbb{R}.$$

In order to use the Fourier transform in time, we first prolongate u by zero for negative times. Since the initial conditions are zero for $x \geq 0$, the prolongated function $\tilde{u}$ still satisfies the homogeneous wave equation

$$\frac{\partial^2 \tilde{u}}{\partial t^2} - c^2 \left(\frac{\partial^2 \tilde{u}}{\partial x^2} + \frac{\partial^2 \tilde{u}}{\partial y^2} \right) = 0, \quad x > 0, \quad t, y \in \mathbb{R}.$$

We take the Fourier transform in t and y (dual variables are ω and k, respectively) of this equation

$$-\omega^2 \hat{\tilde{u}} - c^2 \frac{\partial^2 \hat{\tilde{u}}}{\partial x^2} + c^2 k^2 \hat{\tilde{u}} = 0.$$

For a given (ω, k), it is an ordinary differential equation whose solution may be sought in the form $\hat{\tilde{u}}(\omega, x, k) = \sum_j \alpha_j(\omega, k) exp(\lambda_j(\omega, k)x)$. We have

$$\lambda^2 = k^2 - \frac{\omega^2}{c^2}.$$

We have two possibilities:

- *evanescent waves*, $|k| > \frac{\omega}{c}$: $\lambda^{\pm} = \pm\sqrt{k^2 - \frac{\omega^2}{c^2}}$
- *propagative waves*, $|k| < \frac{\omega}{c}$: $\lambda^{\pm} = \pm i \frac{\omega}{c}\sqrt{1 - \frac{c^2 k^2}{\omega^2}}$ $(i^2 = -1)$

By using inverse Fourier transform, the general form for $\tilde{u}$ is

$$\tilde{u}(t, x, y) = \int \int_{|k| > \frac{\omega}{c}} \alpha_-(\omega, k) e^{-\sqrt{k^2 - \frac{\omega^2}{c^2}}x} e^{i(\omega t + ky)} dk\, d\omega \tag{6.2}$$

$$+ \int \int_{|k| < \frac{\omega}{c}} \alpha_-(\omega, k) e^{i\left(\omega t - \frac{\omega}{c}\sqrt{1 - \frac{c^2 k^2}{\omega^2}}x + ky\right)} dk\, d\omega \tag{6.3}$$

$$+ \int \int_{|k| > \frac{\omega}{c}} \alpha_+(\omega, k) e^{\sqrt{k^2 - \frac{\omega^2}{c^2}}x} e^{i(\omega t + ky)} dk\, d\omega \tag{6.4}$$

$$+ \int \int_{|k| < \frac{\omega}{c}} \alpha_+(\omega, k) e^{i\left(\omega t + \frac{\omega}{c}\sqrt{1 - \frac{c^2 k^2}{\omega^2}}x + ky\right)} dk\, d\omega. \tag{6.5}$$

Next, we show that $\alpha_+(\omega, k) \equiv 0$ for all (ω, k). In (6.4) indeed, the term $e^{\sqrt{k^2-\frac{\omega^2}{c^2}}x}$ tends to infinity as x tends to infinity. Since the solution $\tilde{u}$ is bounded, we must have $\alpha_+(\omega, k) = 0$ for $|k| > \frac{\omega}{c}$. Moreover, the integral (6.5) corresponds to a linear combination of waves propagating in the direction of negative x. It is impossible that waves originate from infinity, only the boundary Γ creates waves. As a result, we have $\alpha_+(\omega, k) = 0$ for $|k| < \frac{\omega}{c}$. Thus, we have

$$
\begin{aligned}
\tilde{u}(t,x,y) = &\int \int_{|k|>\frac{\omega}{c}} \alpha_-(\omega,k) e^{-\sqrt{k^2-\frac{\omega^2}{c^2}}x} e^{i(\omega t+ky)} dk\, d\omega \\
&+ \int \int_{|k|<\frac{\omega}{c}} \alpha_-(\omega,k) e^{i\left(\omega t - \frac{\omega}{c}\sqrt{1-\frac{c^2k^2}{\omega^2}}x + ky\right)} dk\, d\omega.
\end{aligned} \tag{6.6}
$$

By taking $x = 0$ in this expression, we see that $\alpha_-(\omega, k) = \hat{\tilde{u}}(\omega, 0, k)$.

By differentiating (6.6) with respect to x, we have thus at $x = 0$

$$
\frac{\partial \tilde{u}}{\partial x}(t,0,y) - \int\int \hat{\tilde{u}}(\omega,0,k)\lambda^-(\omega,k)e^{i(\omega t+ky)} dk\, d\omega = 0. \tag{6.7}
$$

where

$$
\lambda^-(\omega,k) = \begin{cases} -\sqrt{k^2 - \dfrac{\omega^2}{c^2}} & \text{for } |k| > \frac{\omega}{c} \\ -i\dfrac{\omega}{c}\sqrt{1 - \dfrac{c^2k^2}{\omega^2}} & \text{for } |k| < \frac{\omega}{c}. \end{cases}
$$

We assume that relation (6.7) can be used as a boundary condition. It is then an exact ABC since solution u to problem (6.1) has a unique solution and satisfies (6.7) as well.

2.2 Approximation of the exact ABC

The exact ABC (6.7) involves inverse and direct Fourier transforms in both time t and y through the multiplication by a nonpolynomial function in the Fourier variables ω and k. Due to the square root in the formula for λ^-, the inverse Fourier transform of (6.7) is not a differential equation. Its implementation would thus be costly. Moreover, for a varying velocity field $c(x, y)$ or a curved boundary, the previous computation is not valid and the analytic form for the exact absorbing boundary condition is not known. For these reasons, integrodifferential equation (6.7) is approximated by a partial differential equation. They are easy to implement and have a straightforward extension to variable coefficients and/or curved boundaries.

In the Fourier space, (6.7) reads

$$
\frac{\partial \hat{\tilde{u}}}{\partial x}(\omega,0,k) - \lambda^-(\omega,k)\hat{\tilde{u}}(\omega,0,k) = 0. \tag{6.8}
$$

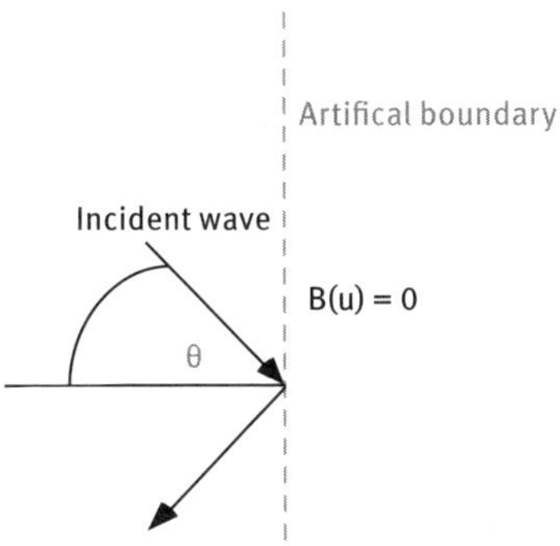

Figure 6.2: Incident wave with an angle θ with respect to the normal to the artificial boundary.

For reasons that will be given later (see Section 3), small values of $c\frac{k}{\omega}$ correspond to waves whose propagation direction is close to the normal to the artificial boundary, see Figure 6.2. We approximate (6.8) by a Taylor expansion in $c\frac{k}{\omega}$ of order zero and centered at zero:

$$\frac{\partial \hat{\tilde{u}}}{\partial x}(\omega,0,k) + i\frac{\omega}{c}\sqrt{1-\frac{c^2k^2}{\omega^2}}\hat{\tilde{u}}(\omega,0,k) = 0 \approx \frac{\partial \hat{\tilde{u}}}{\partial x}(\omega,0,k) + i\frac{\omega}{c}\hat{\tilde{u}}(\omega,0,k) = 0.$$

The inverse Fourier transform yields a zero-order approximation to the exact ABC (6.7):

$$\frac{\partial u}{\partial t}(t,0,y) + c\frac{\partial u}{\partial n}(t,0,y) = 0 \quad \text{(zero-order ABC)} \tag{6.9}$$

where we replaced x by the outward normal to the boundary n in order to have a more intrinsic form. This is a partial differential equation that enables the dropping of the superscript $\tilde{}$. It is easy to check that (6.9) is an exact ABC for the one dimensional constant coefficient wave equation.

A higher-order ABC is obtained with a Taylor expansion of order 2 ($\sqrt{1-\epsilon^2} \approx 1-\frac{\epsilon^2}{2}$):

$$\frac{\partial \hat{\tilde{u}}}{\partial x}(\omega,0,k) + i\frac{\omega}{c}\sqrt{1-\frac{c^2k^2}{\omega^2}}\hat{\tilde{u}}(\omega,0,k) = 0 \tag{6.10}$$

$$\approx \frac{\partial \hat{\tilde{u}}}{\partial x}(\omega,0,k) + i\frac{\omega}{c}\hat{\tilde{u}}(\omega,0,k) + c\frac{k^2}{2i\omega}\hat{\tilde{u}}(\omega,0,k) = 0. \tag{6.11}$$

The term $c\frac{k^2}{2i\omega}$ is not polynomial in ω and thus does not correspond (under inverse Fourier transform) to a partial differential equation on the artificial boundary. In order to fix this difficulty, we have two possibilities. The first one consists simply in multiplying relation (6.11) by $i\omega$ before taking an inverse Fourier transform:

$$\frac{1}{c}\frac{\partial^2 u}{\partial t \partial n} + \frac{1}{c^2}\frac{\partial^2 u}{\partial t^2} - \frac{1}{2}\frac{\partial^2 u}{\partial \tau^2} = 0 \quad \text{(Order 2)} \tag{6.12}$$

where τ is the tangential direction on the artificial boundary. The second possibility is to introduce an auxiliary function $\hat{\phi} = c\frac{k^2}{2i\omega}\hat{\hat{u}}(\omega, 0, k)$ or equivalently, $i\omega\hat{\phi} = c\frac{k^2}{2}\hat{\hat{u}}(\omega, 0, k)$. We take the inverse Fourier transform and get

$$\begin{cases} \frac{1}{c}\frac{\partial u}{\partial t} + \frac{\partial u}{\partial n} + \phi = 0 \\ -\frac{1}{2}\frac{\partial^2 u}{\partial \tau^2} = \frac{1}{c}\frac{\partial \phi}{\partial t}. \end{cases} \quad \text{(Order 2)} \tag{6.13}$$

For $\phi = 0$, this is equivalent to condition (6.9). Condition (6.13) is thus an incremental modification of (6.9) and is easier to implement than (6.12), see [12] for more details.

In order to get higher-order ABCs a natural idea consists in using Taylor expansions of higher and higher orders. For instance, one could use a fourth-order expansion $\sqrt{1-\epsilon^2} \approx 1 - \frac{\epsilon^2}{2} - \frac{\epsilon^4}{8}$. The corresponding ABC would read in Fourier space:

$$\frac{\partial \hat{\hat{u}}}{\partial x}(\omega, 0, k) + i\frac{\omega}{c}\hat{\hat{u}}(\omega, 0, k) + c\frac{k^2}{2i\omega}\hat{\hat{u}}(\omega, 0, k) + \frac{c^3k^4}{8i\omega^3}\hat{\hat{u}}(\omega, 0, k) = 0. \tag{6.14}$$

In fact, the boundary value problem for the wave equation with the inverse Fourier transform of this boundary condition is not well posed. It does not satisfy the Kreiss condition [13] or the Shapiro–Lopatinski condition. In order to formulate stable high-order ABCs it was proposed in [7] to use Padé approximations:

$$\sqrt{1-\epsilon^2} \approx \frac{1 - \frac{3\epsilon^2}{4}}{1 - \frac{\epsilon^2}{4}},$$

see [12] as well for more details.

3 Plane waves analysis of an ABC

The analysis consists in computing for the various interface conditions the reflection of a plane wave as a function of its angle of incidence with respect to the normal to the interface. This gives an intuitive insight into the behavior of the ABCs we have introduced in the previous section. It enables us to quantify the superiority of the ABC of order 2 (6.12) over the zeroth order one (6.9). It is also justified by the fact that both the incident wave and the scattered one can be decomposed into a linear combination of plane waves as in formula (6.2)–(6.5). The same kind of analysis will be performed in Section 5 for the PML method. The incident wave is written as follows:

$$u^I = e^{-i\left(\omega t - \frac{\omega}{c}\sqrt{1-\frac{c^2k^2}{\omega^2}}x + ky\right)}.$$

The angle of incidence θ with respect to the normal of the artificial boundary is such that

$$\tan\theta = \frac{\frac{ck}{\omega}}{\sqrt{1-\frac{c^2k^2}{\omega^2}}} \quad \text{and} \quad \cos\theta = \sqrt{1-\frac{c^2k^2}{\omega^2}}.$$

The reflected wave reads

$$u^R = R(\theta)e^{-i\left(\omega t + \frac{\omega}{c}\sqrt{1-\frac{c^2k^2}{\omega^2}}x + ky\right)}.$$

Function $u^I + R(\theta)u^R$ solves the homogeneous wave equation. The reflection coefficient $R(\theta)$ is computed by inserting the expression of $u^I + R(\theta)u^R$ in condition (6.9) or (6.12).

For the zeroth-order ABC (6.9), we have

$$R_0(\theta) = \frac{\cos\theta - 1}{\cos\theta + 1}; \quad R < 10^{-2} \text{ for } |\theta| \le 11.45^\circ$$

and for the second-order ABC (6.12):

$$R_2(\theta) = -\left(\frac{\cos\theta - 1}{\cos\theta + 1}\right)^2; \quad R < 10^{-2} \text{ for } |\theta| \le 35^\circ .$$

The reflection coefficient for the first-order ABC is smaller than one percent for a cone of angle 11.45° whereas the reflection of the second order reaches this magnitude for a larger cone of angle 35°. The latter ABC is sometimes referred to as a wide angle ABC.

4 Perfectly matched layers

Historically, the first idea was to border the computational domain by a dissipative zone where the waves are damped. The dissipative equations were imposed a priori. The problem with this approach lies in the interface between the physical zone and the "sponge" layer. It leads to reflection. This matching problem was first solved by Berenger in 1994 [5]. He introduced perfectly matched layers (PML) for use with Maxwell's equations, and since that time there have been several related reformulations of PML for both Maxwell's equations and for other wave equations. Berenger's original formulation is based on a splitting of the electromagnetic fields into two unphysical fields in the PML region. In [6] PMLs were shown to correspond to a coordinate transformation in which the coordinate normal to the artificial boundary is mapped to complex numbers. This is actually an analytic continuation of the wave equation into complex coordinates, replacing propagating (oscillating) waves by exponentially decaying waves. The idea also consists in imposing the behavior of the solution in the PML and then retrieve the corresponding equations and interface conditions. In this way, it is possible to design a reflectionless PML. Thus the key property of a PML that distinguishes it from an ordinary absorbing material is that it is designed so that waves incident upon the PML from a non-PML medium do not reflect at the interface. This property allows the PML to strongly absorb outgoing waves from the interior of a computational region without reflecting them back into the interior.

4.1 Helmholtz equation

Let $f : \mathbb{R}^2 \to \mathbb{C}$ be a function with a compact support in the half-space of negative x, $c > 0$ the speed of sound and ω a nonzero real number. We start with a Helmholtz problem defined in the whole space with f as the right-hand side:

$$(-\omega^2 - c^2\Delta)(u) = f \quad \text{in } \mathbb{R}^2$$

with Sommerfeld radiation condition at infinity. We consider a PML that will be located in the positive x half-space. Let u_0 be the solution at $x = 0$, $u_0(y) := u(0, y)$. We know from formula (6.6) that the solution to the Helmholtz equation in the right half-space $x > 0$ has the form

$$u(x, y) = \int_{\mathbb{R}} \hat{u}_0(k) \exp\left(-i\frac{\omega}{c}\sqrt{1 - \frac{c^2k^2}{\omega^2}}\, x\right) \exp(iky)\, dk. \tag{6.15}$$

The PML must damp the propagative modes of the solution which correspond to the zone $c|k| \le |\omega|$. For this, we first consider an analytic continuation of the solution with respect to the x coordinate. The complex variable will be denoted by $z \in \mathbb{C}$. The analytic continuation is formally defined as

$$\tilde{u}(z, y) = \int_{\mathbb{R}} \hat{u}_0(k) \exp\left(-i\frac{\omega}{c}\sqrt{1 - \frac{c^2k^2}{\omega^2}}\, z\right) \exp(iky)\, dk. \tag{6.16}$$

We now make a change of coordinate with respect to the first coordinate. Let $\sigma(x)$ be a function such that $\sigma(x) \ge 0$ for $x \ge 0$. We define a path in the complex plane $s : \mathbb{R} \to \mathbb{C}$:

$$s(x) := x + \frac{c}{i\omega}\int_0^x \sigma(a)\, da \tag{6.17}$$

and the function $U : \mathbb{R}^2 \to \mathbb{C}$:

$$U(x, y) := \begin{cases} u(x, y) & \text{if } x < 0 \\ \tilde{u}(s(x), y) & \text{if } x > 0. \end{cases}$$

We first investigate the behavior of $U(x, y)$ for positive x. Let

$$\hat{u}(x, k) := \hat{u}_0(k) \exp\left(-i\frac{\omega}{c}\sqrt{1 - \frac{c^2k^2}{\omega^2}}\, x\right).$$

We have with this notation:

$$U(x, y) = \int_{\mathbb{R}} \hat{u}(x, k) \exp\left(-\sqrt{1 - \frac{c^2k^2}{\omega^2}} \int_0^x \sigma(a)\, da\right) \exp(iky)\, dk.$$

We see that U is exponentially decreasing for $c|k| < |\omega|$ as x goes to infinity. The propagative modes of u have been turned into vanishing modes. As for the vanishing modes of u ($c|k| > |\omega|$), they are still exponentially decreasing as x goes to infinity although now with a sinusoidal modulation. In summary, U contains only vanishing modes as $x \to \infty$.

We now have to find out the equations satisfied by U. First of all, since $U(x, y) = u(x, y)$ for $x < 0$, we have

$$(-\omega^2 - c^2\Delta)(U) = f \quad \text{if } x < 0 . \tag{6.18}$$

We now determine the equation satisfied by u in the right half-space $x > 0$. We have

$$\frac{\partial U}{\partial x}(x, y) = s'(x)\frac{\partial \tilde{u}}{\partial z}(s(x), y) .$$

This yields

$$\frac{1}{s'(x)}\frac{\partial U}{\partial x}(x, y) = \frac{\partial \tilde{u}}{\partial z}(s(x), y) .$$

Differentiating again with respect to x, we get

$$\frac{1}{s'(x)}\frac{\partial}{\partial x}\left(\frac{1}{s'(x)}\frac{\partial U}{\partial x}\right)(x, y) = \frac{\partial^2 \tilde{u}}{\partial z^2}(s(x), y) .$$

As for the y derivatives, we have for any integer m:

$$\frac{\partial^m U}{\partial y^m}(x, y) = \frac{\partial^m \tilde{u}}{\partial y^m}(s(x), y) .$$

From formula (6.16), $\tilde{u}$ satisfies the equation

$$\left(-\omega^2 - c^2\frac{\partial^2}{\partial z^2} - c^2\frac{\partial^2}{\partial y^2}\right)(\tilde{u}) = 0 .$$

Therefore, the last three equations yield the equation for U on $x > 0$:

$$\left(-\omega^2 - c^2\frac{1}{s'(x)}\frac{\partial}{\partial x}\left(\frac{1}{s'(x)}\frac{\partial}{\partial x}\right) - c^2\frac{\partial^2}{\partial y^2}\right)(U) = 0 \quad \text{if } x > 0 . \tag{6.19}$$

We introduce the x-pml derivative:

$$\partial_x^{\text{pml}} := \frac{1}{s'(x)}\frac{\partial}{\partial x} = \frac{i\omega}{i\omega + c\sigma(x)}\frac{\partial}{\partial x} .$$

We now turn to the interface conditions at $x = 0$. Since U satisfies a second-order equation on both half-spaces, we need two matching conditions on the interface $x = 0$. By definition of U and since $s(0) = 0$, we have that U is continuous along the interface $x = 0$:

$$U(0^-, y) = U(0^+, y) . \tag{6.20}$$

We consider now the derivative with respect to x. We know that

$$\partial_x^{\text{pml}} U(0, y) = \frac{\partial \tilde{u}}{\partial z}(s(0), y).$$

By formula (6.16), we have $\frac{\partial \tilde{u}}{\partial z}(s(0), y) = \frac{\partial u}{\partial x}(0, y)$. Finally, since $s(0) = 0$, we get

$$\frac{\partial u}{\partial x}(0, y) = \partial_x^{\text{pml}} U(0, y),$$

which yields the interface condition at $x = 0$ for U

$$\partial_x U(0^-, y) = \partial_x^{\text{pml}} U(0^+, y). \tag{6.21}$$

In practice, it is necessary to consider a bounded PML zone, for instance a vertical slab $(0, \delta) \times \mathbb{R}$. On the new boundary $x = \delta$, we have to impose a boundary condition, for instance a homogeneous boundary condition. Finally, we get the following boundary value problem:

$$\begin{aligned}
&(-\omega^2 - c^2\Delta)(v) = f \quad \text{in } \mathbb{R}_- \times \mathbb{R} \\
&\left(-\omega^2 - c^2 \frac{i\omega}{i\omega + c\sigma(x)} \frac{\partial}{\partial x}\left(\frac{i\omega}{i\omega + c\sigma(x)} \frac{\partial}{\partial x}\right) - c^2 \frac{\partial^2}{\partial y^2}\right)(v) = 0, \quad 0 < x < \delta \\
&v(0^-, y) = v(0^+, y) \quad \forall y \in \mathbb{R} \\
&\partial_x(v)(0^-, y) = \frac{i\omega}{i\omega + c\sigma(x)} \frac{\partial}{\partial x}(v)(0^+, y) \quad \forall y \in \mathbb{R} \\
&v(\delta, y) = 0 \quad \forall y \in \mathbb{R}.
\end{aligned} \tag{6.22}$$

4.2 The wave equation

In order to design a PML for the wave equation, we should take the inverse Fourier transform of the PML of the Helmholtz equation (6.22). The terms in $(i\omega + c\sigma(x))^{-1}$ correspond to nondifferential operators. The way to bypass this difficulty is, similarly to what is done in (6.13), to introduce two auxiliary unknowns in the PML region: $\hat{w}(\omega, x, y)$ and $\hat{\phi}(\omega, x, y)$ defined as

$$\begin{aligned}
\hat{w}(\omega, x, y) &= \frac{i\omega}{i\omega + c\sigma(x)} \frac{\partial \hat{u}}{\partial x}(\omega, x, y) \\
\hat{\phi}(\omega, x, y) &= \frac{i\omega}{i\omega + c\sigma(x)} \frac{\partial \hat{w}}{\partial x}(\omega, x, y)
\end{aligned} \tag{6.23}$$

or equivalently

$$\begin{aligned}
(i\omega + c\sigma(x))\hat{w}(\omega, x, y) &= i\omega \frac{\partial \hat{u}}{\partial x}(\omega, x, y) \\
(i\omega + c\sigma)\hat{\phi}(\omega, x, y) &= i\omega \frac{\partial \hat{w}}{\partial x}(\omega, x, y)
\end{aligned} \tag{6.24}$$

At this point, an inverse Fourier transform in time yields partial differential equations. The time-dependent counterpart of equation (6.22) reads

$$
\begin{aligned}
\frac{\partial^2 v}{\partial t^2} - c^2(x,y)\Delta v &= f(t,x,y), && t>0, x<0, y\in\mathbb{R}\\
\frac{\partial w}{\partial t} + c\sigma(x) w &= \frac{\partial^2 v}{\partial t\partial x}, && t>0,\ 0<x<\delta, y\in\mathbb{R}\\
\frac{\partial \phi}{\partial t} + c\sigma(x)\phi &= \frac{\partial^2 w}{\partial t\partial x}, && t>0,\ 0<x<\delta, y\in\mathbb{R}\\
-\frac{1}{c^2(x,y)}\frac{\partial^2 v}{\partial t^2} + \phi - \frac{\partial^2 v}{\partial y^2} &= 0, && t>0,\ 0<x<\delta, y\in\mathbb{R}\\
v(t,0^-,y) &= v(t,0^-,y) && t>0,\ y\in\mathbb{R}\\
\frac{\partial v}{\partial x}(t,0^-,y) &= w(t,0^+,y) && t>0,\ y\in\mathbb{R}
\end{aligned}
\tag{6.25}
$$

together with initial conditions at $t = 0$: $v(0,x,y) = u_0(x,y)$, $\partial v/\partial t(0,x,y) = u_1(x,y)$ and since u_0 has compact support in $\mathbb{R}_- \times \mathbb{R}$, we can take zero initial conditions on the auxiliary unknowns in the PML region w and ϕ.

5 Computation of the reflection coefficient of a PML

The analysis consists in computing the reflection of a plane wave as a function of its angle of incidence with respect to the normal to the interface. This gives an intuitive insight into the behavior of the PML and the impact of the damping parameter σ and the thickness of the PML. It is also justified by the fact that both the incident wave and the scattered one can be decomposed into a linear combination of plane waves as in formula (6.2)–(6.5). The same kind of analysis was performed in Section 3 for the ABCs. It enables us to compare quantitatively ABCs with the PML technique.

In this section, the function σ takes a constant value still denoted by σ. As in (6.22), the Helmholtz domain is the left half-space $x < 0$. The PML is a band of width δ, $(0,\delta)\times\mathbb{R}$. We compute the reflection coefficient for a PML with a homogeneous Dirichlet at the boundary $x = \delta$ of the PML. We make a planar wave analysis by considering homogeneous equations but instead of having a radiation condition at minus infinity, we consider a planar wave "coming" from minus infinity.

Let

$$
\begin{aligned}
\lambda_H &= \frac{i\omega}{c}\sqrt{1-\frac{c^2k^2}{\omega^2}}\\
\lambda_H^{\text{pml}} &= \left(1+\frac{c\sigma}{i\omega}\right)\lambda_H
\end{aligned}
$$

It can be checked that the functions $e^{\pm\lambda_H x}$ are solutions to the homogeneous Helmholtz equations and $e^{\pm\lambda_H^{\text{pml}} x}$ are solutions to the homogeneous PML equations. We shall make use of the following formulas:

$$\begin{cases} \partial_x\left(e^{\lambda_H x}\right) = \lambda_H e^{\lambda_H x} \\ \partial_x^{\text{pml}}\left(e^{\lambda_H^{\text{pml}} x}\right) = \dfrac{i\omega}{i\omega + c\sigma}\lambda_H^{\text{pml}} e^{\lambda_H^{\text{pml}} x} = \lambda_H e^{\lambda_H^{\text{pml}} x}. \end{cases} \tag{6.26}$$

The solution u_H to the homogeneous Helmholtz equation in the left domain has the form

$$\hat{u}_H = \alpha_H e^{\lambda_H x} + \beta_H e^{-\lambda_H x}$$

and in the PML zone:

$$\hat{u}_{\text{pml}} = \alpha_{\text{pml}} e^{\lambda_H^{\text{pml}} x} + \beta_{\text{pml}} e^{-\lambda_H^{\text{pml}} x}.$$

Using (6.26), the coupling conditions (6.20) and (6.21) at the interface between the Helmholtz and the PML yield

$$\begin{cases} \alpha_H + \beta_H = \alpha_{\text{pml}} + \beta_{\text{pml}} \\ \lambda_H(\alpha_H - \beta_H) = \lambda_H(\alpha_{\text{pml}} - \beta_{\text{pml}}). \end{cases} \tag{6.27}$$

The homogeneous Dirichlet boundary condition in the PML at $x = \delta$ yields

$$\alpha_{\text{pml}} e^{\lambda_H^{\text{pml}} \delta} + \beta_{\text{pml}} e^{-\lambda_H^{\text{pml}} \delta} = 0. \tag{6.28}$$

We study the damping of the PML by considering β_H to be given. This corresponds to an ingoing wave from the physical (in space) medium and moving towards the interface between the physical medium and the PML medium. The three other quantities $(\alpha_H, \alpha_{\text{pml}}, \beta_{\text{pml}})$ are determined by the interface conditions. The reflection caused by a finite width PML is given by the ratio $R := |\alpha_H/\beta_H|$. A direct computation shows that

$$R = e^{-2\sqrt{1-\frac{c^2k^2}{\omega^2}}\sigma\delta}. \tag{6.29}$$

Using the physical meaning of $\sqrt{1 - \frac{c^2k^2}{\omega^2}}$ as the cosine of the angle θ between the ingoing wave and the normal to the boundary, R reads

$$R = (e^{-2\sigma\delta})^{\cos(\theta)}. \tag{6.30}$$

We see here the main advantage of the PML approach over the ABC approach. In order to decrease the reflection coefficient, it is sufficient to increase the width of the PML and/or to take a large damping coefficient σ whereas for the ABC approach, it is necessary to implement more and more complex absorbing boundary conditions. At this point, one could think that it is enough to take a very large damping parameter σ. Actually, it creates a numerical reflection that is proportional to $\sigma\Delta x$. A remedy consists in taking a variable damping $\sigma(x) = Cx^2/\delta$ and tune the parameters C and δ. In practice, δ is roughly a wavelength $2\pi c/\omega$ and C is 20–40. Of course, taking a smaller Δx (finer mesh) will improve the results as well since we come closer to the continuous limit.

6 Conclusion

We have designed absorbing boundary conditions and perfectly matched layers for some scalar second-order partial differential operators: the Helmholtz equation and the wave equation. The construction of the ABC is based on Fourier symbol computations of an exact ABC which is then approximated by a rational fraction. By introducing auxiliary unknowns, this approximation yields, in the physical space, a system of partial differential equations. This technique is quite general and can be applied to systems of partial differential equations, like the isotropic or anisotropic elasticity system with or without anisotropy, Oseen equations (linearized Navier–Stokes equations), Maxwell system, ... As for PML, their designs rely on a complex coordinate coordinate change that will be stable in the sense that propagative modes are turned into evanescent ones and that evanescent modes remain evanescent. For complex systems of PDEs with anisotropy, it is not always easy to find such a change of coordinates. Thus, although PMLs are usually better fitted to numerical computations than ABCs, their designs are not always feasible and are still an active area of research.

We have considered the design of absorbing boundary conditions and perfectly matched layers in a common Fourier framework and thus restricted ourselves to rectilinear boundaries for scalar equations. Actually, these concepts are also valid for curved boundaries and/or systems of partial differential equations, see for example [2, 3, 9–11, 14] and references therein.

An appealing feature of ABCs and PMLs is that these notions are building blocks for apparently unrelated applications such as domain decomposition methods [4], approximate factorizations preconditioners for Helmholtz-type problems [8] or inverse problems via time reversal techniques [1]. From a general point of view, it is because ABCs or PMLs are cheap approximations to nonlocal operators.

References

[1] Franck Assous, Marie Kray, Frédéric Nataf, and Eli Turkel, Time Reversed Absorbing Condition: Application to inverse problem, *Inverse Problems* 27 (2011).

[2] Hélène Barucq, Julien Diaz, and Véronique Duprat, Micro-differential boundary conditions modelling the absorption of acoustic waves by 2D arbitrarily-shaped convex surfaces, *Commun. Comput. Phys.* 11 (2012), 674–690.

[3] Alvin Bayliss and Eli Turkel, Radiation boundary conditions for wave-like equations, *Comm. Pure Appl. Math.* 33 (1980), 707–725.

[4] Jean-David Benamou and Bruno Després, A domain decomposition method for the Helmholtz equation and related optimal control problems, *J. Comput. Phys.* 136 (1997), 68–82.

[5] J. P. Berenger, A perfectly matched layer for the absorption of electromagnetic waves, *J. Comput. Phys.* 114 (1994), 185–200.

[6] W. C. Chew and W. H. Weedon, A 3D perfectly matched medium from modified Maxwell's equations with stretched coordinates, *IEEE Trans. Microwave Opt. Technol. Lett.* 7 (1994), 599–604.

[7] B. Engquist and A. Majda, Absorbing Boundary Conditions for the Numerical Simulation of Waves, *Math. Comp.* 31 (1977), 629–651.

[8] Björn Engquist and Lexing Ying, Sweeping preconditioner for the Helmholtz equation: moving perfectly matched layers, *Multiscale Model. Simul.* 9 (2011), 686–710.

[9] Dan Givoli, *Numerical methods for problems in infinite domains*, Studies in Applied Mechanics 33, Elsevier Scientific Publishing Co., Amsterdam, 1992.

[10] Marcus J. Grote and Joseph B. Keller, Nonreflecting boundary conditions for Maxwell's equations, *J. Comput. Phys.* 139 (1998), 327–342.

[11] Marcus J. Grote and Joseph B. Keller, Exact nonreflecting boundary condition for elastic waves, *SIAM J. Appl. Math.* 60 (2000), 803–819 (electronic).

[12] Robert L. Higdon, Absorbing boundary conditions for difference approximations to the multidimensional wave equation, *Math. Comp.* 47 (1986), 437–459.

[13] Heinz-Otto Kreiss, Initial boundary value problems for hyperbolic systems, *Comm. Pure Appl. Math.* 23 (1970), 277–298.

[14] S. V. Tsynkov, Artificial boundary conditions for the numerical simulation of unsteady acoustic waves, *J. Comput. Phys.* 189 (2003), 626–650.

Roland W. E. Potthast

Dynamic inverse scattering

Abstract: Inverse scattering theory is concerned with the reconstruction of scatterers and their physical properties from remote measurement of the scattered fields. Methods and tools from inverse scattering in acoustics and electromagnetics are employed in diverse areas such as medical imaging, nondestructive testing, geophysical exploration, or meteorology. Today, the use of inverse methods often includes a dynamic component, i.e. the waves that are used to investigate some scatterer are pulses in the time domain or the phenomena or scatterers are time dependent. The need to develop dynamic methods has led to various activity in the field over the past years, including the development of methods for object reconstruction in the time domain. The goal of this chapter is the development of cycled probing and sampling methods which are addressing dynamic scattering problems (with moving objects) using time-domain wave scattering. In particular, we investigate a time-domain probe method (TDPM) and orthogonality sampling (OS) within a cycled data assimilation framework for dynamic reconstructions. A cycled TDPM and a cycled OS method are described. Finally, we demonstrate the feasibility of the ideas by numerical results for a cycled OS method in two dimensions.

Keywords: inverse scattering, data assimilation, cycled probe methods, time-domain probe method, orthogonality sampling

2010 Mathematics Subject Classification: 34L25, 31A25, 31B20, 35R30, 45Q05, 78A46, 86A22

Roland W. E. Potthast: Deutscher Wetterdienst, Frankfurter Str. 135, 63067 Offenbach, Germany; and Department of Mathematics, University of Reading Whiteknights, PO Box 220 Berkshire, RG6 6AX, United Kingdom, r.w.e.potthast@reading.ac.uk

1 Introduction

The goal of this chapter is to study *dynamic inverse scattering problems* using techniques from the area of *sampling and probe methods* in combination with *data assim-*

Part of this research has been carried out under support of the EPSRC with a "Bridging the Gaps" award (University of Reading, UK) and supported by the Special Research Programme funded by the German Transport Ministery BMVBS, for Data Assimilation at Deutscher Wetterdienst.
The work has been carried out at two funded research stays at the Radon Institute, Linz, Austria, in Autumn 2011. The author would like to thank the organizers of the Special Semester on Multiscale Simulation and Analysis in Energy and the Environment for their invitation and support.

ilation techniques which are typically employed in the framework of prediction and forecasting of dynamical systems.

Often, inverse scattering problems are studied in the framework of time-harmonic waves, compare [3, 6, 18]. More recently, dynamic aspects have been increasingly investigated, see for example [21, 33]. The use of multi-frequency data has been pursued by Chen [5], Bao and Triki [1], see also [34], where several linear inverse problems at different frequencies are solved successively to fully recover an unknown object.

There are two main directions of study, the *first* is concerned with the dynamic nature of the waves which are used to probe the unknown scatterers. The *second* is studying the dynamic nature of the scatterer itself, its changing location, orientation and shape. Both directions of study are often referred to as *dynamic*. The investigation of dynamic scatterers is strongly related to the task of determining the current state of a dynamical system, usually called data assimilation in geophysical sciences [16, 38]. Here, the state of the dynamical system is the location, orientation, and shape of the scattering object.

Data assimilation techniques have been intensively studied in applications. This involves variational techniques [15, 16, 24, 36, 37] as well as intensive activity on ensemble Kalman filters [7, 12]. The basic idea behind data assimilation is to employ the knowledge about previous reconstructions for calculating a first guess for the current state of the system. Then, measurements are used to combine the first guess into an up-to-date reconstruction. When ill-posed observation operators are involved, this includes regularization in every reconstruction step (also denoted as *analysis* step) of the method. *Cycling* refers to the process of reconstruction of the current state and its propagation to the next time step where a new reconstruction needs to be carried out. The cycling adds knowledge, but can also increase the ill-posedness of the problem. Recently, the instability of cycled regularization for inverse problems has been investigated in [22, 23].

Sampling and probing methods have been very popular since the introduction of the linear sampling method by Colton and Kirsch 1996 and the probing methods by Ikehata and Potthast in 1998/1999. The basic idea of the schemes is to define an indicator functional, which provides knowledge about the location and shape of the object in the sense that it is large at the boundary of the unknown scatterer.

More recently, a *time-domain probe method* (TDPM) has been suggested by Burkard et al. [2]. The basic idea here is to employ causality of the time-dependent scattered field to detect the location and shape of the unknown object. It is based on the time-domain point source method of Luke et al. [21], based on ideas of Chandler–Wilde and Lines [4]. Orthogonality sampling (OS) has been first suggested in [31] with significant progress on the theory by Griesmaier [10]. The OS has also been discovered independently by Ito et al. [13, 14], usually denoted as *direct sampling method*. For a survey about sampling and probe methods we refer to [3, 11, 30].

Here, our goal is to develop an *integrated* approach, where dynamic inverse scattering methods are employed within a *data assimilation framework* for reconstructing moving objects. To this end, we will employ a temporal *scale separation* in the sense that we assume that the dynamical change of the object is slow compared to the dynamic pulses used for probing the objects. In this case, we can employ time-domain methods for object identification at discrete time intervals $I_k := [t_k, t_k + \triangle t]$, while the objects are considered to be invariant in I_k, for $k = 1, 2, 3, \ldots$. In particular, we will describe a cycled TDPM following Burkhard et al. or alternatively a cycled OS based on Potthast, Griesmaier, Ito, Zou for dynamic object reconstruction. For methods that go beyond the range of scale separation and study the full integrated problem we refer to [9, 35].

We study scattering of acoustic waves by some scatterer $D \subset \mathbb{R}^m$, $m = 2, 3$, in two or three dimensions. Here, we assume D to be bounded and its surface is Lipschitz continuous and piecewise C^2 smooth. In general, D will consist of several connected components. We assume that the exterior D^e of D is connected. Further, assume that we are given a time-dependent pulse

$$U^i(t, x, d) = \int_{-\infty}^{\infty} e^{i\kappa t} \rho(\kappa) e^{i\kappa x \cdot d} d\kappa, \quad t \in (-\infty, \infty), \ x \in \mathbb{R}^m, \tag{7.1}$$

with direction of incidence $d \in \mathbb{S}$ and $\rho \in L^2(\mathbb{R})$, i.e. U^i is a pulse which is a Fourier transform of a plane wave $e^{i\kappa x \cdot d}$, $x \in \mathbb{R}^m$ with the direction of incidence d and a frequency spectrum given by $\rho(\kappa)$. Then, the scattered field $U^s(t, x, d)$ is given by a corresponding superposition

$$U^s(t, x, d) = \int_{-\infty}^{\infty} e^{i\kappa t} \rho(\kappa) u^s(\kappa, x, d) d\kappa, \quad t \in (-\infty, \infty), \ x \in \mathbb{R}^m \setminus D, \tag{7.2}$$

where $u^s(\kappa, x, d)$ solves the time-harmonic scattering problem, compare [21]. Its far field pattern is given by

$$U^\infty(t, \hat{x}, d) = \int_{-\infty}^{\infty} e^{i\kappa t} \rho(\kappa) u^\infty(\kappa, \hat{x}, d) d\kappa, \quad t \in (-\infty, \infty), \ \hat{x} \in \mathbb{S}. \tag{7.3}$$

Our task is to solve the inverse problem where we know the time-dependent far field pattern $U^s(t, x, d)$ for $t \in [t_k, t_k + \triangle t]$ and $x \in \mathbb{S}_R$ on the boundary of some ball of radius R in every time step, where U^s is supported in this interval. Then, by Fourier transform and standard integral equation methods [6] we can retrieve $u^\infty(\kappa, \ldots, d)$ for the directions of incidence $d \in \mathbb{S}$ and all wave numbers $\kappa > 0$ in every reconstruction step.

We will provide a concise introduction into the time-domain point source method in Section 2. It is used as a key ingredient for the TDPMs described in Section 3. Orthogonality sampling is described in Section 4. Then, the methods are integrated into a *cyclic data assimilation scheme* in Section 5, where we compare our approach to standard three-dimensional variational data assimilation (3dVar). We show numerical experiments based on the cyclic OS in our last Section 6, proving the feasibility of the cycled probing and sampling method.

2 Reconstruction of time-dependent pulses by the point-source method

The point-source method as suggested in [27, 28] and [20] is based on Green's formula for field representation and an approximation of the fundamental solution (the *point-source*) by a superposition of incident plane- or spherical waves. Chandler–Wilde and Lines [4] and Luke and Potthast [21] develop a version which is capable of field reconstructions in the time domain. It has been used practically, for example, by Fazi, Nelson et al. [32] for field splitting and source separation in acoustics.

Here, we will base the TDPM on the time-domain pulse reconstruction by the point-source method. To this end here we provide a concise introduction and error estimate, where we carry out our derivation in three dimensions $m = 3$. Let D be a scatterer with Dirichlet boundary condition. Then, cf. [6], the scattered field can be represented as

$$u^s(x) = \int_{\partial D} \Phi(x,y) \frac{\partial u}{\partial \nu}(y)\, ds(y), \quad x \in \mathbb{R}^m \setminus \overline{D}. \tag{7.4}$$

The far field pattern of u^s is given by

$$u^\infty(\hat{x}) = \frac{1}{4\pi} \int_{\partial D} e^{-i\kappa \hat{x}\cdot y} \frac{\partial u}{\partial \nu}(y)\, ds(y), \quad \hat{x} \in \mathbb{S}. \tag{7.5}$$

Let us now assume that we have the approximation

$$\Phi(x,y) \approx \int_{\mathbb{S}} e^{i\kappa y\cdot\theta} g_x(\theta)\, ds(\theta), \quad y \in \overline{D}, \tag{7.6}$$

in the sense that

$$\left| \Phi(x,y) - \int_{\mathbb{S}} e^{i\kappa y\cdot\theta} g_x(\theta)\, ds(\theta) \right| \le \epsilon \tag{7.7}$$

for all $y \in \overline{D}$ and $x \in \mathbb{R}^m \setminus \overline{D}$. Then, we can use the approximation (7.6) in (7.4) and obtain

$$\begin{aligned} u^s(x) &= \int_{\partial D} \Phi(x,y) \frac{\partial u}{\partial \nu}(y)\, ds(y) \\ &\approx \int_{\partial D} \int_{\mathbb{S}} e^{i\kappa y\cdot\theta} g_x(\theta)\, ds(\theta) \frac{\partial u}{\partial \nu}(y)\, ds(y) \\ &= \int_{\mathbb{S}} \Big(\int_{\partial D} e^{i\kappa y\cdot\theta} \frac{\partial u}{\partial \nu}(y)\, ds(y) \Big) g_x(\theta)\, ds(\theta) \\ &= \int_{\mathbb{S}} u^\infty(\kappa, -\theta) g_x(\theta)\, ds(\theta) \end{aligned} \tag{7.8}$$

in the sense that

$$\left| u^s(x) - \int_{\mathbb{S}} u^\infty(\kappa, -\hat{x}) g_x(\theta)\, ds(\theta) \right| \le |\partial D|\, \epsilon \left\| \frac{\partial u}{\partial \nu} \right\|_\infty . \tag{7.9}$$

The time-dependent field U^s is a weighted superposition of u^s with respect to the wave number κ. This leads to

$$\begin{aligned} U^s(x,t) &= \int_{-\infty}^{\infty} e^{i\kappa t} \rho(\kappa) u^s(\kappa, x)\, d\kappa \\ &\approx \int_{-\infty}^{\infty} e^{i\kappa t} \rho(\kappa) \Big(\int_{\mathbb{S}} u^\infty(\kappa, -\theta) g_x(\kappa, \theta)\, ds(\theta) \Big)\, d\kappa \\ &= \int_{\mathbb{S}} \int_{-\infty}^{\infty} e^{i\kappa t} \left(\rho(\kappa) g_x(\kappa,\theta)\right)\, u^\infty(\kappa, -\theta)\, d\kappa\, ds(\theta) \\ &= \int_{\mathbb{S}} \left(G_x(\theta, \cdot) * U^\infty(\cdot, -\theta)\right)(t)\, ds(\theta), \end{aligned} \tag{7.10}$$

where $*$ denotes convolution with respect to time and the time-dependent kernel G_x is given by

$$G_x(\theta, t) := \int_{-\infty}^{\infty} e^{i\kappa t} \rho(\kappa) g_x(\kappa, \theta)\, d\kappa. \tag{7.11}$$

Here, the error estimate for (7.10) is

$$\left\| U^s(x,\cdot) - \int_{\mathbb{S}} G_x(\theta,\cdot) * U^\infty(\cdot, -\theta)\, ds(\theta) \right\|_2 \le c\, \epsilon\, \|\rho\|_2 \tag{7.12}$$

with $c = |\partial D| \cdot \sup_\kappa \|\partial u(\kappa, \cdot)/\partial \nu\|_\infty$.

The main computational effort of the point-source method is the calculation of the reconstruction kernel g_x or G_x, respectively. When the background is homogeneous, as assumed in our work here, we can use simple translations and rotations of some domain of approximation to obtain a highly efficient method for the calculation of g_x. Choose a reference domain Ω_x with $\partial D \subset \Omega_x$ and $x \notin \Omega$ and solve (7.6) on the boundary $\partial\Omega_x$ of Ω_x. If $\overline{D} \subset \Omega_x$, the approximation of $\Phi(x,\cdot)$ on $\partial\Omega_x$ leads to an approximation of $\Phi(x,\cdot)$ on ∂D, where the error estimate now involved a constant C which arises from estimating the norm of the solution of an interior Dirichlet problem in Ω_x with respect to its Dirichlet boundary values. Now, translations of Ω_x into Ω_{x+v} with a translation vector $v \in \mathbb{R}^m$ lead to kernels $\tilde{g}_x(\theta) = e^{-i\kappa v\cdot\theta} g_x(\theta)$, $\theta \in \mathbb{S}$ with the kernel g_x calculated for Ω_x. In the time domain by standard properties of the Fourier transform this translates into a translation in time. More details on the approximation procedure for the frequency domain can be found in [28] or [29].

3 Time-domain probe method (TDPM)

The basic idea of probe methods is to generate particular incident fields for which the scattered fields provide information about the location and shape of a scatterer. This is of course a very old principle, already employed by radar and tomography. But radar and tomography traditionally work in the linear regime of optics, where rays are clearly defined and where the nonlinearities of the full scattering process are negligible.

Probing methods in the frequency domain have been suggested, for example, by Ikehata and Potthast, compare the survey [30] with many citations. The basic feature of all these methods is the fact that from measurement the response $\mu(x)$ of the scatterer with respect to some particular incident field is constructed, which for example has a singularity in some point x. When x is moved towards the boundary of the scatterer, the response $\mu(x)$ tends to infinity or has some other characteristic behaviour. This can be used to locate scatterers or singularities of their boundary and to reconstruct their shape or physical properties.

More recently, Burkard et al. [2] suggested a TDPM, which is built on causality rather than singular behaviour of incident fields. The method is based on measurements of some time-dependent scattered pulse, or alternatively on measuring the scattered field or far field pattern for a wide range of wave numbers κ.

Let us study the time-dependent field U^s in $\mathbb{R}^m \setminus \overline{D}$. Then, we know that U^s is subject to the causality. Initially, i.e. for time $t = -\infty$, the scattered field U^s will be zero. Let us assume that we have some incident field $U^i(x,t)$ which is zero in x until it reaches the point at time $T(x)$. Then, it is nonzero in some interval $[T, T+s]$ with $s > 0$. The field U^s is created by an interaction of the incident field with the scatterer, i.e. we can apply Huygens' principle for a point $x \in \mathbb{R}^m \setminus \overline{D}$. This means that the interaction of the primary incident field with a point $y \in \partial D$ on the surface of the scatterer leads to a singular scattered field arising in y. This scattered field will spread and can interact with other parts of the scatterer as secondary incident field, leading to the well-known complex scattering patterns. However, the secondary scattered field will never be faster than the original incident field to reach a point x in our domain of consideration.

The TDPM is now built on the following observations. Let $T(x)$ be the *first hitting time* of the primary incident field in x.

(i) the scattered field $U^s(x,t)$ will be zero as long as x is not reached by the primary incident field $U^i(x,t)$, i.e. for $t < T(x)$.

(ii) If x is on the surface ∂D of the scatterer D, there will be a scattered field arising with $U^i(x)$. That means for every $t \in [T, T+s]$ with some $s > 0$ there will be some scattered field $|U^s(x,t)| > 0$.

(iii) If $x \notin \overline{D}$, then there is an interval $[T, T+s]$ with $s > 0$ in which the scattered field satisfies $U^s(x,t) = 0$, $t \in [T, T+s]$.

The points (i)–(iii) can be used to detect the location and shape of some object as follows:

- We follow the level curve Γ_t of first arrival times, i.e. the surface given by

$$\begin{aligned}\Gamma_t := \{x \in \mathbb{R}^m :\ & U^i(x,s) = 0\ \forall s < t,\\ & U^i(x,s) \neq 0 \text{ in an interval } [t, t+s] \text{ with some } s > 0\}.\end{aligned} \tag{7.13}$$

- For each $x \in \Gamma_t$ we reconstruct $U^s(x)$ from the measured remote scattered field or its far field pattern using the point-source method above. Here, we can in fact use Γ_t as a part of the boundary of our approximation domain.
- For each $x \in \Gamma_t$ we test whether $|U^s(x,s)| > 0$ for all s in an interval $[t, t+\epsilon]$ for some $\epsilon > 0$. If this is positive, then $x \in \partial D$. If $U^s(x,s) \equiv 0$ in $[t, t+\epsilon]$ for some $\epsilon > 0$, then $x \notin \partial D$.

The above approach can find all points $x \in \partial D$ which are not in the shadow region of the scatterer for the given incident pulse U^i, where we define the shadow region Ω_S by the condition

$$\min_s \{|U^i(x,s)| > 0\} = \min_s \left\{|U^i(x,s) + U^s(x,s)| > 0\right\}. \tag{7.14}$$

The condition (7.14) is important, since in the time domain the arrival time of U^i in the shadow does not fit any more to the time where the total field reaches a point x, so the total field remains zero, when our virtual incident field has already reached the point x under consideration.

The treatment of shadow regions is more complicated and beyond the scope of this chapter, since one has to use the total field arising from earlier time $s < t$ as new incident field. But for scatterers that have only nonshadow regions the method is straightforward, as formulated by Burkard et al. [2] in the case of rough surface scattering.

4 Orthogonality sampling

Orthogonality sampling is a method to find the location and shape of an unknown scatterer, when the scattered field $u^s(\kappa, y)$ or far-field pattern $u^\infty(\kappa, \theta)$ is given for one to many different wave numbers κ and possibly for one or several directions of incidence $d \in \mathbb{S}$.

The idea of OS is to study scalar products, i.e. *orthogonality*, of particular functions with a measured scattered field or far field pattern, for example

$$\mu(\kappa, x) := \left| \left\langle e^{-i\kappa x \cdot \theta}, u^\infty(\theta) \right\rangle_{L^2(\mathbb{S})} \right| = \left| \int_{\mathbb{S}} e^{-i\kappa x \cdot \theta} u^\infty(\kappa, \theta)\, ds(\theta) \right|, \tag{7.15}$$

or

$$\mu(x) := \int_{\kappa_1}^{\kappa_2} \left| \int_{\mathbb{S}} e^{-i\kappa x \cdot \theta} u^\infty(\kappa, \theta)\, ds(\theta) \right| d\kappa \tag{7.16}$$

As shown in [31] the indicator $\mu(\kappa, x)$ is equal to the modulus of the *reduced scattered field*

$$u^s_{\mathrm{red}}(\kappa, x) := \int_{\partial D} j_0(\kappa|x - y|) \frac{\partial u(y)}{\partial \nu}\, ds(y), \quad x \in \mathbb{R}^m \setminus \overline{D} \tag{7.17}$$

with j_0 denoting the spherical Bessel function of order zero. It is demonstrated in [31] or [10] that the functional $\mu(\kappa, x)$ defined by (7.15) and $\mu(x)$ given by (7.16) provides a lot of information about the location and shape of the unknown scatterer. In particular, [10] provides a convergence theory for small scatterers, showing that indeed the indicator function is large at the location of the scatterer. The general case remains still open, and in particular in the case of the Neumann boundary condition the behaviour of the indicator function is quite involved and not simple at all, showing a significant drop to zero at the boundary, but larger values in a neighbourhood of the unknown boundary.

The same idea has been employed recently by Ito et al. [13, 14], usually denoted as *direct sampling method*, with similar numerical results.

5 Dynamic inversion via data assimilation techniques

Now, we consider a scatterer D which depends on time, i.e. $D = D(t)$ for $t \in \mathbb{R}$. In the simplest case $D(t)$ is a fixed-shape scatterer D_0 which is moved around in $\mathbb{R}^m$, i.e. we have a time-dependent translation vector $v(t)$ and a rotation matrix $R(t)$ such that

$$D(t) = R(t)D_0 + v(t), \quad t \in \mathbb{R}. \tag{7.18}$$

We will call this setup the *moving object case*.

Now, assume we measure the scattered field for some incident acoustic wave. The wave will usually be a time-dependent pulse $U^i(x,t)$ which is reflected from $D(t)$. The oldest and best-known technique for inversion is RADAR, originally introduced for electromagnetic waves as an abbreviation of *radio detection and ranging*. Radar sends out pulses in different directions and calculates distances of a reflecting body by travel-time measurements. This technique relies on the geometrical optics approximation of the field propagation. It can find the approximate location of the scatterer, but not its shape.

For shape reconstruction we need to work with the full scattering process, i.e. we need to take into account the nonlinear interaction of the incoming wave U^i with the scattering surface $\partial D(t)$. We will assume that we can solve the corresponding scattering problem, for which techniques like FDTD, FEM, or BEM have been developed. For our numerical tests we have used a Nyström method based on second-kind boundary integral equations as used in [19, 21].

Here, we assume to measure at least a part of the cross-section of the scattered field $U^s(x,t)$, i.e. we need many receivers with synchronized recording. Here, we will assume that we can measure $U^s(x,t)$ on a subset Λ of some large ball or its far-field pattern $U^\infty(\hat{x},t)$ on a subset $\Lambda \subset \mathbb{S}$ of the unit sphere. Λ can be an open subset, but also a discrete set with measurements at some points.

Let us assume that we send a pulse at times t_k for $k \in \mathbb{N}$ and measure the reflected pulse during some interval $I_k = [t_k, t_k + \Delta t]$. We assume that the scattering process is fast compared to the change of location and rotation of the scatter, such that we separate the time-scales and consider a sequence of *static* inverse problems. Then, the task is to reconstruct $D(t_k)$ from the measurement of $U^s(y,s)$ for $y \in \Lambda$ and $s \in [t_k, t_k + \Delta t]$.

Usually, we have some knowledge about the dynamics given by the speed $v(t)$ and rotation $R(t)$ of the scatterer. We write this in the form M_k and call it the *model dynamics* or just *the model*

$$D(t_k) = M_k(D(t_{k-1})), \quad k \in \mathbb{N}, \tag{7.19}$$

which means that the model takes a scatterer $D(t_{k-1})$ at time t_{k-1} and maps it into a moved scatterer $D(t_k)$ at time t_k. Given a reconstruction $D_{k-1}^{(a)}$ of the scatterer $D(t)$ at time t_{k-1}, we can calculate a *first guess*, also called *background* (indicated by the superscript *b*)

$$D_k^{(b)} := M_k(D_{k-1}^{(a)}), \quad k \in \mathbb{N}. \tag{7.20}$$

Definition 7.1 (Dynamic Inverse Scattering Problem). The task is, given the initial guess (or background) $D_k^{(b)}$ calculated iteratively by (7.20), to generate a series of reconstructions (the analysis) $D_k^{(a)}$ of $D(t_k)$ at time t_k from the scattered field U_k^s or far-field pattern U_k^∞, respectively, of a scattered pulse U_k^i, where the index k denotes the pulse sent at time t_k.

The typical way to study the above problem in the area of data assimilation is to introduce an observation operator H, which maps the unknown state onto the measurements. Here, this is the mapping of the unknown boundary $\partial D(t)$ onto the scattered field $U^s(y, s)$, $y \in \Lambda$, $s \in [t_k, t_k + \Delta t]$, i.e.

$$H : X \to Y : \quad H(D) = U^s. \tag{7.21}$$

Here, X is the state space that contains the domains D under consideration and Y is the measurement space, which we can choose to be either $L^2(\Lambda, [t_k, t_k + \Delta t])$ or $L^2(\Lambda, \mathbb{R})$.

We have reduced the above problem to the following *data assimilation problem*, which we formulate in general form.

Definition 7.2 (Data Assimilation Problem). Let X, Y be Hilbert spaces and $H : X \to Y$ some *observation operator*. Further, assume that we have some model dynamics $M_k : X \to X$ for $k \in \mathbb{N}$ describing the dynamical evolution of a state from time t_k to t_{k+1}. Given measurements $f_k \in Y$ of some true measured data $f_k^{(\mathrm{true})} := H(\varphi_k^{(\mathrm{true})})$ for some true state $\varphi_k^{(\mathrm{true})} \in X$ and some initial guess φ_0 at time t_0, calculate an analysis $\varphi_k^{(a)} \in X$ at time t_k, which takes into account both the measurements f_ξ for $\xi = 1, \ldots, k$, the initial guess φ_0 and the dynamics M_k, $k \in \mathbb{N}$.

There is a wide range of methods to approach the above data assimilation problem, which has been developed mostly in geophysical and meteorological applications, where operational systems are providing analysis and forecast on a regular basis, p.e. every three hours. Here, we will present a dynamic inverse method that shares some features with *three-dimensional variational assimilation*, but replaces the minimization step by modern probe or sampling approaches. To fully discuss the differences and advantages of our approach, we will describe 3dVar in more detail first and then present our dynamic approach based on TDPM or OS.

5.1 Three-dimensional variational data assimilation

The basic idea behind variational data assimilation is the well-known approach to reformulate the task as a minimization problem. It can be seen as a *maximum likelihood estimator* (MAP) for Gaussian densities in a Bayesian framework [16], which is a very popular viewpoint in the Bayesian community. The use of data and the knowledge from previous analysis steps is achieved by a joint minimization of the data discrepancy $\|f_k - H(\varphi)\|$ and the background discrepancy $\|\varphi - \varphi_k^{(b)}\|$ by

$$J_k(\varphi) := \|\varphi - \varphi_k^{(b)}\|^2 + \|f_k - H(\varphi)\|^2, \quad \varphi \in X. \tag{7.22}$$

Usually weighted norms are used that incorporate the knowledge about the data error, the model error, and corrollations between the state variables under consideration. In a finite-dimensional framework, we can use invertible weight-matrices B for the background and R for the data weights and define scalar product on X by

$$\langle \varphi, \tilde{\varphi} \rangle_{B^{-1}} := \langle \varphi, B^{-1} \tilde{\varphi} \rangle, \quad \varphi, \tilde{\varphi} \in X \tag{7.23}$$

and on Y by

$$\langle \psi, \tilde{\psi} \rangle_{R^{-1}} := \langle \psi, R^{-1} \tilde{\psi} \rangle, \quad \psi, \tilde{\psi} \in Y \tag{7.24}$$

with norms $\| \cdot \|_{B^{-1}}$ included by (7.23) and $\| \cdot \|_{R^{-1}}$ induced by (7.24). In this case the 3dVar-functional is given by

$$J_k(\varphi) := \|\varphi - \varphi^{(b)}\|^2_{B^{-1}} + \|f_k - H(\varphi)\|^2_{R^{-1}}, \quad \varphi \in X. \tag{7.25}$$

Definition 7.3 (3dVar for Dynamic Inverse Scattering). Consider the data assimilation problem given by Definition 7.2. We define our starting analysis by $\varphi_0^{(a)} := \varphi_0$. Then, the 3dVar for dynamic inverse scattering consists of the following two steps, which are *cycled* for $k = 1, 2, \ldots$.

(i) Calculate a *first guess* $\varphi_k^{(b)} := M_{k-1}(\varphi_{k-1}^{(a)})$ propagating the previous analysis $\varphi_{k-1}^{(a)}$ using the model dynamics M_{k-1}.

(ii) Calculate an analysis $\varphi_k^{(a)}$ by minimizing J_k as defined in (7.22) or (7.25).

We can understand 3dVar as a *cycling of Tikhonov regularization*, where the result of the previous cycling step is used as the background information for the next cycling step. Here, 3dVar has some joint features with iteratively regularized Gauss–Newton methods for nonlinear inverse problems. Stability of cycled Tikhonov regularization and 3dVar has been investigated recently by Marx et al. [22], see also Moodey et al. [23].

The operator H is nonlinear, such that we need to solve a nonlinear optimization problem in each assimilation step. This is usually carried out by an iterative method such as a CG or gradient method, i.e. we employ an interior loop in the cycling loop of Definition 7.3.

The linearization of H for the case of obstacle scattering is given by the Fréchet derivative of the scattering problem with respect to the boundary ∂D. The derivative has been calculated for example by Kirsch [17] by variational methods or in [25, 26] using boundary integral equations. Then, under appropriate smoothness assumptions an approximation

$$H(\varphi) = H(\tilde{\varphi}) + H'(\tilde{\varphi})(\varphi - \tilde{\varphi}) + O(\|\varphi - \tilde{\varphi}\|^2) \tag{7.26}$$

is obtained. Now, an interior loop is achieved by iterative minimization of the quadratic minimization problem

$$\tilde{J}_{k,\xi}(\varphi) := \left\|\varphi - \tilde{\varphi}^{(\xi)}\right\|^2_{B^{-1}} + \left\|\left(f_k - H(\tilde{\varphi}^{(\xi)})\right) - H'(\tilde{\varphi}^{(\xi)})\left(\varphi - \tilde{\varphi}^{(\xi)}\right)\right\|^2_{R^{-1}} \tag{7.27}$$

over $\varphi \in X$ for $\xi = 1, 2, \ldots$, starting with $\tilde{\varphi}^{(0)} := \varphi^{(b)}$ and iteratively setting $\tilde{\varphi}^{(\xi)}$ as the minimizer of $\tilde{J}_{k,\xi-1}(\varphi)$.

The use of the full Fréchet derivative is useful for the case where the shape ∂D is unknown and changing in time. Here, however, we want to focus on the case of a rigid object, whose shape is known to the observer. The unknown quantities are location and rotation of the object. In this case it is possible to use easier and faster methods for reconstruction. Here, we employ the TDPM described in Section 3 or the OS of Section 4.

5.2 Cycled probing and sampling method

Many methods need some kind of first guess to be efficient. This is true for iterative schemes, for which convergence holds locally only. It is also true for probe or sampling methods, since one has to evaluate some characteristic function in a region of space. We need to have some a priori information where to search, to make the evaluation efficient. For probe methods, it is well known (c.f. [8, 30]) that some a priori knowledge about the size of the objects helps to make approximations much better and save time for the reconstruction.

When we study inverse scattering methods in the framework of dynamics, the first guess is obtained from a propagation of the analysis of the previous time-step. Its quality is mainly depending on the quality of the dynamical model M which is employed for the reconstruction. But for probe methods we do not need a very good guess. Depending on our search strategy and the particular method, we need to identify some *outer domain*, i.e. one which contains the unknown scatterer in its interior. Then, the search is carried out within the outer domain.

We now describe a cycled probing or sampling method, where the reconstruction in the analysis step is carried out by a combination of a probe method (which in its present form can detect only the visible part of the object) and a *matching scheme*, which fits a know shape into its partial reconstruction.

Definition 7.4 (Cycled Probing/Sampling Method (cycP)). Consider the data assimilation problem given by Definition 7.2 for the dynamic inverse scattering problem described in Definition 7.1. We define our starting analysis by $D_0^{(a)} := D_0$. Also, assume that the time-dependent pulse U^i has direction of incidence $d \in \mathbb{S}$. Then, the cycled probing method (cycP) for dynamic inverse scattering consists of the following steps, which are *cycled* for $k = 1, 2, \ldots$.

(i) Calculate a *first guess* $D_k^{(b)} := M_{k-1}(D_{k-1}^{(a)})$ propagating the previous analysis $D_{k-1}^{(a)}$ using the model dynamics M_{k-1}.
(ii) Using some error estimate for the dynamics determine some outer domain G around the first guess which contains the unknown scatterer.

(iii) (a) *Time-domain probing method.* Evaluate the probe functional approaching the unknown object from its visible side, i.e. we evaluate $U^s(y, T(y))$ for $y = x + hd$ for each x on the visible side of ∂G and for $h = j \cdot \Delta h$, $j = 0, 1, 2, \ldots$ until $U^s(y, T(y)) > 0$. The minimal $h = h_{min}(x)$ for which this is the case define a surface $\Gamma = \{x + h_{min}(x) \cdot d\}$ which is an approximation to the visible side of the unknown scatterer.
(b) *Orthogonality sampling.* Evaluate the orthogonality functional (7.16) on G. Then functional is large on the visible boundary of the unknown scatterer.

(iv) Calculate an analysis $D_k^{(a)}$ by matching the domain D_0 into the indicator function provided by sampling or probing. This is carried out by constructing a translation v and rotation R such that $D_k^{(a)} \subset RD_{k-1}^{(a)} + v$.

Remark 7.5. For OS, numerical experiments [31] show that there will be artifacts at focal points of subsets of the scatterer.

We need to specify the *matching algorithm* for step (iv) of Definition 7.3. Given some subset Γ of a domain $D = RD_0 + v$ with an unknown rotation R and an unknown translation v, the task is to determine D. Here, we will not discuss the uniqueness question. Of course, depending on the particular domain D and the reconstructed part Γ the task can be uniquely solvable or it can show nonuniqueness when parts Γ of ∂D could be found in different places. For the moment, we assume that the problem is uniquely solvable in the sense that the domain D_0 has the property that the visible side of D for the given $d \in \mathbb{S}$ uniquely determines D.

5.3 Partial reconstruction matching scheme

Assume that for some $k \in \mathbb{N}$ we are given some surface patch Γ and a domain D_{k-1}, where $\Gamma \subset RD_{k-1} + v$ with some orthonormal matrix R in the special orthogonal group (i.e. with determinant 1) and a vector $v \in \mathbb{R}^m$. Let us first study the two-dimensional case and set $k = 1$ for simplicity.

We first need to describe our discrepancy function, i.e. how we calculate the distance between two domains. We could, of course, start with some joint parametrization and then calculate the distance of the parametrizations in some metric space. This is, however, less useful if you are given only a subset of a surface and want to match it to a full surface. Here, we use the following measure:

$$d(\Gamma, \partial D) := 1 - \frac{1}{|\Gamma|} \int_{\Gamma} \chi(x) \, ds(x), \tag{7.28}$$

where

$$\chi(x) = \begin{cases} 0, & \text{if } x \notin \partial D, \\ 1, & \text{if } x \in \partial D. \end{cases} \tag{7.29}$$

and $|\Gamma| := \int_\Gamma ds$ is the arc length. The distance $d(\Gamma, \partial D)$ will be 0 if $\Gamma \subset \partial D$. Under uniqueness assumptions this determines D from Γ. It will be larger than 0 if $\Gamma \not\subset \partial D$. In a discrete environment we define $\chi(x)$ by checking whether ∂D intersects the pixel containing x.

An efficient way to calculate the matching is obtained by trying to move Γ such that it is a part of ∂D (instead of moving ∂D to fit Γ). A rotation and translation can be completely described by using two points $x_1 \neq x_2 \in \Gamma$ on the surface patch Γ. We now carry out a search by evaluating the discrepancy functional on a one-dimensional interval $[0, 2\pi]$ as follows.

(i) The surface ∂D_0 can be parameterized over $[0, 2\pi]$ by a function y and we can map x_1 onto each point $y(s) \in \partial D_0$ for $s \in [0, 2\pi]$ by a translation T.

(ii) Then, we calculate the intersections of the circle with radius $\rho := |x_2 - x_1|$ around $Tx_1 = y(s)$ with ∂D. There might be several solutions $x^{(\xi)}$, $\xi = 1, 2, \dots, m$.

(iii) For each solution we apply a rotation R around x_1 of the translated surface patch $T\Gamma$ by an angle which maps x_2 onto $x^{(\xi)}$. We calculate $d(RT\Gamma, \partial D)$ by (7.28) for each of the solutions $x^{(\xi)}$, $\xi = 1, \dots, m$ and then select the minimal value $d(s)$ of the discrepancy. For each minimal solution we store the point $x^{(\xi)}$.

(iv) The matching translation and rotation is found by minimizing the one-dimensional function d over $[0, 2\pi]$.

Under the condition that there is only one unique solution to the matching problem, there will be only one translation T and rotation R for which $d(RT\Gamma, \partial D) = 0$.

In $\mathbb{R}^m$ the rotation and translation are given by three points $x_1, x_2, x_3 \in \Gamma$ which are not on a line. Again, we choose a translation of x_1 onto ∂D, where now ∂D is a two-dimensional surface, such that the search will me more costly. Then, we test rotations that map x_2 and x_3 onto ∂D. A discrepancy function is calculated as described in (7.28).

We remark that the matching algorithm consists of calculating integrals over Γ for every or at least many points $x \in \partial D$. The computational effort is proportional to the product of the discretizations m_Γ and $m_{\partial D}$ which are chosen on the two boundaries. Of course, instead of calculating d on the full interval $[0, 2\pi]$, we could carry out a gradient algorithm for $d(s) = 0$ directly. But there will be many local maxima and minima depending on the particular shape of ∂D, such that a full search might be needed anyway.

In any case, the matching scheme (i)–(iv) is much less costly than solving one direct scattering problem, which is proportional to $m_{\partial D}^3$ if we use a direct inversion method for a typical boundary integral equation or $\ell \cdot m_{\partial D}^2$ for iterative methods with ℓ iteration steps employed for calculating the inverse of the boundary integral equation. Even if we use fast methods which are proportional to $\ell \cdot m_{\partial D} \log m_{\partial D}$ for ℓ iteration steps we can expect our matching to be feasible within the calculation time of one forward problem, such that in general the method is comparable to or quicker than iterative schemes.

6 Numerical examples

The task of this final section is to demonstrate the feasibility of a cycled orthogonality scheme by numerical experiments. We will show that OS is feasible for the dynamic reconstruction of their location and shape.

For the following tests, we have employed a boundary-integral approach following [19, 21] for calculating the time-dependent scattered field of an incident pulse. The scheme is based on the Fourier transform and a time-harmonic boundary integral equation following Brackhage–Werner as presented in [6] for calculating the far field pattern of a scattered time-harmonic wave in the frequency domain.

We now sent pulses at times $t = t_k$ for $k = 1, 2, 3, \dots$ toward the moving object. As described, we assume that the time-scale for scattering is much smaller than the time-scale of the object speed, such that every pulse is basically scattered from a translated and rotated fixed object. We obtain a sequence of time-dependent scattered fields U_k^s, $k = 1, 2, 3, \dots$ on some surface Λ. For our experiments, we used $\Lambda = \mathbb{S}$ and then studied the far field pattern of the fields as in (7.3). We then reconstruct the objects $D = D_j$ at time t_k, $k = 1, 2, 3, \dots$. The error for the calculation of U^s by the Nyström method within $N = 200$ discretization points is approximately 1%, which then is the *data error* for the inverse problem.

For reconstructions, we employ the *cycled orthogonality sampling scheme* as described in Definition 7.4. The sampling domain is determined from the previous step by tracking the centre of the previous reconstruction and then choosing a square large enough to contain the domain in its interior when it has some maximal anticipated speed. By this strategy, we can allow arbitrary changes of direction within the movement of the object. But of course we need to sample on an area larger than if we would also estimate and use the direction of the movement.

Results of the dynamic reconstructions are shown in Figures 7.2 and 7.3. The figures display the result of the sampling function as well as the reconstruction which is provided by domain matching as described in (7.28). We have tested the matching scheme introduced in the previous section and, also, tested the calculation of (7.28) by precomputing a *dictionary* of vectors which correspond to a location and rotation of the domain $RD_0 + v$ over some limited set of rotations and translations around some centre v_0. For each of these domains we calculate a mask over a grid of points in a rectangle

$$Q := [-a, a] \times [-b, b] \tag{7.30}$$

with $a = b = 5$. A demonstration of such a mask is shown in Figure 7.1, where the mask is 1 on points touching the boundary of the domain and 0 otherwise. Then,

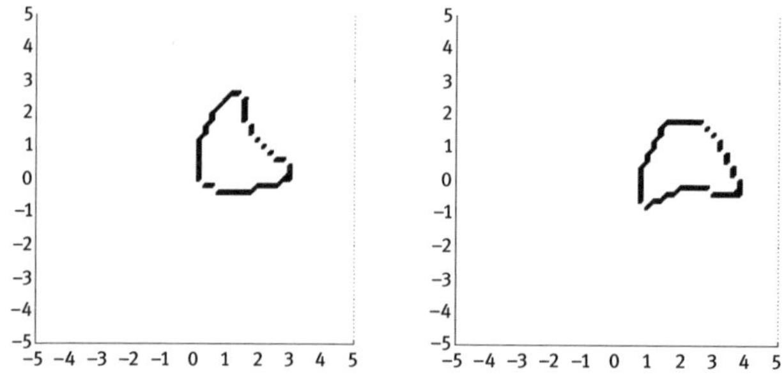

Figure 7.1: We demonstrate the mask that has been used to carry out the domain matching according to (7.28).

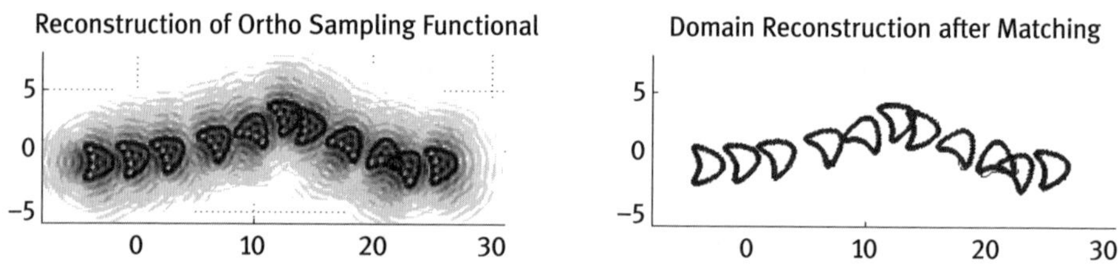

Figure 7.2: We demonstrate the cycled OS for tracking and reconstruction of a moving object, here with many pulses from different directions and $\kappa = 5$ for calculating the OS functional. Left: We plot all indicator functions in one plot. Right: Original domain with thick line, reconstruction with thin line. The feasibility of the approach is clearly demonstrated.

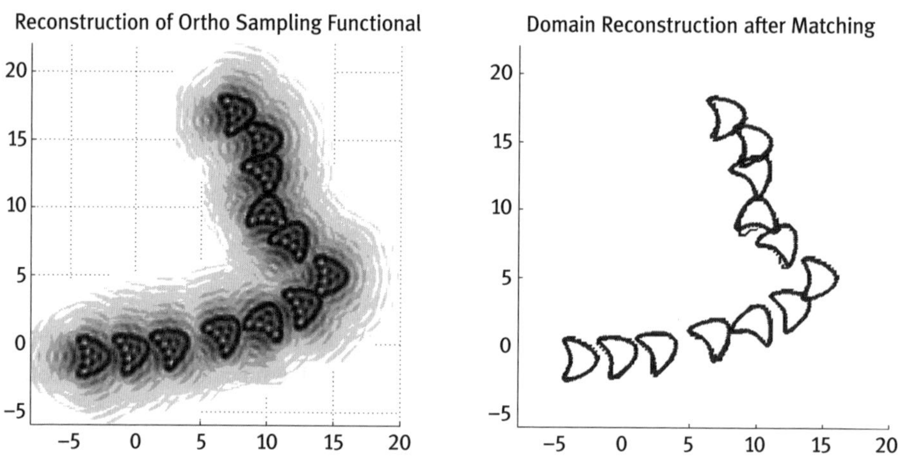

Figure 7.3: We demonstrate the cycled OS for tracking and reconstruction of a moving object, here with many pulses from different directions and $\kappa = 5$ for calculating the OS functional. Left: We plot all indicator functions in one plot. Right: Original domain with thick line, reconstruction with thin line. The feasibility of the approach is clearly demonstrated.

each image is stored as a column vector in a matrix A. A matching can now be carried out by a matrix–vector multiplication of the transpose A^T of the matrix A with the vector μ of reconstructed OS functional. The matrix–vector multiplication carries out the calculation of the metric (7.28) for all test domains stored in A. The maximal entry of the vector $\rho := A^T \circ \mu$ gives the best match.

After the matrix A is precomputed once, the matching is very efficiently carried out by $A^T \circ \mu$ for every reconstruction of the orthogonality functional on a shifted version of the rectangle Q, compare Figures 7.2 and 7.3.

For the reconstructions in Figures 7.2 and 7.3 we have used wave numbers $\kappa \in [4.5, 5]$ and 16 directions of incoming pulses. Then, the matching is very stable and provides very good reconstructions. When we use less pulses, the OS functional will give maxima only at a part of the boundary and then we partially run into nonuniqueness effects for the matching part. We demonstrate this in Figure 7.4, where the matching is partly wrong. Note that this *nonuniqueness* cannot be fully resolved, if one does allow arbitrary rotations from one observation time-step to the next.

In [31] it has been shown by a numerical study that OS has a resolution of π/κ, such that we need to use relatively high wave number for reconstructions compared to the size of D, to make the matching step work. Low wave numbers will provide a functional which has a broad maximum where the unknown domain is located. The functional for higher wave numbers is large at the boundary of the unknown scatterer (since we work with a Dirichlet boundary condition).

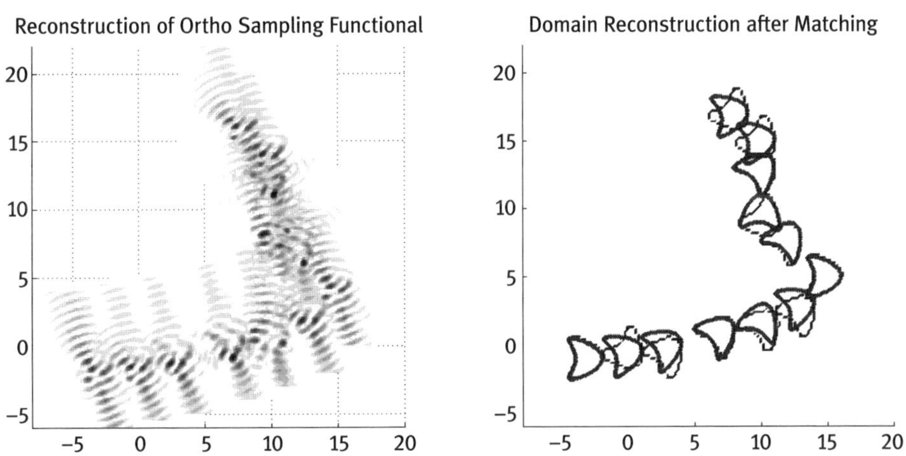

Figure 7.4: We demonstrate the cycled OS for tracking and reconstruction of a moving object, where we used just *one* incident pulse with direction $[-0.31, 0.95]$ and $\kappa = 5$ for calculating the orthogonality sampling functional. Left: We plot all indicator functions in one plot. Right: Original domain with thick line, reconstruction with thin line. Here, nonuniqueness in some cases can be clearly seen by the wrong rotation of the reconstructed object.

References

[1] Gang Bao and Faouzi Triki, Error estimates for the recursive linearization of inverse medium problems, *Journal of Computational Mathematics* 28 (2010), 725.

[2] C. Burkard and R. Potthast, A time-domain probe method for three-dimensional rough surface reconstructions, *Inverse Problems and Imaging (IPI)* 3 (2) (2009), 259 – 274.

[3] F. Cakoni and D. Colton, *Qualitative methods in inverse scattering theory*, Springer, 2006.

[4] S. N. Chandler-Wilde and C. D. Lines, A Time Domain Point Source Method for Inverse Scattering by Rough Surfaces, *Computing* 75 (2005), 157–180.

[5] Yu Chen, Inverse scattering via Heisenberg's uncertainty principle, *Inverse Problems* 13 (1997), 253.

[6] David Colton and Rainer Kress, *Inverse acoustic and electromagnetic scattering theory*, second ed, Applied Mathematical Sciences 93, Springer-Verlag, Berlin, 1998.

[7] John W. Dower, P. L. Houtekamer, and H. L. Mitchell, Data assimilation using an ensemble Kalman filter technique., *Monthly Weather Review* 126 (1998), 796–811.

[8] Klaus Erhard, *Point Source Approximation Methods in Inverse Obstacle Reconstruction Poblems*, Ph.D. thesis, University of Göttingen, 2005, p. 205.

[9] Gregory Eskin and James Ralston, The determination of moving boundaries for hyperbolic equations, *Inverse Problems* 26 (2010), 015001.

[10] Roland Griesmaier, Multi-frequency orthogonality sampling for inverse obstacle scattering problems., *Inverse Problems* 27 (2012), doi:10.1088/0266–5611/27/8/085005.

[11] N. Grinberg and A. Kirsch, The Factorization Method for Inverse Problems, *Oxford Lecture Series in Mathematics and Its Applications* 36 (2008).

[12] P. L. Houtekamer and H. L. Mitchell, A sequential ensemble Kalman Filter technique., *Monthly Weather Review* 129 (2001), 123–137.

[13] Kazufumi Ito, Bangti Jin, and Jun Zou, A Two-stage Method for Inverse Medium Scattering., *http://arxiv.org/abs/1205.4277*.

[14] Kazufumi Ito, Bangti Jin, and Jun Zou, A direct sampling method to an inverse medium scattering problem., *Inverse Problems* 28 (2012), doi:10.1088/0266–5611/28/2/025003.

[15] S. Lakshmivarahan, J. M. Lewis, and S. Dhall, *Dynamic Data Assimilation: A Least Squares Approach*, Cambridge University Press, 2006.

[16] Jari P. Kaipio and Erkki Somersalo, *Computational and Statistical Methods for Inverse Problems*, 160, Springer-Verlag, 2004.

[17] A. Kirsch, The domain derivative and two applications in inverse scattering theory, *Inverse Problems* 9 (1993), 81–96.

[18] A. Kirsch, *An introduction to the mathematical theory of inverse problems*, Applied Mathematical Sciences 120, Springer-Verlag, New York, 1996.

[19] R. Kress, *Linear Integral Equations*, Springer, Berlin, 1989.

[20] J. J. Liu, G. Nakamuray, and R. Potthast, A new approach and improved error analysis for reconstructing the scattered wave by the point source method, (2003).

[21] D. Russell Luke and Roland Potthast, The point source method for inverse scattering in the time domain, *Math. Meth. Appl. Sci.* 29, Issue 13 (2006), 1501–1521.

[22] Boris Marx and Roland W. E. Potthast, On Instabilities for Data Assimilation Algorithms, *GEM - International Journal on Geomathematics* DOI: 10.1007/s13137-012-0034-5. (2012).

[23] Alexander J. F. Moodey, Amos S. Lawless, Roland W. E. Potthast, and Peter Jan van Leeuwen, Nonlinear error dynamics for cycled data assimilation methods, *Inverse Problems* (2012).

[24] Seon K. Park and Liang Xu, *Data Assimilation for Atmospheric, Oceanic and Hydrologic Applications*, Springer, 2009.

[25] R. W. E. Potthast, Frèchet differentiability of boundary integral operators in inverse acoustic scattering, *Inverse Problems* 10 (1994), 431–447.

[26] R. W. E. Potthast, Domain derivatives in electromagnetic scattering, *Math. Methods Appl. Sci.* 19 (1996), 1157–1175.

[27] R. W. E. Potthast, A fast new method to solve inverse scattering problems, *Inverse Problems* 12 (1996), 731–742.

[28] R. W. E. Potthast, A point source method for inverse acoustic and electromagnetic obstacle scattering problems, *IMA J. Appl. Math.* 61 (1998), 119–140.

[29] R. W. E. Potthast, *Point sources and multipoles in inverse scattering theory*, Chapman & Hall/CRC Research Notes in Mathematics 427, Chapman & Hall/CRC, Boca Raton, FL, 2001.

[30] R. W. E. Potthast, A survey on sampling and probe methods for inverse problems, *Inverse Problems* 22 (2006), R1–R47.

[31] R. W. E. Potthast, A Study of Orthogonality Sampling for Object Identifcation, *Inverse Problems* (2010).

[32] R. W. E. Potthast, F. M. Fazi, and P. A. Nelson, Source splitting methods via the point source method, *Inverse Problems* (2010).

[33] A. Lechleiter Q. Chen, H. Haddar, and P. Monk, A sampling method for inverse scattering in the time domain, *Inverse Problems* 26 (2010).

[34] Mourad Sini and Nguyen Trung Thanh, Inverse acoustic obstacle scattering problems using multifrequency measurements, *Inverse Problems and Imaging* 6 (2012), 749 – 773.

[35] P. D. Stefanov, Inverse scattering problem for moving obstacles, *Math. Zeitschr.* 207 (1991), 461–480.

[36] A M Stuart, Inverse problems: a Bayesian perspective, *Acta Numerica* 19 (2010), 451–559.

[37] Thomas Tomkins Warner, *Numerical Weather and Climate Prediction*, Cambridge University Press, 2011.

[38] Richard Swinbank William Lahoz, and Boris Khattatov, *Data Assimilation: Making Sense of Observations*, Springer, 2010.

Olaf Steinbach

Boundary integral equations for Helmholtz boundary value and transmission problems

Abstract: In this chapter, we review, analyse and discuss several boundary integral formulations for a stable solution of exterior boundary value and transmission problems for the Helmholtz equation. Based on the characterisation of spurious modes, which correspond to eigensolutions of the interior Laplace equation, direct, indirect, combined and regularised boundary integral equations are considered for the solution of the exterior Dirichlet boundary value problem. In addition to established approaches such as Burton/Miller, Brakhage/Werner, or the combined Helmholtz integral equation formulation (CHIEF) method, we also include a discussion of more recent results that can also be applied in the more general case of Lipschitz domains. For a stable boundary integral formulation for the solution of transmission problems, we rely on the use of both boundary integral equations of the direct approach and on suitable linear combinations. Here we restrict ourselves to single trace formulations, including the rather standard Steklov–Poincaré operator formulation. This contribution reviews the mathematical analysis of stable boundary integral formulations for the solution of Helmholtz boundary value and transmission problems, and it will provide a foundation for the error and stability analysis of related Galerkin boundary element methods.

Keywords: Helmholtz equation, transmission problem, boundary integral equations

2010 Mathematics Subject Classification: 65N38

Olaf Steinbach: Institut für Numerische Mathematik, TU Graz, Steyrergasse 30, 8010 Graz, Austria, o.steinbach@tugraz.at

1 Introduction

Boundary integral equation methods and related boundary element methods are well-established tools for the solution of direct and inverse acoustic and electromagnetic scattering problems. There is a variety of several boundary integral formulations around, e.g. direct or indirect approaches, and first or second-kind Fredholm integral equations, just to name a few. Moreover, although exterior boundary value problems for the Helmholtz equation admit, due to the imposed radiation condition, unique solutions, a naive use of boundary integral equations may result in formulations which may not be solvable or whose solution may not be unique. These so–called spurious modes are related to eigensolutions of the, in the case of acoustics, interior Laplace equation. Stable boundary integral formulations which are robust for all wave numbers are based on complex linear combinations of the standard boundary integral equations, e.g. the direct approach of Burton and Miller [5], the indirect formulation of Brakhage and Werner [2], or the combined Helmholtz integral equation formulation of Schenck [26]. While in most cases the classical analysis, either in spaces of continuous or Hölder continuous functions, or for square integrable functions, relies on the compactness of certain boundary integral operators, sufficiently smooth surfaces have to be assumed. More recent works, see, e.g. [3, 4, 6, 10, 11], on regularised combined boundary integral equations also allow for the consideration of Lipschitz surfaces. While the focus of this contribution is on the formulation of boundary integral equations which are stable for all wave numbers, we do not analyse the behaviour of solutions in the high-frequency regime, see [6] for the related state-of-the-art. In addition to exterior boundary value problems we also consider the formulation of stable boundary integral equations for the solution of transmission problems, see, e.g. [15], by appropriate combinations of boundary integral equations for both the interior and exterior subproblem. The resulting formulations can be classified into single and multiple trace formulations [7, 24], which also include the minimal coupling formulation of Mitzner [19]. Probably more common, and similar to the Laplace case, is the Steklov–Poincaré operator formulation which is based on the solution of local Dirichlet boundary value problems to enhance the Neumann transmission condition.

In this chapter, we aim to present a unified approach to the analysis of boundary integral equations for the solution of acoustic boundary value and transmission problems. In Section 2 we recall the definition and we summarise the mapping properties of all boundary integral operators under consideration. A particular focus is on the characterisation of the kernels and their relations with eigensolutions of the interior Laplace equation. Stable boundary integral formulations for the solution of the exterior Dirichlet boundary value problem are considered in Section 3. We start with the direct approach, where the standard boundary integral equations turn out to be solvable for all wave numbers, but the solution may not be unique. If the eigensolution is known, an appropriate orthogonality condition results in a stabilised variational formulation. Otherwise one may consider the evaluation of the interior representation formula to describe the unique solution of the exterior problem, what is known as combined Helmholtz integral equation formulation (CHIEF). Next, complex linear combinations of both boundary integral equations of the direct approach are considered, which correspond to the method of Burton and Miller, originally designed for the exterior Neumann boundary value problem. Finally, we consider the boundary integral equation system of the exterior Calderon projector which turns out to be injective in the unknown Neumann datum. In fact, the symmetric representation of the exterior Steklov–Poincaré operator is well defined for all wave numbers. In addition to direct formulations we also discuss the use of indirect methods, and several complex linear combinations such as the approach of Brakhage and Werner, and regularised boundary integral equations, which are applicable also in the case of Lipschitz surfaces. In Section 4 we consider the solution of free-space transmission problems. As a model problem we consider the scattering at an interface between two media of different density. It turns out that in total four boundary integral equations and two transmission conditions can be used to determine the four unknown Cauchy data of the direct approach. The elimination of the Cauchy data of the exterior problem results in single trace formulations, while the elimination of the local Neumann data results in Steklov–Poincaré operator formulations, as known from the Laplace equation. Finally, we discuss certain combined formulations including the minimal coupling formulation of Mitzner. We do not discuss multiple trace formulations, for this we refer to [7].

This chapter may provide a foundation for the error and stability analysis of Galerkin boundary element methods and for the construction and analysis of preconditioned iterative solution procedures for an efficient solution of boundary value and transmission problems for the Helmholtz equation with moderate wave numbers. However, almost all approaches and methodologies as presented here can be carried over to the case of electromagnetic scattering problems, to the coupling with different physical fields, and to the stable coupling of finite and boundary element methods.

2 Boundary integral equations

The first part of this section is a general introduction to boundary integral equations and boundary integral operators, see also [9, 14, 18, 25, 27]. In the second part we recall the basic mapping properties of boundary integral operators in the particular case of the Helmholtz equation, see, e.g. [6, 8, 21].

2.1 Boundary integral operators

For a bounded Lipschitz domain $\Omega \subset \mathbb{R}^3$ and for a wave number $\kappa \in \mathbb{R}$ we consider either the interior Helmholtz equation

$$-\Delta u(x) - \kappa^2 u(x) = 0 \quad \text{for } x \in \Omega, \tag{8.1}$$

or the exterior Helmholtz problem

$$-\Delta u(x) - \kappa^2 u(x) = 0 \quad \text{for } x \in \Omega^c := \mathbb{R}^3 \backslash \overline{\Omega}, \tag{8.2}$$

where for the exterior problem we have to include the Sommerfeld radiation condition

$$\lim_{r \to \infty} \int\limits_{|x|=r} \left| \frac{\partial}{\partial n_x} u(x) - i\kappa u(x) \right|^2 ds_x = 0, \tag{8.3}$$

and n_x is the exterior normal vector. Recall from [8, Remark 3.4] that any solution u of the Helmholtz equation (8.2) satisfying the radiation condition (8.3) automatically satisfies

$$u(x) = \mathcal{O}\left(\frac{1}{|x|}\right) \quad \text{as } |x| \to \infty.$$

In addition we consider boundary or transmission conditions on $\Gamma = \partial\Omega$.

In the direct approach, any solution of the interior Helmholtz equation (8.1) is given by the representation formula

$$u(x) = \int_\Gamma U_\kappa^*(x,y)\frac{\partial}{\partial n_y}u(y)ds_y - \int_\Gamma \frac{\partial}{\partial n_y}U_\kappa^*(x,y)u(y)ds_y \quad \text{for } x \in \Omega, \tag{8.4}$$

where the Helmholtz fundamental solution is given by

$$U_\kappa^*(x,y) = \frac{1}{4\pi}\frac{e^{i\kappa|x-y|}}{|x-y|}.$$

In the representation formula (8.4) we have used the single-layer potential

$$(\widetilde{V}_\kappa w)(x) = \int_\Gamma U_\kappa^*(x,y)w(y)ds_y \quad \text{for } x \in \mathbb{R}^3\backslash\Gamma, \quad \widetilde{V}_\kappa : H^{-1/2}(\Gamma) \to H^1(\Omega),$$

and the double-layer potential

$$(W_\kappa v)(x) = \int_\Gamma \frac{\partial}{\partial n_y}U_\kappa^*(x,y)v(y)ds_y \quad \text{for } x \in \mathbb{R}^3\backslash\Gamma, \quad W_\kappa : H^{1/2}(\Gamma) \to H^1(\Omega).$$

By definition, both the single- and double-layer potentials are solutions of the interior and exterior Helmholtz equation, i.e. for $x \in \mathbb{R}^3\backslash\Gamma$ we have

$$\Delta(\widetilde{V}_\kappa w)(x) + \kappa^2(\widetilde{V}_\kappa w)(x) = 0, \quad \Delta(W_\kappa v)(x) + \kappa^2(W_\kappa v)(x) = 0.$$

For the interior and exterior Dirichlet traces of the single-layer potential $\widetilde{V}_\kappa w$ we find

$$\gamma_0^{\text{int}}(\widetilde{V}_\kappa w)(x) = \gamma_0^{\text{ext}}(\widetilde{V}_\kappa w)(x) = \int_\Gamma U_\kappa^*(x,y)w(y)ds_y =: (V_\kappa w)(x) \quad \text{for } x \in \Gamma,$$

where $V_\kappa := \gamma_0\widetilde{V}_\kappa : H^{-1/2}(\Gamma) \to H^{1/2}(\Gamma)$ is the single-layer boundary integral operator. For the interior and exterior Neumann traces of the single-layer potential $\widetilde{V}_\kappa w$ we obtain, in the sense of $H^{-1/2}(\Gamma)$, i.e. for all $v \in H^{1/2}(\Gamma)$,

$$\left\langle \gamma_1^{\text{int}}\widetilde{V}_\kappa w, v\right\rangle_\Gamma = \left\langle \frac{1}{2}w + K_\kappa' w, v\right\rangle_\Gamma, \quad \left\langle \gamma_1^{\text{ext}}\widetilde{V}_\kappa w, v\right\rangle_\Gamma = \left\langle -\frac{1}{2}w + K_\kappa' w, v\right\rangle_\Gamma. \tag{8.5}$$

Here, $K'_\kappa : H^{-1/2}(\Gamma) \to H^{-1/2}(\Gamma)$ is the adjoint double-layer boundary integral operator

$$(K'_\kappa w)(x) = \int_\Gamma \frac{\partial}{\partial n_x} U^*_\kappa(x,y)w(y)ds_y \quad \text{for } x \in \Gamma.$$

In particular, from (8.5) we conclude the jump relation of the normal derivative of the single-layer potential $\widetilde{V}_\kappa w$, i.e.

$$\left[\gamma_1 \widetilde{V}_\kappa w\right]_{|\Gamma} := \gamma_1^{\text{ext}} \widetilde{V}_\kappa w - \gamma_1^{\text{int}} \widetilde{V}_\kappa w = -w \quad \text{in } H^{-1/2}(\Gamma). \tag{8.6}$$

For the interior and exterior Dirichlet traces of the double-layer potential $W_\kappa v$ we find for almost all $x \in \Gamma$

$$\gamma_0^{\text{int}}(W_\kappa v)(x) = -\frac{1}{2}v(x) + (K_\kappa v)(x), \quad \gamma_0^{\text{ext}}(W_\kappa v)(x) = \frac{1}{2}v(x) + (K_\kappa v)(x), \tag{8.7}$$

where $K_\kappa : H^{1/2}(\Gamma) \to H^{1/2}(\Gamma)$ is the double-layer boundary integral operator

$$(K_\kappa v)(x) = \int_\Gamma \frac{\partial}{\partial n_y} U^*_\kappa(x,y)v(y)ds_y \quad \text{for } x \in \Gamma.$$

From (8.7) we now conclude the jump relation of the double-layer potential $W_\kappa v$, i.e.

$$[\gamma_0 W_\kappa v]_{|\Gamma} := \gamma_0^{\text{ext}}(W_\kappa v)(x) - \gamma_0^{\text{int}}(W_\kappa v)(x) = v(x) \quad \text{for } x \in \Gamma.$$

Finally, the interior and exterior Neumann traces of the double-layer potential $W_\kappa v$ define the hypersingular boundary integral operator $D_\kappa : H^{1/2}(\Gamma) \to H^{-1/2}(\Gamma)$,

$$(D_\kappa v)(x) = -\frac{\partial}{\partial n_x} \int_\Gamma \frac{\partial}{\partial n_y} U^*_\kappa(x,y)v(y)ds_y \quad \text{for } x \in \Gamma.$$

By using integration by parts we obtain a representation of the hypersingular boundary integral operator D_κ which involves tangential rather than normal derivatives [17, 19], i.e. for $x \in \Gamma$ we have

$$\begin{aligned}(D_\kappa v)(x) = -\int_\Gamma \Big[&(n_x \times \nabla_x U^*_\kappa(x,y)) \cdot (n_y \times \nabla v(y)) \\ &+\kappa^2 (n_x \cdot n_y) U^*_\kappa(x,y) v(y)\Big] ds_y.\end{aligned} \tag{8.8}$$

Note that $n_y \times \nabla v(y)$, $y \in \Gamma$, describes the surface curl of a function given on the boundary. By considering the weak form of the relation (8.8), and using once again integration by parts, the bilinear form of the hypersingular boundary integral operator D_κ can be written as [20]

$$\begin{aligned}\langle D_\kappa u, v\rangle_\Gamma = {}& \frac{1}{4\pi}\int_\Gamma\int_\Gamma \frac{e^{i\kappa|x-y|}}{|x-y|}\Big(n_y \times \nabla u(y)\Big) \cdot (n_x \times \nabla v(x))\, ds_y ds_x \\ & - \frac{\kappa^2}{4\pi}\int_\Gamma\int_\Gamma \frac{e^{i\kappa|x-y|}}{|x-y|} u(y)v(x)\,(n_y \cdot n_x)\, ds_y ds_x.\end{aligned}$$

From the representation formula (8.4) to describe solutions of the interior Helmholtz equation (8.1) we find, by considering the interior Dirichlet and Neumann traces of the single- and double-layer potentials, the system of boundary integral equations

$$\begin{pmatrix} u \\ t \end{pmatrix} = \begin{pmatrix} \frac{1}{2}I - K_\kappa & V_\kappa \\ D_\kappa & \frac{1}{2}I + K'_\kappa \end{pmatrix} \begin{pmatrix} u \\ t \end{pmatrix} =: C \begin{pmatrix} u \\ t \end{pmatrix} \quad \text{on } \Gamma. \tag{8.9}$$

Note that we have used $t(x) := \frac{\partial}{\partial n_x} u(x)$, $x \in \Gamma$. From the projection property $C = C^2$ of the interior Calderon operator C as defined in (8.9) we find, as in the case of the Laplace equation [27], the well-known relations [5]

$$K_\kappa V_\kappa = V_\kappa K'_\kappa, \quad V_\kappa D_\kappa = \frac{1}{4}I - K_\kappa^2. \tag{8.10}$$

2.2 Coercivity of boundary integral operators

The mapping properties of all boundary integral operators as introduced above are well established, see, e.g. [8, 9, 14, 18, 21, 25, 27]. Recall that we consider the general case when $\Gamma = \partial\Omega$ is a Lipschitz boundary.

Lemma 8.1. *The single-layer boundary integral operator $V_\kappa : H^{-1/2}(\Gamma) \to H^{1/2}(\Gamma)$ is coercive, i.e. there is a compact operator $C_V : H^{-1/2}(\Gamma) \to H^{1/2}(\Gamma)$ such that*

$$\langle (V_\kappa + C_V)w, w\rangle_\Gamma \geq c_1^V \|w\|^2_{H^{-1/2}(\Gamma)} \quad \textit{for all } w \in H^{-1/2}(\Gamma). \tag{8.11}$$

Proof. Let $V = V_0$ denote the Laplace single-layer boundary integral operator which is $H^{-1/2}(\Gamma)$-elliptic, see, e.g. [27, Section 6.6.1]. Moreover, $C_V := V - V_\kappa : H^{-1/2}(\Gamma) \to H^{1/2}(\Gamma)$ is compact, see, for example, [27, Section 6.9]. Now the coercivity follows from

$$\langle (V_\kappa + C_V)w, w\rangle_\Gamma = \langle Vw, w\rangle_\Gamma \geq c_1^V \|w\|^2_{H^{-1/2}(\Gamma)} \quad \text{for all } w \in H^{-1/2}(\Gamma). \qquad \square$$

For $t, \tau \in H^{-1/2}(\Gamma)$ we obtain

$$\begin{aligned} \langle V_\kappa t, \tau\rangle_\Gamma &= \int_\Gamma (V_\kappa t)(x)\overline{\tau(x)}\, ds_x \\ &= \frac{1}{4\pi}\int_\Gamma\int_\Gamma \frac{e^{i\kappa|x-y|}}{|x-y|} t(y) ds_y\, \overline{\tau(x)}\, ds_x \\ &= \frac{1}{4\pi}\int_\Gamma t(y) \overline{\int_\Gamma \frac{e^{-i\kappa|x-y|}}{|x-y|}\tau(y) ds_x}\, ds_y \\ &= \int_\Gamma t(y)\overline{(V_{-\kappa}\tau)(y)}\, ds_y = \langle t, V_{-\kappa}\tau\rangle_\Gamma. \end{aligned} \tag{8.12}$$

Moreover, for $\kappa \in \mathbb{R}$ we find that, see, e.g. [10, Lemma 3.1],

$$\Im(\langle V_\kappa \tau, \tau\rangle_\Gamma) \geq 0 \quad \text{for all } \tau \in H^{-1/2}(\Gamma).$$

Lemma 8.2. *The hypersingular boundary integral operator $D_\kappa : H^{1/2}(\Gamma) \to H^{-1/2}(\Gamma)$ is coercive, i.e. there is a compact operator $C_D : H^{1/2}(\Gamma) \to H^{-1/2}(\Gamma)$ such that*

$$\langle (D_\kappa + C_D)v, v\rangle_\Gamma \geq c_1^D \|v\|^2_{H^{1/2}(\Gamma)} \quad \textit{for all } v \in H^{1/2}(\Gamma).$$

Proof. Let $\widetilde{D} : H^{1/2}(\Gamma) \to H^{-1/2}(\Gamma)$ denote the stabilised Laplace hypersingular boundary integral operator defined as [22]

$$\langle \widetilde{D}u, v\rangle_\Gamma := \langle D_0 u, v\rangle_\Gamma + \langle u, V^{-1}1\rangle_\Gamma \overline{\langle v, V^{-1}1\rangle_\Gamma} \quad \text{for } u, v \in H^{1/2}(\Gamma). \tag{8.13}$$

Since $C_D := \widetilde{D} - D_\kappa : H^{1/2}(\Gamma) \to H^{-1/2}(\Gamma)$ is compact, the assertion follows from

$$\langle (D_\kappa + C_D)v, v\rangle_\Gamma = \langle \widetilde{D}v, v\rangle_\Gamma \geq c_1^D \|v\|^2_{H^{1/2}(\Gamma)} \quad \text{for all } v \in H^{1/2}(\Gamma).$$ □

For the statement of some coercivity results for the double-layer boundary integral operator $\frac{1}{2}I - K_\kappa$ we first recall that the inverse Laplace single-layer boundary integral operator V induces an equivalent norm $\|\cdot\|_{V^{-1}} = \sqrt{\langle V^{-1}\cdot, \cdot\rangle_\Gamma}$ in $H^{1/2}(\Gamma)$.

Lemma 8.3. *The boundary integral operator $\frac{1}{2}I - K_\kappa : H^{1/2}(\Gamma) \to H^{1/2}(\Gamma)$ is coercive, i.e. there is a compact operator $C_K : H^{1/2}(\Gamma) \to H^{1/2}(\Gamma)$ such that*

$$\left\langle \left[\left(\frac{1}{2}I - K_\kappa\right) + C_K\right] v, v\right\rangle_{V^{-1}} \geq (1 - c_K)\|v\|^2_{V^{-1}} \quad \textit{for all } v \in H^{1/2}(\Gamma). \tag{8.14}$$

Here, $c_K < 1$ is the contraction constant of $\frac{1}{2}I + K_0 : H^{1/2}(\Gamma) \to H^{1/2}(\Gamma)$.

Proof. Let $K = K_0$ denote the Laplace double-layer boundary integral operator for which we can use the contraction property [29]

$$\left\|\left(\frac{1}{2}I + K\right) v\right\|_{V^{-1}} \leq c_K \|v\|_{V^{-1}} \quad \text{for all } v \in H^{1/2}(\Gamma), \quad c_K < 1.$$

Moreover, $C_K := K_\kappa - K : H^{1/2}(\Gamma) \to H^{1/2}(\Gamma)$ is compact, and hence we obtain

$$\left\langle \left[\left(\frac{1}{2}I - K_\kappa\right) + C_K\right] v, v\right\rangle_{V^{-1}} = \left\langle \left(\frac{1}{2}I - K\right) v, v\right\rangle_{V^{-1}} \geq (1 - c_K)\|v\|^2_{V^{-1}}$$

for all $v \in H^{1/2}(\Gamma)$. □

In a similar way, we conclude that $\frac{1}{2}I - K'_\kappa : H^{-1/2}(\Gamma) \to H^{-1/2}(\Gamma)$ is coercive, i.e.

$$\left\langle \left[\left(\frac{1}{2}I - K'_\kappa\right) + C_{K'}\right] \tau, \tau\right\rangle_V = \left\langle \left(\frac{1}{2}I - K'\right) \tau, \tau\right\rangle_V \geq (1 - c_K)\|\tau\|^2_V \tag{8.15}$$

for all $\tau \in H^{-1/2}(\Gamma)$, where $\|\cdot\|_V = \sqrt{\langle V\cdot, \cdot\rangle_\Gamma}$ is an equivalent norm in $H^{-1/2}(\Gamma)$ which is induced by the Laplace single-layer boundary integral operator V. Note that $C_{K'} := K'_\kappa - K' : H^{-1/2}(\Gamma) \to H^{-1/2}(\Gamma)$ is compact. Moreover, as for the single-layer boundary integral operator V_κ, see (8.12), we conclude

$$\langle K_\kappa v, \tau\rangle = \langle v, K'_{-\kappa}\tau\rangle_\Gamma \quad \text{for all } v \in H^{1/2}(\Gamma), \tau \in H^{-1/2}(\Gamma). \tag{8.16}$$

2.3 Injectivity of boundary integral operators

Since all boundary integral operators are coercive, we can apply Fredholm's alternative to investigate the unique solvability of boundary integral equations arising in certain transmission and boundary value problems for the Helmholtz equation. It remains to consider the injectivity of the involved boundary integral operators. This is related to interior eigenvalue problems of the Laplace operator.

The next two results are direct consequences of the boundary integral equations (8.9) which hold for any solution of the interior Helmholtz equation.

Proposition 8.4. *Let $(\lambda, u_\lambda) \in \mathbb{R}\times H^1(\Omega)$, $\lambda > 0$, be a solution of the interior Dirichlet eigenvalue problem*

$$-\Delta u_\lambda(x) = \lambda u_\lambda(x) \quad \text{for } x \in \Omega, \quad u_\lambda(x) = 0 \quad \text{for } x \in \Gamma. \tag{8.17}$$

For $0 < \kappa := \sqrt{\lambda} \in \mathbb{R}$ and $t_\lambda := \dfrac{\partial}{\partial n_x} u_\lambda \in H^{-1/2}(\Gamma)$ we then conclude for $x \in \Gamma$

$$\begin{aligned} (V_\kappa t_\lambda)(x) &= (V_{-\kappa} t_\lambda)(x) = 0, \\ \left(\frac{1}{2}I - K'_\kappa\right) t_\lambda(x) &= \left(\frac{1}{2}I - K'_{-\kappa}\right) t_\lambda(x) = 0. \end{aligned} \tag{8.18}$$

In particular, the single-layer boundary integral operator V_κ and the adjoint double-layer boundary integral operator $\frac{1}{2}I - K'_\kappa$ are not injective if κ^2 corresponds to an eigenvalue λ of the interior Dirichlet eigenvalue problem (8.17).

Proposition 8.5. *Let $(\mu, u_\mu) \in \mathbb{R} \times H^1(\Omega)$, $\mu \geq 0$, be a solution of the interior Neumann eigenvalue problem*

$$-\Delta u_\mu(x) = \mu u_\mu(x) \quad \text{for } x \in \Omega, \quad t_\mu(x) := \frac{\partial}{\partial n_x} u_\mu(x) = 0 \quad \text{for } x \in \Gamma. \tag{8.19}$$

For $0 \leq \kappa := \sqrt{\mu} \in \mathbb{R}$ we then conclude for $x \in \Gamma$

$$\begin{aligned} \left(\frac{1}{2}I + K_\kappa\right) u_\mu(x) &= \left(\frac{1}{2}I + K_{-\kappa}\right) u_\mu(x) = 0, \\ (D_\kappa u_\mu)(x) &= (D_{-\kappa} u_\mu)(x) = 0. \end{aligned} \tag{8.20}$$

Hence, the hypersingular boundary integral operator D_κ and the double-layer boundary integral operator $\frac{1}{2}I + K_\kappa$ are not injective if κ^2 corresponds to an eigenvalue μ of the interior Neumann eigenvalue problem (8.19).

To characterise the null spaces of all boundary integral operators as discussed above we need to analyse the equivalence of the boundary integral operator eigenvalue problems (8.18) and (8.20) with the eigenvalue problems (8.17) and (8.19). This has to be done in several steps, first we consider the single-layer boundary integral operator V_κ.

Proposition 8.6. *Let $(\kappa, t) \in \mathbb{R} \times H^{-1/2}(\Gamma)$ be an eigensolution of the single-layer boundary integral operator eigenvalue problem $V_\kappa t = 0$. Then,*

$$\lambda := \kappa^2, \quad u_\lambda(x) := \int_\Gamma U_\kappa^*(x,y)t(y)ds_y \quad \text{for } x \in \Omega$$

defines an eigensolution of the interior Dirichlet eigenvalue problem (8.17). *Moreover,*

$$t = \frac{\partial}{\partial n_x} u_\lambda \quad \text{on } \Gamma.$$

Proof. Since u_λ is defined as a single-layer potential, by definition, u_λ is a solution of the interior Helmholtz equation, i.e.

$$-\Delta u_\lambda(x) - \kappa^2 u_\lambda(x) = 0 \quad \text{for } x \in \Omega.$$

Moreover,

$$u_\lambda(x) = \int_\Gamma U_\kappa^*(x,y)t(y)ds_y = (V_\kappa t)(x) = 0 \quad \text{for } x \in \Gamma.$$

Hence, u_λ is a solution of the interior Dirichlet eigenvalue problem (8.17).

Since the single-layer potential $u_\lambda = \widetilde{V}_\kappa t$ is also a solution of the exterior Helmholtz problem (8.2) satisfying the radiation condition (8.3), the homogeneous Dirichlet boundary condition $u_\lambda = 0$ on Γ implies $u_\lambda \equiv 0$ in Ω^c, and the jump relation (8.6) of the single-layer potential gives

$$\gamma_1^{\text{ext}} \widetilde{V}_\kappa t - \gamma_1^{\text{int}} \widetilde{V}_\kappa t = -t, \quad \text{i.e. } \frac{\partial}{\partial n_x} u_\lambda = \gamma_1^{\text{int}} \widetilde{V}_\kappa t = t \quad \text{on } \Gamma. \qquad \square$$

Remark 8.7. In addition to real eigenvalues $\kappa \in \mathbb{R}$ there also exist complex solutions $\kappa \in \mathbb{C}$ of the single-layer boundary integral operator eigenvalue problem $V_\kappa t = 0$ which correspond to the scattering poles, and where the associated single-layer potential is zero inside Ω. For an approximate solution of the single-layer boundary integral operator eigenvalue problem $V_\kappa t = 0$ by using boundary element methods, see, e.g. [28, 33].

Corollary 8.8. *Since any eigensolution $(\kappa, t) \in \mathbb{R} \times H^{-1/2}(\Gamma)$ of the single-layer boundary integral operator eigenvalue problem $V_\kappa t = 0$ induces an eigensolution $(\lambda, u_\lambda) \in \mathbb{R} \times H^1(\Omega)$ of the interior Dirichlet eigenvalue problem* (8.17) *satisfying*

$$\frac{\partial}{\partial n_x} u_\lambda = t \quad \text{on } \Gamma,$$

Proposition 8.4 *implies*

$$V_{-\kappa} t = 0, \quad \left(\frac{1}{2}I - K'_\kappa\right) t = 0, \quad \left(\frac{1}{2}I - K'_{-\kappa}\right) t = 0.$$

Proposition 8.9. *Let $t_i \in H^{-1/2}(\Gamma)$, $i = 1, 2$, be a pair of linear independent eigensolutions for an eigenvalue $\kappa \in \mathbb{R}$, in particular we have $V_\kappa t_i = 0$, and $\alpha_1 t_1 + \alpha_2 t_2 = 0$ implies $\alpha_1 = \alpha_2 = 0$. Then the associated eigensolutions*

$$u_i(x) := \int_\Gamma U_\kappa^*(x, y) t_i(y) ds_y \quad \text{for } x \in \Omega$$

of the interior Dirichlet eigenvalue problem (8.17) *are linear independent.*

Proof. Consider $u := \alpha_1 u_1 + \alpha_2 u_2$ which is a solution of the interior Dirichlet eigenvalue problem (8.17) with the Neumann datum

$$\frac{\partial}{\partial n_x} u = \alpha_1 t_1 + \alpha_2 t_2 \quad \text{on } \Gamma.$$

Interpreting the interior eigenvalue problem as a Poisson equation with the right-hand side $\kappa^2 u$, this is equivalent to the boundary integral equation

$$V[\alpha_1 t_1 + \alpha_2 t_2](x) = -\kappa^2 \int_\Omega U^*(x, y)[\alpha_1 u_1(y) + \alpha_2 u_2(y)] dy \quad \text{for } x \in \Gamma.$$

Hence, $\alpha_1 u_1 + \alpha_2 u_2 \equiv 0$ implies $V[\alpha_1 t_1 + \alpha_2 t_2] \equiv 0$, and since the Laplace single-layer boundary integral operator V is invertible, $\alpha_1 t_1 + \alpha_2 t_2 = 0$ follows. Moreover, since t_1 and t_2 are assumed to be linear independent, we conclude $\alpha_1 = \alpha_2 = 0$. □

A similar result as given in Proposition 8.6 for the single-layer boundary integral operator V_κ also holds for the hypersingular boundary integral operator D_κ.

Proposition 8.10. *Let $(\kappa, u) \in \mathbb{R} \times H^{1/2}(\Gamma)$ be an eigensolution of the hypersingular boundary integral operator eigenvalue problem $D_\kappa u = 0$. Then,*

$$\mu := \kappa^2, \quad u_\mu(x) := -\int_\Gamma \frac{\partial}{\partial n_y} U_\kappa^*(x, y) u(y) ds_y \quad \text{for } x \in \Omega$$

defines an eigensolution of the interior Neumann eigenvalue problem (8.19) *satisfying $u_\mu = u$ on Γ. Moreover, we also have*

$$D_{-\kappa} u = 0, \quad \left(\frac{1}{2} I + K_\kappa\right) u = 0, \quad \left(\frac{1}{2} I + K_{-\kappa}\right) u = 0.$$

More involved is the consideration of the adjoint double-layer boundary integral operator eigenvalue problem $(\frac{1}{2} I - K'_\kappa) t = 0$.

Lemma 8.11. *Let $(\kappa, t) \in \mathbb{R} \times H^{-1/2}(\Gamma)$ be an eigensolution of the adjoint boundary integral operator eigenvalue problem $(\frac{1}{2} I - K'_\kappa) t = 0$. Then,*

$$\lambda := \kappa^2, \quad u_\lambda(x) := \int_\Gamma U_\kappa^*(x, y) t(y) ds_y \quad \text{for } x \in \Omega$$

defines an eigensolution of the interior Dirichlet eigenvalue problem (8.17) *satisfying*

$$\frac{\partial}{\partial n_x} u_\lambda = t \quad \text{on } \Gamma.$$

Moreover, we have

$$\left(\frac{1}{2}I - K'_{-\kappa}\right) t = 0, \quad V_\kappa t = 0, \quad V_{-\kappa} t = 0.$$

Proof. By definition, u_λ is a solution of the interior Helmholtz equation. For the Dirichlet trace of u_λ we obtain

$$u_\lambda(x) = (V_\kappa t)(x) \quad \text{for } x \in \Gamma,$$

and therefore, by using (8.10),

$$\left(\frac{1}{2}I - K_\kappa\right) u_\lambda(x) = \left(\frac{1}{2}I - K_\kappa\right)(V_\kappa t)(x) = V_\kappa \left(\frac{1}{2}I - K'_\kappa\right) t(x) = 0 \quad \text{for } x \in \Gamma$$

follows. On the other hand, by considering the Neumann trace of u_λ we obtain

$$\frac{\partial}{\partial n_x} u_\lambda(x) = \left(\frac{1}{2}I + K'_\kappa\right) t(x) = t(x) - \left(\frac{1}{2}I - K'_\kappa\right) t(x) = t(x) \quad \text{for } x \in \Gamma,$$

and by using the second boundary integral equation of (8.9)

$$(D_\kappa u_\lambda)(x) = \left(\frac{1}{2}I - K'_\kappa\right) t(x) = 0 \quad \text{for } x \in \Gamma$$

follows. Since u_λ implies a solution of the Neumann eigenvalue problem (8.19), see Proposition 8.10, we further conclude, by using Proposition 8.5, that

$$\left(\frac{1}{2}I + K_\kappa\right) u_\lambda(x) = 0 \quad \text{for } x \in \Gamma,$$

and therefore

$$u_\lambda(x) = (V_\kappa t)(x) = 0 \quad \text{for } x \in \Gamma$$

follows. □

By using Proposition 8.4, Proposition 8.6, Corollary 8.8, and Lemma 8.11, we finally conclude

$$\begin{aligned} \ker V_\kappa &= \ker V_{-\kappa} = \ker\left(\frac{1}{2}I - K'_\kappa\right) = \ker\left(\frac{1}{2}I - K'_{-\kappa}\right) \\ &= \left\{\frac{\partial}{\partial n_x} u_\lambda \in H^{-1/2}(\Gamma) : -\Delta u_\lambda = \kappa^2 u_\lambda \text{ in } \Omega, u_\lambda = 0 \text{ on } \Gamma\right\}. \end{aligned} \tag{8.21}$$

A similar result as given for the adjoint double-layer boundary integral operator $\frac{1}{2}I - K'_\kappa$ in Lemma 8.11 also holds for the double-layer boundary integral operator $\frac{1}{2}I + K_\kappa$.

Lemma 8.12. *Let $(\kappa, u) \in \mathbb{R} \times H^{1/2}(\Gamma)$ be an eigensolution of the double-layer boundary integral operator eigenvalue problem $(\frac{1}{2}I + K_\kappa)u = 0$. Then,*

$$\mu := \kappa^2, \quad u_\mu(x) := -\int_\Gamma \frac{\partial}{\partial n_y} U_\kappa^*(x,y)u(y)ds_y \quad \textit{for } x \in \Omega$$

defines an eigensolution of the interior Neumann eigenvalue problem (8.19) *satisfying $u_\mu = u$ on Γ, and we have $D_\kappa u = 0$. Moreover,*

$$\begin{aligned} ker D_\kappa &= ker D_{-\kappa} = ker\left(\frac{1}{2}I + K_\kappa\right) = ker\left(\frac{1}{2}I + K_{-\kappa}\right) \\ &= \left\{u_{\mu|\Gamma} \in H^{1/2}(\Gamma) : -\Delta u_\mu = \kappa^2 u_\mu \textit{ in } \Omega, \frac{\partial}{\partial n_x} u_\mu = 0 \textit{ on } \Gamma\right\}. \end{aligned}$$

2.4 Interior Robin boundary value problem

From Propositions 8.4 and 8.5 we conclude that the solution of the interior Helmholtz equation (8.1) with either Dirichlet or Neumann boundary conditions is not unique when κ^2 corresponds to either a Dirichlet or a Neumann eigenvalue, respectively. Note that in these cases one has to assume certain solvability conditions to ensure existence of a solution. But when considering the interior Helmholtz equation (8.1) with a Robin-type boundary condition this admits a unique solution for all wave numbers.

Theorem 8.13. *Let $\Omega \subset \mathbb{R}^3$ be a bounded Lipschitz domain, and let $\kappa \in \mathbb{R}$. Let $f \in \widetilde{H}^{-1}(\Omega)$ and $g \in H^{-1/2}(\Gamma)$ be given. Then there exists a unique solution of the interior Robin boundary value problem, $\eta \in \mathbb{R}$, $\eta \neq 0$,*

$$\begin{aligned} -\Delta u(x) - \kappa^2 u(x) &= f(x) \quad \textit{for } x \in \Omega, \\ \frac{\partial}{\partial n_x} u(x) - i\eta u(x) &= g(x) \quad \textit{for } x \in \Gamma. \end{aligned} \tag{8.22}$$

Proof. The variational formulation of the Robin boundary value problem (8.22) is to find $u \in H^1(\Omega)$ such that

$$\begin{aligned} &\int_\Omega \nabla u(x) \cdot \nabla v(x) dx - \kappa^2 \int_\Omega u(x)v(x)dx - i\eta \int_\Gamma u(x)v(x)ds_x \\ &= \int_\Omega f(x)v(x)dx + \int_\Gamma g(x)v(x)ds_x \end{aligned}$$

is satisfied for all $v \in H^1(\Omega)$. Since the sesquilinear form

$$a(u,v) := \int_\Omega \nabla u(x) \cdot \overline{\nabla v(x)} dx - \kappa^2 \int_\Omega u(x)\overline{v(x)}dx - i\eta \int_\Gamma u(x)\overline{v(x)}ds_x$$

is bounded for all $u, v \in H^1(\Omega)$, and since it satisfies a Gårding inequality, i.e.

$$\begin{aligned}\Re(a(v,v)) &= \int_\Omega |\nabla v(x)|^2 dx + \int_\Omega |v(x)|^2 dx - (1+\kappa^2)\int_\Omega |v(x)|^2 dx \\ &= \|v\|^2_{H^1(\Omega)} - c(v,v) \quad \text{for all } v \in H^1(\Omega)\end{aligned}$$

with a compact bilinear form $c(\cdot,\cdot)$, it remains to prove injectivity.

Let $w \in H^1(\Omega)$ be a solution of the homogeneous Robin boundary value problem

$$-\Delta w(x) - \kappa^2 w(x) = 0 \quad \text{for } x \in \Omega, \qquad \frac{\partial}{\partial n_x} w(x) - i\eta w(x) = 0 \quad \text{for } x \in \Gamma.$$

Then we have

$$\int_\Omega \nabla w(x) \cdot \nabla v(x) dx - \kappa^2 \int_\Omega w(x) v(x) dx - i\eta \int_\Gamma w(x) v(x) ds_x = 0$$

for all $v \in H^1(\Omega)$. In particular for $v = \overline{w}$ this gives

$$\int_\Omega |\nabla w(x)|^2 dx - \kappa^2 \int_\Omega |w(x)|^2 dx - i\eta \int_\Gamma |w(x)|^2 ds_x = 0,$$

and by considering the imaginary part

$$\int_\Gamma |w(x)|^2 ds_x = 0$$

follows due to $\kappa \in \mathbb{R}$. Hence we conclude $w = 0$ and $\frac{\partial}{\partial n} w = 0$ on Γ and therefore $w = 0$ in Ω follows when considering the representation formula (8.4). □

Remark 8.14. The unique solvability of the interior Robin boundary value problem (8.22) and the proof as presented are the essential tools to ensure unique solvability of combined boundary integral equations as to be discussed in the subsequent sections. Obviously, the proof of Theorem 8.13 remains true when considering complex wave numbers $\kappa \in \mathbb{C}$ when assuming $\eta > 0$ for $\Re(\kappa) \cdot \Im(\kappa) > 0$ and $\eta < 0$ for $\Re(\kappa) \cdot \Im(\kappa) < 0$. Moreover, we may replace the Robin boundary condition in (8.22) by the regularised boundary condition

$$\frac{\partial}{\partial n_x} u(x) - i\eta(Bu)(x) = g(x) \quad \text{for } x \in \Gamma$$

where $B : H^{1/2}(\Gamma) \to H^{-1/2}(\Gamma)$ is some self-adjoint and $H^{1/2}(\Gamma)$-elliptic operator such that

$$\langle Bw, w\rangle_\Gamma > 0 \quad \text{for all } w \in H^{1/2}(\Gamma).$$

Remark 8.15. While Theorem 8.13 ensures unique solvability of the interior Robin boundary value problem (8.22) it is possible to bound the energy norm of the solution u by an expression which is explicit in the wave number κ, see [12, Theorem 2.4].

2.5 Boundary integral equations for exterior boundary value problems

As for the interior Helmholtz equation (8.1) we can describe the solution of the exterior Helmholtz problem (8.2) that satisfies the radiation condition (8.3), by the representation formula

$$u(x) = -\int_\Gamma U_\kappa^*(x,y)t(y)ds_y + \int_\Gamma \frac{\partial}{\partial n_y} U_\kappa^*(x,y)u(y)ds_y \quad \text{for } x \in \Omega^c. \tag{8.23}$$

By considering the exterior Dirichlet and Neumann traces of the representation formula (8.23) we obtain the related system of boundary integral equations

$$\begin{pmatrix} u \\ t \end{pmatrix} = \begin{pmatrix} \frac{1}{2}I + K_\kappa & -V_\kappa \\ -D_\kappa & \frac{1}{2}I - K_\kappa' \end{pmatrix} \begin{pmatrix} u \\ t \end{pmatrix}. \tag{8.24}$$

In addition, by applying Green's formula with respect to the interior domain Ω, this gives

$$-\int_\Gamma U_\kappa^*(x,y)t(y)ds_y + \int_\Gamma \frac{\partial}{\partial n_y} U_\kappa^*(x,y)u(y)ds_y = 0 \quad \text{for } x \in \Omega \tag{8.25}$$

for the solution of the exterior Helmholtz equation.

Recall that the exterior Helmholtz problem (8.2) with either Dirichlet or Neumann boundary conditions admits in all cases, due to the Sommerfeld radiation condition (8.3), a unique solution.

3 Exterior Dirichlet boundary value problem

As a model problem we consider the exterior Dirichlet boundary value problem

$$-\Delta u(x) - \kappa^2 u(x) = 0 \quad \text{for } x \in \Omega^c, \quad u(x) = g(x) \quad \text{for } x \in \Gamma, \tag{8.26}$$

together with the Sommerfeld radiation condition (8.3).

In what follows we will review and discuss different direct and indirect formulations of boundary integral equations to describe solutions of the exterior Dirichlet boundary value problem (8.26).

Note that the exterior Dirichlet boundary value problem (8.26) serves as a model problem, all further considerations can be done analogously when considering either an exterior Neumann or Robin boundary value problem.

3.1 Direct boundary integral equations

In the direct approach, the solution of the exterior Dirichlet boundary value problem (8.26) is given by the representation formula (8.23), i.e.

$$u(x) = -\int_\Gamma U_\kappa^*(x,y)t(y)ds_y + \int_\Gamma \frac{\partial}{\partial n_y} U_\kappa^*(x,y)g(y)ds_y \quad \text{for } x \in \Omega^c.$$

Hence it remains to find the yet unknown Neumann datum $t \in H^{-1/2}(\Gamma)$ as the solution of one of the boundary integral equations as given in (8.24). In fact, we can use either the first-kind boundary integral equation

$$(V_\kappa t)(x) = \left(-\frac{1}{2}I + K_\kappa\right) g(x) \quad \text{for } x \in \Gamma, \tag{8.27}$$

or the second-kind boundary integral equation

$$\left(\frac{1}{2}I + K_\kappa'\right) t(x) = -(D_\kappa g)(x) \quad \text{for } x \in \Gamma. \tag{8.28}$$

To investigate the unique solvability of the boundary integral equation (8.27) we first combine the results on coercivity and injectivity of the single-layer boundary integral operator V_κ.

Corollary 8.16. *The single-layer boundary integral operator* $V_\kappa : H^{-1/2}(\Gamma) \to H^{1/2}(\Gamma)$ *is coercive and satisfies the Gårding inequality* (8.11). *Moreover, if* κ^2 *does not coincide with an eigenvalue* λ *of the interior Dirichlet eigenvalue problem* (8.17)*, then* V_κ *is also injective, and the boundary integral equation* (8.27) *admits, for any* $g \in H^{1/2}(\Gamma)$*, a unique solution* $t \in H^{-1/2}(\Gamma)$ *satisfying the variational problem*

$$\langle V_\kappa t, \tau\rangle_\Gamma = \left\langle \left(-\frac{1}{2}I + K_\kappa\right) g, \tau \right\rangle_\Gamma \quad \textit{for all } \tau \in H^{-1/2}(\Gamma).$$

Although the single-layer boundary integral operator V_κ is not injective when $\lambda = \kappa^2$ is an eigenvalue of the interior Dirichlet eigenvalue problem (8.17), the boundary integral equation (8.27) of the direct approach is solvable. Since V_κ is Fredholm, we can apply the closed range theorem, i.e. we have $\operatorname{Im} V_\kappa = (\ker V_{-\kappa})^0$ where the polar space is given as

$$(\ker V_{-\kappa})^0 := \left\{ v \in H^{1/2}(\Gamma) : \langle v, t_\lambda\rangle_\Gamma = 0 \text{ for all } t_\lambda \in \ker V_{-\kappa} \right\}.$$

By using $\ker V_\kappa = \ker\left(\frac{1}{2}I - K_\kappa'\right)$ for all $\kappa \in \mathbb{R}$, see (8.21), we obtain for the right-hand side of the boundary integral equation (8.27)

$$\left\langle \left(-\frac{1}{2}I + K_\kappa\right) g, t_\lambda \right\rangle_\Gamma = -\left\langle g, \left(\frac{1}{2}I - K_{-\kappa}'\right) t_\lambda \right\rangle_\Gamma = 0,$$

and therefore $(-\frac{1}{2}I + K_\kappa)g \in \operatorname{Im} V_\kappa$ follows. In particular, the boundary integral equation (8.27) of the direct approach is solvable, but the solution is not unique. Hence one needs to introduce some suitable constraint for scaling.

A natural choice for doing so would be to require

$$t \in H_\lambda^{-1/2}(\Gamma) := \left\{ w \in H^{-1/2}(\Gamma) : \langle Vw, t_\lambda \rangle_\Gamma = 0 \quad \text{for all } t_\lambda \in \ker V_\kappa \right\},$$

i.e. to select a solution which is orthogonal to all eigensolutions $t_\lambda \in \ker V_\kappa$, where the orthogonality in $H^{-1/2}(\Gamma)$ is realized by using an inner product which is induced by the Laplace single-layer boundary integral operator V. Then we can consider an extended variational problem to find $t \in H^{-1/2}(\Gamma)$ such that

$$\langle V_\kappa t, \tau \rangle_\Gamma + \sum_{t_{\lambda,\ell} \in \ker V_\kappa} \langle Vt, t_{\lambda,\ell} \rangle_\Gamma \overline{\langle V\tau, t_{\lambda,\ell} \rangle_\Gamma} = \left\langle \left(-\frac{1}{2}I + K_\kappa \right) g, \tau \right\rangle_\Gamma \tag{8.29}$$

is satisfied for all $\tau \in H^{-1/2}(\Gamma)$. Note that the sum is taken over a finite set of linear independent eigensolutions $t_{\lambda,\ell}$. Since the sesquilinear form of the extended variational problem (8.29) is coercive and injective, unique solvability follows.

Obviously, the extended variational problem (8.29) is only of practical interest when the eigensolutions $t_{\lambda,\ell}$ are known, which restricts the applicability of this approach.

Instead one may combine the boundary integral equation (8.27) with the interior representation formula (8.25), see [26], i.e. we consider the problem to find $t \in H^{-1/2}(\Gamma)$ such that

$$\langle V_\kappa t, \tau \rangle_\Gamma = \left\langle \left(-\frac{1}{2}I + K_\kappa \right) g, \tau \right\rangle_\Gamma \quad \text{for all } \tau \in H^{-1/2}(\Gamma), \tag{8.30}$$

and

$$\int_\Gamma U_\kappa^*(x,y) t(y) ds_y = \int_\Gamma \frac{\partial}{\partial n_y} U_\kappa^*(x,y) g(y) ds_y \quad \text{for all } x \in \Omega. \tag{8.31}$$

Lemma 8.17. *For all wave numbers $\kappa \in \mathbb{R}$ there exists a unique solution $t \in H^{-1/2}(\Gamma)$ of the boundary integral formulation* (8.30) *which also satisfies* (8.31).

Proof. Since the variational formulation (8.30) admits a unique solution when κ^2 is not an eigenvalue of the interior Dirichlet eigenvalue problem (8.17), it is sufficient to consider the remaining case only. If $\lambda = \kappa^2$ is an eigenvalue of the interior Dirichlet problem (8.17), the variational problem (8.30) is solvable, but the solution is not unique.

Let $t_i \in H^{-1/2}(\Gamma)$, $i = 1, 2$, denote two solutions of the system (8.30) and (8.31), i.e.

$$V_\kappa t_i = \left(-\frac{1}{2}I + K_\kappa \right) g \quad \text{on } \Gamma,$$

$$\int_\Gamma U_\kappa^*(x,y) t_i(y) ds_y = \int_\Gamma \frac{\partial}{\partial n_y} U_\kappa^*(x,y) g(y) ds_y \quad \text{in } \Omega.$$

For the difference $t_0 := t_1 - t_2$ we therefore conclude

$$V_\kappa t_0 = 0 \quad \text{on } \Gamma, \qquad \int_\Gamma U_\kappa^*(x,y)t_0(y)ds_y = 0 \quad \text{in } \Omega.$$

From the first equation we find

$$t_0 = \sum_{t_{\lambda,\ell}\in\ker V_\kappa} \alpha_\ell t_{\lambda,\ell}, \quad \alpha_\ell \in \mathbb{C}$$

with linearly independent eigensolutions $t_{\lambda,\ell}$, while the second equation gives

$$0 = \sum_{t_{\lambda,\ell}\in\ker V_\kappa} \alpha_\ell \int_\Gamma U_\kappa^*(x,y)t_{\lambda,\ell}(y)ds_y = \sum_{t_{\lambda,\ell}\in\ker V_\kappa} \alpha_\ell u_{\lambda,\ell}(x) \quad \text{for all } x \in \Omega.$$

Since the eigensolutions $t_{\lambda,\ell} \in \ker V_\kappa$ are linearly independent, the induced eigensolutions $u_{\lambda,\ell}$ of the interior Dirichlet eigenvalue problem (8.17) are also linear independent, see Proposition 8.9, and therefore, $\alpha_\ell = 0$ follows for all ℓ. □

The discretisation of the coupled formulation (8.30) and (8.31) by using either a collocation or Galerkin scheme for (8.30), and choosing a finite set of interior nodes for (8.31), results in an overdetermined system of linear equations to be solved. This approach is known as *Combined Helmholtz Integral Equation Formulation* (CHIEF) [26], see also [1] for a further discussion.

In any case, the boundary integral equation (8.27) admits a unique solution $t \in H^{-1/2}(\Gamma)$ when κ^2 is not an eigenvalue of the Dirichlet eigenvalue problem (8.17), or a unique solution $t \in H_\lambda^{-1/2}(\Gamma)$ when $\lambda = \kappa^2$ is an eigenvalue of the Dirichlet eigenvalue problem (8.17). In the latter case, the general solution of the boundary integral equation (8.27) is given by

$$t = -V_\kappa^{-1}\left(\frac{1}{2}I - K_\kappa\right)g + \sum_{t_{\lambda,\ell}\in\ker V_\kappa} \alpha_\ell t_{\lambda,\ell}, \quad \alpha_\ell \in \mathbb{C},$$

where the application of V_κ^{-1} acts between appropriate factor spaces. When using, in addition, the second boundary integral equation as given by the exterior Calderon projection (8.24), this gives

$$\begin{aligned}
t &= -D_\kappa g + \left(\frac{1}{2}I - K_\kappa'\right)t \\
&= -D_\kappa g + \left(\frac{1}{2}I - K_\kappa'\right)\left[-V_\kappa^{-1}\left(\frac{1}{2}I - K_\kappa\right)g + \sum_{t_{\lambda,\ell}\in\ker V_\kappa} \alpha_\ell t_{\lambda,\ell}\right] \\
&= -\left[D_\kappa + \left(\frac{1}{2}I - K_\kappa'\right)V_\kappa^{-1}\left(\frac{1}{2}I - K_\kappa\right)\right]g + \sum_{t_{\lambda,\ell}\in\ker V_\kappa} \alpha_\ell \left(\frac{1}{2}I - K_\kappa'\right)t_{\lambda,\ell} \\
&= -S_\kappa^{\text{ext}} g,
\end{aligned}$$

i.e. the symmetric representation of the exterior Steklov–Poincaré operator

$$S_\kappa^{\text{ext}} := D_\kappa + \left(\frac{1}{2}I - K_\kappa'\right) V_\kappa^{-1} \left(\frac{1}{2}I - K_\kappa\right), \tag{8.32}$$

which is related to the exterior Dirichlet boundary value problem, and which is well defined for all wave numbers $\kappa \in \mathbb{R}$. Note that the non-symmetric representation

$$S_\kappa^{\text{ext}} = V_\kappa^{-1} \left(\frac{1}{2}I - K_\kappa\right) \tag{8.33}$$

is not well defined when $\lambda = \kappa^2$ is an eigenvalue of the Dirichlet eigenvalue problem (8.17). However, the products $V_\kappa S_\kappa^{\text{ext}}$ and $(\frac{1}{2}I - K_\kappa')S_\kappa^{\text{ext}}$ are well defined for all wave numbers $\kappa \in \mathbb{R}$ also for the non-symmetric representation (8.33).

Instead of the first-kind boundary integral equation (8.27) we may also use the second-kind boundary integral equation (8.28) to describe the solution of the exterior Dirichlet boundary value problem (8.26). To investigate the unique solvability of the boundary integral equation (8.28) we first recall that the boundary integral operator $\frac{1}{2}I + K_\kappa'$ is not injective when $\frac{1}{2}I + K_\kappa$ is not injective, and vice versa. Moreover, it is possible to characterise the null space of $\frac{1}{2}I + K_\kappa'$. As a motivation we first recall the situation when considering the Laplace equation.

Remark 8.18. In the case of the Laplace equation we have $(\frac{1}{2}I+K)u_0 = 0$ for $u_0 \equiv 1$. Since the Laplace single-layer boundary integral operator V is bijective, this implies

$$V^{-1}\left(\frac{1}{2}I + K\right) u_0 = \left(\frac{1}{2}I + K'\right) V^{-1}u_0 = \left(\frac{1}{2}I + K'\right) t_{\text{eq}} = 0,$$

where $t_{\text{eq}} = V^{-1}u_0$ is the natural density.

The relation as just described for the Laplace equation can be generalised to the case of the Helmholtz equation.

Lemma 8.19. *Let $\mu = \kappa^2$ be an eigenvalue of the Neumann eigenvalue problem* (8.19). *Any eigensolution $u_\mu \in \ker(\frac{1}{2}I + K_\kappa)$ implies an eigensolution $t_\mu \in \ker(\frac{1}{2}I + K_\kappa')$ via $u_\mu = V_\kappa t_\mu$ and vice versa.*

Proof. Let $t \in \ker(\frac{1}{2}I + K_\kappa')$, i.e. $(\frac{1}{2}I + K_\kappa')t = 0$. Define $u = V_\kappa t$ which turns out to be non-trivial. Indeed, in the case $u \equiv 0$, $V_\kappa t = 0$ implies, by using Corollary 8.8, $(\frac{1}{2}I - K_\kappa')t = 0$, and therefore $t \equiv 0$ follows. Hence we have that $u = V_\kappa t$ is non-trivial satisfying, by using (8.10),

$$\left(\frac{1}{2}I + K_\kappa\right) u = \left(\frac{1}{2}I + K_\kappa\right) V_\kappa t = V_\kappa \left(\frac{1}{2}I + K_\kappa'\right) t = 0.$$

Vice versa, let $u \in \ker(\frac{1}{2}I + K_\kappa)$, i.e. $(\frac{1}{2}I + K_\kappa)u = 0$. It remains to prove that there exists a $t \in H^{-1/2}(\Gamma)$ satisfying $V_\kappa t = u$. This is trivial when V_κ is injective. In this case we further conclude

$$\left(\frac{1}{2}I + K_\kappa\right) u = \left(\frac{1}{2}I + K_\kappa\right) V_\kappa t = V_\kappa \left(\frac{1}{2}I + K_\kappa'\right) t = 0,$$

and therefore

$$\left(\frac{1}{2}I + K'_\kappa\right) t = 0, \quad t = V_\kappa^{-1} u.$$

In the case that V_κ is not injective, let $t_\lambda \in \ker V_\kappa = \ker\left(\frac{1}{2}I - K'_{-\kappa}\right)$, see (8.21). From

$$\langle u, t_\lambda \rangle_\Gamma = \left\langle \left(\frac{1}{2}I + K_\kappa\right) u, t_\lambda \right\rangle_\Gamma + \left\langle u, \left(\frac{1}{2}I - K'_{-\kappa}\right) t_\lambda \right\rangle_\Gamma = 0$$

we conclude, by the Fredholm alternative, $u \in \operatorname{Im} V_\kappa$. Hence there exists a unique solution $t \in H^{-1/2}(\Gamma)$ satisfying

$$V_\kappa t = u, \quad \langle Vt, t_\lambda \rangle_\Gamma = 0 \quad \text{for all } t_\lambda \in \ker V_\kappa.$$

Moreover, again by using (8.10), we obtain

$$0 = \left(\frac{1}{2}I + K_\kappa\right) u = \left(\frac{1}{2}I + K_\kappa\right) V_\kappa t = V_\kappa \left(\frac{1}{2}I + K'_\kappa\right) t.$$

In the case

$$\left(\frac{1}{2}I + K'_\kappa\right) t = \sum_{t_{\lambda,\ell} \in \ker V_\kappa} \alpha_\ell t_{\lambda,\ell} \in \ker V_\kappa$$

for some $\alpha_\ell \in \mathbb{C}$ we further conclude

$$t = \sum_{t_{\lambda,\ell} \in \ker V_\kappa} \alpha_\ell t_{\lambda,\ell} + \left(\frac{1}{2}I - K'_\kappa\right) t,$$

which is obviously satisfied for

$$t = \sum_{t_{\lambda,\ell} \in \ker V_\kappa} \alpha_\ell t_{\lambda,\ell}.$$

The orthogonality

$$0 = \langle Vt, t_{\lambda,j} \rangle_\Gamma = \sum_{t_{\lambda,\ell} \in \ker V_\kappa} \alpha_\ell \langle V t_{\lambda,\ell}, t_{\lambda,j} \rangle_\Gamma \quad \text{for all } t_{\lambda,j} \in \ker V_\kappa$$

gives $\alpha_\ell = 0$ for all ℓ, and finally we obtain

$$\left(\frac{1}{2}I + K'_\kappa\right) t = 0. \qquad \square$$

Now we are in a position to investigate the unique solvability of the second-kind boundary integral equation (8.28). Since the boundary integral operator $\frac{1}{2}I + K'_\kappa$ may not be injective, in particular when $\mu = \kappa^2$ is an eigenvalue of the interior Neumann eigenvalue problem (8.19), we introduce the polar space

$$\left(\ker\left(\frac{1}{2}I + K_{-\kappa}\right)\right)^0 := \Big\{ w \in H^{-1/2}(\Gamma) : \langle w, u_\mu \rangle_\Gamma = 0 \\ \text{for all } u_\mu \in \ker\left(\frac{1}{2}I + K_{-\kappa}\right) \Big\},$$

and by the Fredholm alternative we have

$$\operatorname{Im}\left(\frac{1}{2}I + K'_\kappa\right) = \left(\ker\left(\frac{1}{2}I + K_{-\kappa}\right)\right)^0 .$$

Indeed, for $u_\mu \in \ker(\frac{1}{2}I + K_{-\kappa}) = \ker D_{-\kappa}$ we have

$$\langle D_\kappa g, u_\mu\rangle_\Gamma = \langle g, D_{-\kappa} u_\mu\rangle_\Gamma = 0,$$

and therefore, $D_\kappa g \in \operatorname{Im}(\frac{1}{2}I + K'_\kappa)$. In fact, the boundary integral equation (8.28) is solvable, but the solution may not be unique. Since $\frac{1}{2}I + K'_\kappa : H^{-1/2}(\Gamma) \to H^{-1/2}(\Gamma)$ is coercive and satisfies a Gårding inequality similar to (8.15), we can formulate the following result.

Corollary 8.20. *If κ^2 does not coincide with an eigenvalue μ of the interior Neumann eigenvalue problem* (8.19), *then $\frac{1}{2}I + K'_\kappa$ is injective, and the boundary integral equation* (8.28) *admits, for any $g \in H^{1/2}(\Gamma)$, a unique solution $t \in H^{-1/2}(\Gamma)$. Moreover, if $\mu = \kappa^2$ is an eigenvalue of the interior Neumann eigenvalue problem* (8.19), *the boundary integral equation* (8.28) *is solvable, but the solution is only unique up to eigensolutions $t_\lambda \in ker(\frac{1}{2}I + K'_\kappa)$.*

When the solution of the boundary integral equation (8.28) is not unique, we may proceed as in the previous case of the boundary integral equation (8.27) to introduce an appropriate scaling condition. In the case that the eigensolutions of the Neumann eigenvalue problem (8.19) are known, we can define a suitable factor space. In the other case, as in the CHIEF method, we may use the interior representation formula (8.25) for scaling.

Although both boundary integral equations (8.27) and (8.28) are solvable for all wave numbers $\kappa \in \mathbb{R}$, the solutions are not unique when κ^2 corresponds either to a Dirichlet or to a Neumann eigenvalue of the interior Laplace equation. Even when the Dirichlet and Neumann eigenvalues coincide, the underlying eigensolutions of the single-layer boundary integral operator V_κ and of the adjoint double-layer boundary integral operator $\frac{1}{2}I + K'_\kappa$ are different in general. This motivates to use suitable linear combinations of the boundary integral equations (8.27) and (8.28) to derive combined boundary integral equations which admit unique solutions for all wave numbers.

Following the approach of Burton and Miller [5] in the case of an exterior Neumann boundary value problem, in the case of the exterior Dirichlet boundary value problem (8.26) a complex linear combination of the boundary integral equations (8.27) and (8.28) results in the combined boundary integral equation, $\eta \in \mathbb{R}$, $\eta \neq 0$,

$$\left(\frac{1}{2}I + K'_\kappa + i\eta V_\kappa\right) t(x) = \left(-D_\kappa + i\eta\left(-\frac{1}{2}I + K_\kappa\right)\right) g(x) \quad \text{for } x \in \Gamma. \tag{8.34}$$

Lemma 8.21. *The boundary integral operator $\frac{1}{2}I + K'_\kappa + i\eta V_\kappa : H^{-1/2}(\Gamma) \to H^{-1/2}(\Gamma)$, $\eta \in \mathbb{R}$, $\eta \neq 0$, is bounded, coercive and injective. Hence, for any $g \in H^{1/2}(\Gamma)$, there exists a unique solution $t \in H^{-1/2}(\Gamma)$ of the boundary integral equation* (8.34).

Proof. The boundedness of $\frac{1}{2}I + K'_\kappa + i\eta V_\kappa : H^{-1/2}(\Gamma) \to H^{-1/2}(\Gamma)$ follows from $\frac{1}{2}I + K_\kappa : H^{-1/2}(\Gamma) \to H^{-1/2}(\Gamma)$ and $V_\kappa : H^{-1/2}(\Gamma) \to H^{1/2}(\Gamma) \subset H^{-1/2}(\Gamma)$. Since $K'_\kappa - K' + i\eta V_\kappa : H^{-1/2}(\Gamma) \to H^{-1/2}(\Gamma)$ is compact, Gårdings inequality follows.

To prove injectivity, let $w \in H^{-1/2}(\Gamma)$ be a solution of the homogeneous equation

$$\left(\frac{1}{2}I + K'_\kappa + i\eta V_\kappa\right) w(x) = 0 \quad \text{for } x \in \Gamma.$$

Define

$$u(x) = \int_\Gamma U^*_\kappa(x,y) w(y) ds_y \quad \text{for } x \in \Omega,$$

which is a solution of the interior Helmholtz equation. Moreover,

$$\frac{\partial}{\partial n_x} u(x) + i\eta u(x) = \left(\frac{1}{2}I + K'_\kappa\right) w(x) + i\eta (V_\kappa w)(x) = 0 \quad \text{for } x \in \Gamma.$$

In particular we conclude that u is a solution of the homogeneous interior Robin boundary value problem

$$\Delta u(x) + \kappa^2 u(x) = 0 \quad \text{for } x \in \Omega, \qquad \frac{\partial}{\partial n_x} u(x) + i\eta u(x) = 0 \quad \text{for } x \in \Gamma.$$

By using Theorem 8.13 we conclude $u \equiv 0$ in Ω. Then,

$$(V_\kappa w)(x) = 0, \quad \left(\frac{1}{2}I + K'_\kappa\right) w(x) = 0 \quad \text{for } x \in \Gamma.$$

Since w is an eigensolution of the single-layer boundary integral operator V_κ, we conclude, by using Corollary 8.8,

$$\left(\frac{1}{2}I - K'_\kappa\right) w(x) = 0 \quad \text{for } x \in \Gamma,$$

and therefore $w \equiv 0$ follows. □

Remark 8.22. Instead of the exterior Dirichlet boundary value problem (8.26), Burton and Miller [5] have considered the exterior Neumann boundary value problem

$$\Delta u(x) + \kappa^2 u(x) = 0 \quad \text{for } x \in \Omega^c, \qquad \frac{\partial}{\partial n_x} u(x) = 0 \quad \text{for } x \in \Gamma$$

and the radiation condition (8.3), where the total field $u = u_i + u_s$ is decomposed into a given incoming field u_i, and an unknown scattered field u_s. The total field satisfies the modified representation formula for $x \in \Omega^c$

$$u(x) = u_i(x) + \int_\Gamma \left\{ u(y) \frac{\partial}{\partial n_y} U^*_\kappa(x,y) - U^*_\kappa(x,y) \frac{\partial}{\partial n_y} u(y) \right\} ds_y. \tag{8.35}$$

From (8.35) one may conclude either the second-kind boundary integral equation

$$\left(\frac{1}{2}I - K_\kappa\right)u(x) = u_i(x) \quad \text{for } x \in \Gamma, \tag{8.36}$$

or the hypersingular boundary integral equation

$$(D_\kappa u)(x) = \frac{\partial}{\partial n_x} u_i(x) ds_x \quad \text{for } x \in \Gamma \tag{8.37}$$

which are not uniquely solvable when κ^2 corresponds either to a Dirichlet eigenvalue λ or to a Neumann eigenvalue μ, respectively. Hence, Burton and Miller proposed to consider a complex linear combination of the boundary integral equations (8.36) and (8.37), for $\alpha \in \mathbb{C}$ with $\Im(\alpha) \neq 0$,

$$\left(\frac{1}{2}I - K_\kappa + \alpha D_\kappa\right)u(x) = u_i(x) + \alpha\frac{\partial}{\partial n_x}u_i(x) \quad \text{for } x \in \Gamma. \tag{8.38}$$

While in [5] Burton and Miller discussed the uniqueness of the solution of the combined boundary integral equation (8.38) only, they did not comment on an appropriate functional analytic setting. In the case of a smooth surface, Lin [16] gives a rigorous proof on the existence and uniqueness of the solution of (8.38) in Hölder spaces.

Although the alternative representation (8.8) of the hypersingular integral operator D_κ was already given in [5], this was not used for discretisation. Instead, using (8.10), a regularised version of (8.38) was derived, i.e.

$$V\left(\frac{1}{2}I - K_\kappa + \alpha[D_\kappa - D]\right)u + \alpha\left(\frac{1}{4}I - K^2\right)u = V\left(u_i + \alpha\frac{\partial}{\partial n_x}u_i\right).$$

Instead of a complex linear combination of the boundary integral equations (8.27) and (8.28) we may also consider a system of both equations to derive a stable boundary integral formulation of the exterior Dirichlet boundary value problem (8.26). The boundary integral equations of the exterior Calderon projection (8.24) can be written as

$$u = \left(\frac{1}{2}I + K_\kappa\right)u - V_\kappa t = g, \quad D_\kappa u + \left(\frac{1}{2}I + K_\kappa'\right)t = 0.$$

Hence we need to find $(u, t) \in H^{1/2}(\Gamma) \times H^{-1/2}(\Gamma)$ such that

$$\begin{pmatrix} D_\kappa & \frac{1}{2}I + K_\kappa' \\ -\left(\frac{1}{2}I + K_\kappa\right) & V_\kappa \end{pmatrix}\begin{pmatrix} u \\ t \end{pmatrix} = \begin{pmatrix} 0 \\ -g \end{pmatrix}. \tag{8.39}$$

Note that our main interest is in the determination of the unknown Neumann datum t, while $u = g$ is given by the Dirichlet boundary condition.

Lemma 8.23. *The boundary integral operator as considered in* (8.39) *is injective in t. In particular, the homogeneous system*

$$V_\kappa t - \left(\frac{1}{2}I + K_\kappa\right)u = 0, \quad D_\kappa u + \left(\frac{1}{2}I + K_\kappa'\right)t = 0$$

implies $t = 0$ for all wave numbers κ.

Proof. Let $(u, t) \in H^{1/2}(\Gamma) \times H^{-1/2}(\Gamma)$ be a solution of the homogeneous system

$$V_\kappa t - \left(\frac{1}{2}I + K_\kappa\right) u = 0, \quad D_\kappa u + \left(\frac{1}{2}I + K'_\kappa\right) t = 0.$$

When applying the hypersingular boundary integral operator D_κ to the weakly singular boundary integral equation, and using (8.10), this gives

$$\begin{aligned} 0 &= D_\kappa V_\kappa t - D_\kappa \left(\frac{1}{2}I + K_\kappa\right) u \\ &= \left(\frac{1}{2}I + K'_\kappa\right)\left(\frac{1}{2}I - K'_\kappa\right) t - \left(\frac{1}{2}I + K'_\kappa\right) D_\kappa u \\ &= \left(\frac{1}{2}I + K'_\kappa\right)\left(\frac{1}{2}I - K'_\kappa\right) t + \left(\frac{1}{2}I + K'_\kappa\right)\left(\frac{1}{2}I + K'_\kappa\right) t = \left(\frac{1}{2}I + K'_\kappa\right) t, \end{aligned}$$

and therefore,

$$\left(\frac{1}{2}I + K'_\kappa\right) t = 0, \quad D_\kappa u = 0.$$

In the same way, when applying the single-layer boundary integral operator V_κ to the hypersingular boundary integral operator, we have

$$\begin{aligned} 0 &= V_\kappa D_\kappa u + V_\kappa \left(\frac{1}{2}I + K'_\kappa\right) t \\ &= \left(\frac{1}{2}I + K_\kappa\right)\left(\frac{1}{2}I - K_\kappa\right) u + \left(\frac{1}{2}I + K_\kappa\right) V_\kappa t \\ &= \left(\frac{1}{2}I + K_\kappa\right)\left(\frac{1}{2}I - K_\kappa\right) u + \left(\frac{1}{2}I + K_\kappa\right)\left(\frac{1}{2}I + K_\kappa\right) u = \left(\frac{1}{2}I + K_\kappa\right) u, \end{aligned}$$

and therefore,

$$\left(\frac{1}{2}I + K_\kappa\right) u = 0, \quad V_\kappa t = 0.$$

The latter also implies, by using (8.21),

$$\left(\frac{1}{2}I - K'_\kappa\right) t = 0,$$

and therefore, $t = 0$ follows. □

Note that Lemma 8.23 allows no statement concerning u. In particular, the remaining equations

$$D_\kappa u = 0, \quad \left(\frac{1}{2}I + K_\kappa\right) u = 0$$

are both satisfied when $\mu = \kappa^2$ is a Neumann eigenvalue.

Indeed, the first boundary integral equation in (8.39) is always solvable. When κ^2 is not a Neumann eigenvalue, the hypersingular boundary integral operator D_κ is bijective and we obtain

$$u = -D_\kappa^{-1}\left(\frac{1}{2}I + K'_\kappa\right) t.$$

If $\mu = \kappa^2$ is a Neumann eigenvalue, the general solution is given by

$$u = -D_\kappa^{-1}\left(\frac{1}{2}I + K_\kappa'\right)t + \sum_{u_{\mu,\ell}\in\ker D_\kappa} \alpha_\ell u_{\mu,\ell}, \quad \alpha_\ell \in \mathbb{C},$$

where the application of D_κ^{-1} has to be considered between appropriate factor spaces, see also the related discussion for the definition (8.33) of the exterior Steklov–Poincaré operator. However, in both cases we conclude the Schur complement system

$$T_\kappa t := \left[V_\kappa + \left(\frac{1}{2}I + K_\kappa\right) D_\kappa^{-1} \left(\frac{1}{2}I + K_\kappa'\right)\right] t = -g, \tag{8.40}$$

where $T_\kappa : H^{-1/2}(\Gamma) \to H^{1/2}(\Gamma)$ is also coercive, and bijective.

To avoid the use of the inverse hypersingular boundary integral operator D_κ between appropriate factor spaces, in particular when $\mu = \kappa^2$ is an eigenvalue of the interior Neumann eigenvalue problem (8.19), we can modify the system (8.39) as follows. Since $u = g$ is given due to the Dirichlet boundary condition, instead of (8.39) we now consider the modified system, $\eta \in \mathbb{R}$,

$$\begin{pmatrix} D_\kappa + i\eta\widetilde{D} & \frac{1}{2}I + K_\kappa' \\ -\left(\frac{1}{2}I + K_\kappa\right) & V_\kappa \end{pmatrix} \begin{pmatrix} u \\ t \end{pmatrix} = \begin{pmatrix} i\eta\widetilde{D}g \\ -g \end{pmatrix} \tag{8.41}$$

where $\widetilde{D}$ is the stabilised Laplace hypersingular boundary integral operator as given in (8.13). It turns out, see [32], that $D_\kappa + i\eta\widetilde{D} : H^{1/2}(\Gamma) \to H^{-1/2}(\Gamma)$ is invertible for all wave numbers $\kappa \in \mathbb{R}$, and hence, instead of (8.40) we may consider the modified Schur complement system

$$\begin{aligned} &\left[V_\kappa + \left(\frac{1}{2}I + K_\kappa\right)\left(D_\kappa + i\eta\widetilde{D}\right)^{-1}\left(\frac{1}{2}I + K_\kappa'\right)\right] t \\ &= i\eta\left(\frac{1}{2}I + K_\kappa\right)\left(D_\kappa + i\eta\widetilde{D}\right)^{-1}\widetilde{D}g - g. \end{aligned} \tag{8.42}$$

Although both Schur complement systems (8.40) and (8.42) are unique solvable, the modified version allows a direct application of standard arguments to derive a stability and error analysis of related boundary element methods.

3.2 Indirect boundary integral equations

Instead of the representation formula (8.23) of the direct approach one may also consider an indirect approach to use single- and double-layer potentials to describe solutions of the exterior Helmholtz equation (8.2) that satisfy the radiation condition (8.3). In particular, one may use either the single-layer potential ansatz

$$u(x) = \int_\Gamma U_\kappa^*(x,y) w(y) ds_y \quad \text{for } x \in \Omega^c$$

or the double-layer potential ansatz

$$u(x) = \int_\Gamma \frac{\partial}{\partial n_y} U_\kappa^*(x,y) v(y) ds_y \quad \text{for } x \in \Omega^c.$$

By considering the Dirichlet boundary condition $u = g$ on Γ we conclude the boundary integral equations to find $w \in H^{-1/2}(\Gamma)$ such that

$$(V_\kappa w)(x) = g(x) \quad \text{for } x \in \Gamma, \tag{8.43}$$

or to find $v \in H^{1/2}(\Gamma)$ such that

$$\left(\frac{1}{2}I + K_\kappa\right) v(x) = g(x) \quad \text{for } x \in \Gamma. \tag{8.44}$$

Although the boundary integral equation (8.43) of the indirect approach is similar to the boundary integral equation (8.27) of the direct approach, statements on existence and uniqueness of solutions are rather different, in particular when $\lambda = \kappa^2$ corresponds to an eigenvalue of the interior Dirichlet eigenvalue problem (8.17). In this case, V_κ is not injective, and $g \notin \operatorname{Im} V_\kappa$ in general, i.e. the boundary integral equation (8.43) is not solvable in this case. In all other cases, V_κ is injective, and therefore (8.43) admits a unique solution $w \in H^{-1/2}(\Gamma)$.

Analogously, the boundary integral equation (8.44) is in general not solvable when $\mu = \kappa^2$ corresponds to an eigenvalue of the Neumann eigenvalue problem (8.19), and hence the boundary integral operator $\frac{1}{2}I + K_\kappa$ is not injective. In all other cases, the boundary integral operator $\frac{1}{2}I + K_\kappa$ is injective, and since it is also coercive, there exists a unique solution $v \in H^{1/2}(\Gamma)$ of the boundary integral equation (8.44).

To obtain formulations of boundary integral equations which are uniquely solvable for all wave numbers one may consider complex linear combinations of the boundary integral equations (8.43) and (8.44).

As in [2], see also [23], we may consider a complex linear combination of the indirect single- and double-layer potentials to describe a solution of the exterior Helmholtz equation as

$$u(x) = \int_\Gamma v(y) \left(\frac{\partial}{\partial n_y} - i\eta\right) U_\kappa^*(x,y) ds_y \quad \text{for } x \in \Omega^c, \tag{8.45}$$

and from the Dirichlet boundary condition $u = g$ on Γ we conclude the boundary integral equation

$$\left(\frac{1}{2}I + K_\kappa - i\eta V_\kappa\right) v(x) = g(x) \quad \text{for } x \in \Gamma \tag{8.46}$$

to be solved.

Lemma 8.24. *The boundary integral operator $\frac{1}{2}I + K_\kappa - i\eta V_\kappa : H^{1/2}(\Gamma) \to H^{1/2}(\Gamma)$ is bounded, coercive and injective. Hence, for any $g \in H^{1/2}(\Gamma)$, there exists a unique solution $v \in H^{1/2}(\Gamma)$ of the boundary integral equation* (8.46).

Proof. From $\frac{1}{2}I + K_\kappa : H^{1/2}(\Gamma) \to H^{1/2}(\Gamma)$, $V_\kappa : H^{1/2}(\Gamma) \subset H^{-1/2}(\Gamma) \to H^{1/2}(\Gamma)$ we first conclude boundedness. Coercivity of $\frac{1}{2}I + K_\kappa : H^{1/2}(\Gamma) \to H^{1/2}(\Gamma)$ follows as for $\frac{1}{2}I - K_\kappa : H^{1/2}(\Gamma) \to H^{1/2}(\Gamma)$, see Lemma 8.3. Since the contraction property of the Laplace double-layer potential [29]

$$\left\| \left(\frac{1}{2}I - K \right) v \right\|_{V^{-1}} \le c_K \, \|v\|_{V^{-1}}$$

is satisfied only for $v \in H^{1/2}(\Gamma)$ with $\langle v, V^{-1}1 \rangle_\Gamma = 0$, we need to include an appropriate splitting for a general $v \in H^{1/2}(\Gamma)$, we omit the details.

To prove injectivity, let $v \in H^{1/2}(\Gamma)$ be a solution of the homogeneous boundary integral equation

$$\left(\frac{1}{2}I + K_\kappa - i\eta V_\kappa \right) v(x) = 0 \quad \text{for } x \in \Gamma,$$

and we define

$$u(x) = \int_\Gamma \frac{\partial}{\partial n_y} U_\kappa^*(x,y) v(y) ds_y - i\eta \int_\Gamma U_\kappa^*(x,y) v(y) ds_y \quad \text{for } x \in \mathbb{R}^3 \backslash \Gamma.$$

By construction, u is a solution of the exterior Dirichlet boundary value problem

$$-\Delta u(x) - \kappa^2 u(x) = 0 \quad \text{for } x \in \Omega^c, \quad u(x) = 0 \quad \text{for } x \in \Gamma.$$

Since u satisfies the Sommerfeld radiation condition (8.3), $u \equiv 0$ in Ω^c follows. On the other hand, u is also a solution of the interior Helmholtz equation

$$-\Delta u(x) - \kappa^2 u(x) = 0 \quad \text{for } x \in \Omega.$$

For the interior Dirichlet trace of u we further obtain

$$u(x) = \left(-\frac{1}{2}I + K_\kappa \right) v(x) - i\eta (V_\kappa v)(x) = -v(x) \quad \text{for } x \in \Gamma,$$

while for the interior Neumann trace of u we conclude

$$\frac{\partial}{\partial n_x} u(x) = -i\eta v(x) \quad \text{for } x \in \Gamma.$$

Note that we have used the relations of all boundary integral operators involved, and $u \equiv 0$ in Ω^c. Hence we have

$$\frac{\partial}{\partial n_x} u(x) = i\eta u(x) \quad \text{for } x \in \Gamma.$$

Hence, u is a solution of the homogeneous interior Robin boundary value problem

$$-\Delta u(x) - \kappa^2 u(x) = 0 \quad \text{for } x \in \Omega, \quad \frac{\partial}{\partial n_x} u(x) - i\eta u(x) = 0 \quad \text{for } x \in \Gamma.$$

By using Theorem 8.13, $u \equiv 0$ in Ω follows, and by the jump relation of the double-layer potential we finally conclude

$$v(x) = \gamma_0^{\text{ext}} u(x) - \gamma_0^{\text{int}} u(x) = 0 \quad \text{for } x \in \Gamma.$$

□

Remark 8.25. In [2], the authors consider the case of a twice differentiable smooth surface Γ, and the space of continuous functions, i.e. $v \in C(\Gamma)$. In this setting, the operator $K_\kappa - i\eta V_\kappa : C(\Gamma) \to C(\Gamma)$ is compact, and unique solvability of the boundary integral equation (8.46) is a consequence of Fredholm's alternative. Instead, we may also consider the boundary integral equation (8.46) in $L_2(\Gamma)$, but we still need to assume a smooth surface to ensure compactness of $K_\kappa - i\eta V_\kappa : L_2(\Gamma) \to L_2(\Gamma)$.

Instead of the indirect ansatz (8.45) we may also consider the alternative ansatz

$$u(x) = \int_\Gamma v(y) \left(i\eta \frac{\partial}{\partial n_y} + 1 \right) U_\kappa^*(x,y) ds_y \quad \text{for } x \in \Omega^c,$$

which results in the boundary integral equation

$$\left(V_\kappa + i\eta \left(\frac{1}{2} I + K_\kappa \right) \right) v(x) = g(x) \quad \text{for } x \in \Gamma. \tag{8.47}$$

Obviously, unique solvability of the boundary integral equation (8.47) follows as for (8.46). In particular, the equivalence with the boundary integral equation (8.46) ensures unique solvability in $H^{1/2}(\Gamma)$.

When we consider second-kind boundary integral equations in the natural Sobolev spaces, i.e. in $H^{\pm 1/2}(\Gamma)$, we can ensure unique solvability even in case of a general Lipschitz boundary. These results are based on ellipticity estimates of the Laplace double-layer boundary integral operator $\frac{1}{2}I - K$ in $H^{1/2}(\Gamma)$, see [29]. Since a variational formulation in $H^{\pm 1/2}(\Gamma)$ is in general not applicable for a stable and efficient discretisation scheme, alternative formulations are of interest. In most cases we may consider second-kind boundary integral equations in $L_2(\Gamma)$, but in this case we need to assume smooth boundaries to ensure compactness of the Laplace double-layer boundary integral operator. When considering, e.g. the boundary integral equation (8.46) in $L_2(\Gamma)$, we observe a mismatch in the mapping properties of the single-layer boundary integral operator V_κ, and the double-layer boundary integral operator $\frac{1}{2}I + K_\kappa$. This motivates to introduce regularised combined boundary integral equations.

3.3 Regularised combined boundary integral equations

Instead of the combined boundary integral equations (8.46) and (8.47) we now consider regularised boundary integral equations to find either $w \in H^{-1/2}(\Gamma)$ such that

$$\left[\left(\frac{1}{2} I + K_\kappa \right) B - i\eta V_\kappa \right] w(x) = g(x) \quad \text{for } x \in \Gamma, \tag{8.48}$$

or to find $v \in H^{1/2}(\Gamma)$ such that

$$\left[\left(\frac{1}{2}I + K_\kappa\right) - i\eta V_\kappa B^{-1}\right] v(x) = g(x) \quad \text{for } x \in \Gamma.$$

In both cases, $B : H^{1/2}(\Gamma) \to H^{-1/2}(\Gamma)$ is a suitable given operator. Instead of (8.48) we may also consider the equivalent formulation to find $w \in H^{-1/2}(\Gamma)$ such that

$$\left[V_\kappa + i\eta\left(\frac{1}{2}I + K_\kappa\right) B\right] w(x) = g(x) \quad \text{for } x \in \Gamma.$$

In particular, for the Laplace–Beltrami operator $B := V^2 : H^{-1}(\Gamma) \to H^1(\Gamma)$ we can use the compact imbedding $H^1(\Gamma) \hookrightarrow H^{1/2}(\Gamma)$ to prove unique solvability of the regularised boundary integral equation (8.48), see [4, 8]. Another choice is, see [10, 11], $B := \widetilde{D}^{-1}(\frac{1}{2}I + K'_{-\kappa})$, or, as in (8.42), $B = \frac{1}{i\eta}(D_\kappa + i\eta\widetilde{D})^{-1}(\frac{1}{2}I + K'_\kappa)$. For a more detailed discussion of combined boundary integral equations see also [6].

4 Transmission problems

As a model problem we consider the scattering at an interface between two media of different density [15, 19],

$$\Delta u_i(x) + \kappa_i^2 u_i(x) = 0 \quad \text{for } x \in \Omega, \quad \Delta u_e(x) + \kappa_e^2 u_e(x) = 0 \quad \text{for } x \in \Omega^c, \tag{8.49}$$

where the exterior field u_e also satisfies the Sommerfeld radiation condition (8.3). In addition we consider the inhomogeneous transmission conditions for $x \in \Gamma = \partial\Omega$,

$$\varrho_i u_i(x) - \varrho_e u_e(x) = f(x), \quad t_i(x) - t_e(x) := \frac{\partial}{\partial n_i} u_i(x) - \frac{\partial}{\partial n_i} u_e(x) = g(x), \tag{8.50}$$

where the discontinuity in the normal velocity g and in the pressure discontinuity f represent the effect of boundary layer sources, see [19]. Unique solvability of the transmission problem (8.49), (8.50) and (8.3) is based on the following result.

Theorem 8.26. [15, Theorem 3.1] *Assume $\kappa_i, \kappa_e, \varrho_i, \varrho_e \in \mathbb{C}$, $0 \le \arg \kappa_i, \arg \kappa_e < \pi$, and*

$$\varrho := \frac{\varrho_i}{\varrho_e} \frac{\overline{\kappa}_i^2}{\overline{\kappa}_e^2} \in \mathbb{R},$$

where $\varrho \ge 0$ (< 0) if $\Re(\kappa_i), \Re(\kappa_e) \ge 0$ (< 0). Then the only solution of the homogeneous transmission problem is the trivial solution.

In what follows we discuss equivalent reformulations of the transmission boundary value problem (8.49), (8.50), and (8.3) by means of boundary integral equations, in particular for wave numbers $\kappa_i, \kappa_e \in \mathbb{R}$.

Related to the interior Helmholtz equation in (8.49) we obtain the boundary integral equations

$$(V_{\kappa_i} t_i)(x) - \left(\frac{1}{2}I + K_{\kappa_i}\right) u_i(x) = 0 \quad \text{for } x \in \Gamma \tag{8.51}$$

and

$$(D_{\kappa_i} u_i)(x) - \left(\frac{1}{2}I - K'_{\kappa_i}\right) t_i(x) = 0 \quad \text{for } x \in \Gamma, \tag{8.52}$$

while for the exterior Helmholtz problem in (8.49) we conclude the boundary integral equations

$$(V_{\kappa_e} t_e)(x) - \left(-\frac{1}{2}I + K_{\kappa_e}\right) u_e(x) = 0 \quad \text{for } x \in \Gamma \tag{8.53}$$

and

$$(D_{\kappa_e} u_e)(x) + \left(\frac{1}{2}I + K'_{\kappa_e}\right) t_e(x) = 0 \quad \text{for } x \in \Gamma. \tag{8.54}$$

Together with the transmission conditions (8.50) we therefore have six equations to find the four unknowns $(u_i, t_i; u_e, t_e)$. Recall that the boundary integral equations (8.51) and (8.52) as well as (8.53) and (8.54) are not independent of each other. In what follows we may use different combinations of the above boundary integral equations to describe the unique solution of the transmission problem (8.49), (8.50), and (8.3).

4.1 Steklov–Poincaré operator equations

As in the case of the Laplace equation we may use boundary integral operators to define Steklov–Poincaré operators which realize the Dirichlet to Neumann maps for both the interior and exterior boundary value problems. But special care is required when the local wave numbers, κ_i and κ_e, coincide with eigenvalues of either the interior Dirichlet or Neumann eigenvalue problems (8.17) and (8.19), respectively. Since the transmission problem (8.49) and (8.50) admits, by Theorem 8.26, a unique solution, we aim to derive an equivalent boundary integral equation formulation which is stable for all wave numbers.

From the boundary integral equations (8.51) and (8.52) we find the characterisation

$$t_i(x) = (D_{\kappa_i} u_i)(x) + \left(\frac{1}{2}I + K'_{\kappa_i}\right) t_i(x), \quad (V_{\kappa_i} t_i)(x) = \left(\frac{1}{2}I + K_{\kappa_i}\right) u_i(x),$$

while from (8.53) and (8.54) we obtain

$$t_e(x) = -\left[D_{\kappa_e} + \left(\frac{1}{2}I - K'_{\kappa_e}\right) V_{\kappa_e}^{-1} \left(\frac{1}{2}I - K_{\kappa_e}\right)\right] u_e(x) = -\left(S^{\text{ext}}_{\kappa_e} u_e\right)(x),$$

where $S^{\text{ext}}_{\kappa_e}$ is the Steklov–Poincaré operator (8.32) of the exterior Dirichlet boundary value problem, and which is well defined for all wave numbers $\kappa_e \in \mathbb{R}$.

From the Neumann transmission condition in (8.50) we therefore conclude

$$\begin{aligned} g(x) &= t_i(x) - t_e(x) \\ &= (D_{\kappa_i} u_i)(x) + \left(\frac{1}{2}I + K'_{\kappa_i}\right) t_i(x) + \left(S^{\text{ext}}_{\kappa_e} u_e\right)(x), \end{aligned}$$

where, in addition, t_i is a solution of the local boundary integral equation

$$(V_{\kappa_i} t_i)(x) = \left(\frac{1}{2}I + K_{\kappa_i}\right) u_i(x).$$

When inserting the Dirichlet transmission condition, i.e.

$$u_e(x) = \frac{1}{\varrho_e}\left[\varrho_i u_i(x) - f(x)\right] \quad \text{for } x \in \Gamma,$$

we end up with the following system of boundary integral equations:

$$\begin{pmatrix} V_{\kappa_i} & -\left(\frac{1}{2}I + K_{\kappa_i}\right) \\ \frac{1}{2}I + K'_{\kappa_i} & D_{\kappa_i} + \frac{\varrho_i}{\varrho_e} S^{\text{ext}}_{\kappa_e} \end{pmatrix} \begin{pmatrix} t_i \\ u_i \end{pmatrix} = \begin{pmatrix} 0 \\ g + \frac{1}{\varrho_e} S^{\text{ext}}_{\kappa_e} f \end{pmatrix}. \tag{8.55}$$

Lemma 8.27. *Let the assumptions of Theorem* 8.26 *be satisfied. Then the system* (8.55) *is injective.*

Proof. Let $(t_i, u_i) \in H^{-1/2}(\Gamma) \times H^{1/2}(\Gamma)$ be a solution of the homogeneous system of boundary integral equations, i.e.

$$\begin{pmatrix} V_{\kappa_i} & -\left(\frac{1}{2}I + K_{\kappa_i}\right) \\ \frac{1}{2}I + K'_{\kappa_i} & D_{\kappa_i} + \frac{\varrho_i}{\varrho_e} S^{\text{ext}}_{\kappa_e} \end{pmatrix} \begin{pmatrix} t_i \\ u_i \end{pmatrix} = \begin{pmatrix} 0 \\ 0 \end{pmatrix}.$$

We then define

$$U_i(x) = \int_\Gamma U^*_{\kappa_i}(x,y) t_i(y) ds_y - \int_\Gamma \frac{\partial}{\partial n_y} U^*_{\kappa_i}(x,y) u_i(y) ds_y \quad \text{for } x \in \Omega,$$

which is a solution of the interior Helmholtz equation

$$-\Delta U_i(x) - \kappa_i^2 U_i(x) = 0 \quad \text{for } x \in \Omega,$$

and with the Cauchy data

$$\begin{aligned} U_i(x) &= (V_{\kappa_i} t_i)(x) + \left(\frac{1}{2}I - K_{\kappa_i}\right) u_i(x) = u_i(x), \\ \frac{\partial}{\partial n_i} U_i(x) &= \left(\frac{1}{2}I + K'_{\kappa_i}\right) t_i(x) + (D_{\kappa_i} u_i)(x). \end{aligned}$$

Since the Steklov–Poincaré operator $S^{\text{ext}}_{\kappa_e}$ of the exterior Dirichlet boundary value problem is well defined when using the symmetric representation (8.32), we introduce $t_e := -\frac{\varrho_i}{\varrho_e} S^{\text{ext}}_{\kappa_e} u_i$ and we define

$$U_e(x) = -\int_\Gamma U^*_{\kappa_e}(x,y) t_e(y) ds_y + \frac{\varrho_i}{\varrho_e} \int_\Gamma \frac{\partial}{\partial n_i} U^*_{\kappa_e}(x,y) u_i(y) ds_y \quad \text{for } x \in \Omega^c,$$

which is a solution of the exterior Helmholtz equation

$$-\Delta U_e(x) - \kappa_e^2 U_e(x) = 0 \quad \text{for } x \in \Omega^c,$$

and with the Cauchy data

$$\begin{aligned} U_e(x) &= -V_{\kappa_e} t_e(x) + \frac{\varrho_i}{\varrho_e}\left(\frac{1}{2}I + K_{\kappa_e}\right) u_i(x) \\ &= \frac{\varrho_i}{\varrho_e}\left[V_{\kappa_e} S^{\text{ext}}_{\kappa_e} + \left(\frac{1}{2}I + K_{\kappa_e}\right)\right] u_i(x) = \frac{\varrho_i}{\varrho_e} u_i(x), \\ \frac{\partial}{\partial n_i} U_e(x) &= \left(\frac{1}{2}I - K'_{\kappa_e}\right) t_e(x) - \frac{\varrho_i}{\varrho_e}(D_{\kappa_e} u_i)(x) \\ &= -\frac{\varrho_i}{\varrho_e}\left[\left(\frac{1}{2}I - K'_{\kappa_e}\right) S^{\text{ext}}_{\kappa_e} + D_{\kappa_e}\right] u_i(x) = -\frac{\varrho_i}{\varrho_e}\left(S^{\text{ext}}_{\kappa_e} u_i\right)(x). \end{aligned}$$

Note that in the above computations we can use the non-symmetric representation (8.33) of the exterior Steklov–Poincaré operator $S^{\text{ext}}_{\kappa_e}$, since in both cases the boundary integral operators V_{κ_e} and $(\frac{1}{2}I - K'_{\kappa_e})$ eliminate possible eigenfunctions in the case when $\lambda = \kappa_e^2$ is an eigenvalue of the interior Dirichlet eigenvalue problem (8.17).

Hence we conclude that (U_i, U_e) is a solution of the transmission problem

$$-\Delta U_i - \kappa_i^2 U_i = 0 \quad \text{in } \Omega, \qquad -\Delta U_e - \kappa_e^2 U_e = 0 \quad \text{in } \Omega^c$$

with the transmission conditions

$$\varrho_i U_i = \varrho_e U_e, \qquad \frac{\partial}{\partial n_i} U_i = \frac{\partial}{\partial n_i} U_e \quad \text{on } \Gamma.$$

Since U_e also satisfies the radiation condition (8.3), we can apply Theorem 8.26 to conclude $U_i \equiv 0$ in Ω and $U_e \equiv 0$ in Ω^c. From this,

$$V_{\kappa_i} t_i + \left(\frac{1}{2}I - K_{\kappa_i}\right) u_i = 0, \qquad \left(\frac{1}{2}I + K'_{\kappa_i}\right) t_i + D_{\kappa_i} u_i = 0$$

follows. Together with the first equation of the homogeneous system this results in $u_i = 0$, and therefore

$$V_{\kappa_i} t_i = 0, \qquad \left(\frac{1}{2}I + K'_{\kappa_i}\right) t_i = 0$$

follows. Since the first relation implies

$$\left(\frac{1}{2}I - K'_{\kappa_i}\right) t_i = 0,$$

also $t_i = 0$ follows. □

Now we are in a position to state the unique solvability of the system (8.55) of boundary integral equations.

Theorem 8.28. *Let the assumptions of Theorem* 8.26 *be satisfied. Then there exists a unique solution* $(t_i, u_i) \in H^{-1/2}(\Gamma) \times H^{1/2}(\Gamma)$ *of the system* (8.55).

Proof. The system (8.55) of boundary integral equations induces an operator

$$\begin{aligned} \mathcal{A}_\kappa &:= \begin{pmatrix} V_{\kappa_i} & -\left(\frac{1}{2}I + K_{\kappa_i}\right) \\ \frac{1}{2}I + K'_{\kappa_i} & D_{\kappa_i} + \frac{\varrho_i}{\varrho_e} S^{\text{ext}}_{\kappa_e} \end{pmatrix} \\ &= \begin{pmatrix} V & -\left(\frac{1}{2}I + K\right) \\ \frac{1}{2}I + K' & \widetilde{D} + \frac{\varrho_i}{\varrho_e} S^{\text{ext}} \end{pmatrix} + \begin{pmatrix} V_{\kappa_i} - V & -(K_{\kappa_i} - K) \\ K'_{\kappa_i} - K' & D_{\kappa_i} - \widetilde{D} + \frac{\varrho_i}{\varrho_e}\left(S^{\text{ext}}_{\kappa_e} - S^{\text{ext}}\right) \end{pmatrix} \\ &= \mathcal{A} + (\mathcal{A}_\kappa - \mathcal{A}), \end{aligned}$$

where $V : H^{-1/2}(\Gamma) \to H^{1/2}(\Gamma)$ is the Laplace single-layer boundary integral operator, $K : H^{1/2}(\Gamma) \to H^{1/2}(\Gamma)$ is the Laplace double-layer boundary integral operator, and $K' : H^{-1/2}(\Gamma) \to H^{-1/2}(\Gamma)$ its adjoint, and $\widetilde{D} : H^{1/2}(\Gamma) \to H^{-1/2}(\Gamma)$ is the stabilised Laplace hypersingular boundary integral operator. Moreover, S^{ext} is the related Steklov–Poincaré operator of the exterior Dirichlet boundary value problem of the Laplacian. The boundary integral operator $\mathcal{A} : H^{-1/2}(\Gamma) \times H^{1/2}(\Gamma) \to H^{1/2}(\Gamma) \times H^{-1/2}(\Gamma)$ is elliptic, while the operator $\mathcal{A}_\kappa - \mathcal{A}$ is compact. In fact, the boundary integral operator $\mathcal{A}_\kappa$ is coercive, and due to Lemma 8.27, injective. Hence we conclude unique solvability of (8.55) by Fredholm's alternative. □

If the wave number κ_i is not related to an eigenvalue of the interior Dirichlet eigenvalue problem (8.17), then the single-layer boundary integral operator V_{κ_i} is invertible, and we can determine

$$t_i = V_{\kappa_i}^{-1}\left(\frac{1}{2}I + K_{\kappa_i}\right) u_i \in H^{-1/2}(\Gamma).$$

Hence we conclude the Schur complement boundary integral equation

$$\left[S^{\text{int}}_{\kappa_i} + \frac{\varrho_i}{\varrho_e} S^{\text{ext}}_{\kappa_e}\right] u_i(x) = g(x) + \frac{1}{\varrho_e}\left(S^{\text{ext}}_{\kappa_e} f\right)(x) \quad \text{for } x \in \Gamma$$

with the Steklov–Poincaré operator of the interior Dirichlet problem

$$S^{\text{int}}_{\kappa_i} := D_{\kappa_i} + \left(\frac{1}{2}I + K'_{\kappa_i}\right) V_{\kappa_i}^{-1}\left(\frac{1}{2}I + K_{\kappa_i}\right).$$

Although the exterior Steklov–Poincaré operator $S^{\text{ext}}_{\kappa_e}$ is well defined for all wave numbers $\kappa_e \in \mathbb{R}$, alternatively, we may proceed as in (8.41) to end up with a modified system of boundary integral equations. For this, we first rewrite (8.55) as

$$\begin{pmatrix} V_{\kappa_i} & & -\left(\frac{1}{2}I + K_{\kappa_i}\right) \\ & \varrho_e V_{\kappa_e} & -\varrho_i\left(\frac{1}{2}I - K_{\kappa_e}\right) \\ \frac{1}{2}I + K'_{\kappa_i} & & D_{\kappa_i} + \frac{\varrho_i}{\varrho_e} D_{\kappa_e} \end{pmatrix} \begin{pmatrix} t_i \\ t_e \\ u_i \end{pmatrix} = \begin{pmatrix} 0 \\ -\left(\frac{1}{2}I - K_{\kappa_e}\right) f \\ g \end{pmatrix}.$$

Moreover, due to the Neumann transmission condition $t_i - t_e = g$ we may add the equality

$$-i\eta\varrho_e V t_i + i\eta\varrho_e V t_e = -i\eta\varrho_e V g, \quad \eta \in \mathbb{R},$$

to the second equation, to obtain the modified system

$$\begin{pmatrix} V_{\kappa_i} & & -\left(\frac{1}{2}I + K_{\kappa_i}\right) \\ -i\eta\varrho_e V & \varrho_e\left(V_{\kappa_e} + i\eta V\right) & -\varrho_i\left(\frac{1}{2}I - K_{\kappa_e}\right) \\ \frac{1}{2}I + K'_{\kappa_i} & & D_{\kappa_i} + \frac{\varrho_i}{\varrho_e}D_{\kappa_e} \end{pmatrix} \begin{pmatrix} t_i \\ t_e \\ u_i \end{pmatrix}$$

$$= \begin{pmatrix} 0 \\ -\left(\frac{1}{2}I - K_{\kappa_e}\right)f - i\eta\varrho_e V g \\ g \end{pmatrix}.$$

The operator $V_\kappa + i\eta V$ is invertible for all wave numbers, without any additional restriction. Hence we can eliminate t_e, and we can proceed as before to establish unique solvability of the resulting boundary integral equation system.

4.2 Combined boundary integral equations

The boundary integral equation system (8.55) was based on the use of both the interior and exterior Calderon projections (8.9) and (8.24), and on the elimination of the exterior Cauchy data t_e and u_e. Hence, (8.55) can be seen as a *single trace formulation*. Although the system (8.55) is equivalent to the original transmission problem (8.49) and (8.50), and stable for all wave numbers κ_i and κ_e, one may ask for alternative formulations, in particular from a numerical point of view.

By using parameters $\alpha_i, \alpha_e \in \mathbb{C}$ we can write the general linear combination of the boundary integral equations (8.51) and (8.53) as

$$\alpha_i\left[V_{\kappa_i}t_i - \left(\frac{1}{2}I + K_{\kappa_i}\right)u_i\right] + \alpha_e\left[V_{\kappa_e}t_e - \left(-\frac{1}{2}I + K_{\kappa_e}\right)u_e\right] = 0 \quad \text{on } \Gamma,$$

while by using parameters $\beta_i, \beta_e \in \mathbb{C}$ we obtain from (8.52) and (8.54)

$$\beta_i\left[D_{\kappa_i}u_i - \left(\frac{1}{2}I - K'_{\kappa_i}\right)t_i\right] + \beta_e\left[D_{\kappa_e}u_e - \left(-\frac{1}{2}I - K'_{\kappa_e}\right)t_e\right] = 0 \quad \text{on } \Gamma.$$

By using the transmission conditions (8.50) we may insert

$$u_e(x) = \frac{\varrho_i}{\varrho_e}u_i(x) - \frac{1}{\varrho_e}f(x), \quad t_e(x) = t_i(x) - g(x) \quad \text{for } x \in \Gamma$$

to obtain the coupled system, in the most general case,

$$\left[\alpha_i V_{\kappa_i} + \alpha_e V_{\kappa_e}\right]t_i + \frac{1}{2}\left(\alpha_e\frac{\varrho_i}{\varrho_e} - \alpha_i\right)u_i - \left[\alpha_i K_{\kappa_i} + \alpha_e\frac{\varrho_i}{\varrho_e}K_{\kappa_e}\right]u_i$$

$$= \alpha_e\left[V_{\kappa_e}g + \frac{1}{\varrho_e}\left(\frac{1}{2}I - K_{\kappa_e}\right)f\right]$$

and

$$\left[\beta_i D_{\kappa_i} + \beta_e \frac{\varrho_i}{\varrho_e} D_{\kappa_e}\right] u_i + \frac{1}{2}(\beta_e - \beta_i)\, t_i + \left[\beta_i K'_{\kappa_i} + \beta_e K'_{\kappa_e}\right] t_i$$
$$= \beta_e \left[\left(\frac{1}{2} I + K'_{\kappa_e}\right) g + \frac{1}{\varrho_e} D_{\kappa_e} f\right].$$

For the particular choice

$$\alpha_i = 1, \quad \alpha_e = \frac{\varrho_e}{\varrho_i}, \quad \beta_i = 1, \quad \beta_e = 1$$

we then conclude the system of boundary integral equations

$$\left[V_{\kappa_i} + \frac{\varrho_e}{\varrho_i} V_{\kappa_e}\right] t_i - \left[K_{\kappa_i} + K_{\kappa_e}\right] u_i = \frac{\varrho_e}{\varrho_i} V_{\kappa_e} g + \frac{1}{\varrho_i}\left(\frac{1}{2} I - K_{\kappa_e}\right) f, \tag{8.56}$$

$$\left[K'_{\kappa_i} + K'_{\kappa_e}\right] t_i + \left[D_{\kappa_i} + \frac{\varrho_i}{\varrho_e} D_{\kappa_e}\right] u_i = \left(\frac{1}{2} I + K'_{\kappa_e}\right) g + \frac{1}{\varrho_e} D_{\kappa_e} f. \tag{8.57}$$

Note that the system of the boundary integral equations (8.56) and (8.57) is known as *single trace formulation*, see, e.g. [13, 24]. Since the underlying bilinear form can be shown to be coercive on $H^{-1/2}(\Gamma) \times H^{1/2}(\Gamma)$, see, e.g. the proof of Theorem 8.28, it remains to establish injectivity.

Lemma 8.29. *The boundary integral operator*

$$\begin{pmatrix} V_{\kappa_i} + \frac{\varrho_e}{\varrho_i} V_{\kappa_e} & -K_{\kappa_i} - K_{\kappa_e} \\ K'_{\kappa_i} + K'_{\kappa_e} & D_{\kappa_i} + \frac{\varrho_i}{\varrho_e} D_{\kappa_e} \end{pmatrix} : H^{-1/2}(\Gamma) \times H^{1/2}(\Gamma) \to H^{1/2}(\Gamma) \to H^{-1/2}(\Gamma)$$

is injective.

Proof. Let $(t_i, u_i) \in H^{-1/2}(\Gamma) \times H^{1/2}(\Gamma)$ be a solution of the homogeneous system

$$\begin{pmatrix} V_{\kappa_i} + \frac{\varrho_e}{\varrho_i} V_{\kappa_e} & -K_{\kappa_i} - K_{\kappa_e} \\ K'_{\kappa_i} + K'_{\kappa_e} & D_{\kappa_i} + \frac{\varrho_i}{\varrho_e} D_{\kappa_e} \end{pmatrix} \begin{pmatrix} t_i \\ u_i \end{pmatrix} = \begin{pmatrix} 0 \\ 0 \end{pmatrix}.$$

We then define

$$U_i(x) = \int_\Gamma U^*_{\kappa_i}(x,y) t_i(y) ds_y - \int_\Gamma \frac{\partial}{\partial n_y} U^*_{\kappa_i}(x,y) u_i(y) ds_y \quad \text{for } x \in \Omega$$

and

$$U_e(x) = -\int_\Gamma U^*_{\kappa_e}(x,y) t_i(y) ds_y + \frac{\varrho_i}{\varrho_e} \int_\Gamma \frac{\partial}{\partial n_y} U^*_{\kappa_e}(x,y) u_i(y) ds_y \quad \text{for } x \in \Omega^c,$$

which are solutions of the interior and exterior Helmholtz equation, respectively. For the Cauchy data we obtain for $x \in \Gamma$

$$\begin{aligned} U_i(x) &= (V_{\kappa_i} t_i)(x) + \left(\frac{1}{2}I - K_{\kappa_i}\right) u_i(x), \\ \frac{\partial}{\partial n_i} U_i(x) &= \left(\frac{1}{2}I + K'_{\kappa_i}\right) t_i(x) + (D_{\kappa_i} u_i)(x) \end{aligned}$$

and

$$\begin{aligned} U_e(x) &= -(V_{\kappa_e} t_i)(x) + \frac{\varrho_i}{\varrho_e}\left(\frac{1}{2}I + K_{\kappa_e}\right) u_i(x), \\ \frac{\partial}{\partial n_i} U_e(x) &= \left(\frac{1}{2}I - K'_{\kappa_e}\right) t_i(x) - \frac{\varrho_i}{\varrho_e}(D_{\kappa_e} u_i)(x). \end{aligned}$$

Hence we conclude

$$\begin{aligned} U_i - \frac{\varrho_e}{\varrho_i} U_e &= V_{\kappa_i} t_i + \left(\frac{1}{2}I - K_{\kappa_i}\right) u_i - \frac{\varrho_e}{\varrho_i}\left[-V_{\kappa_e} t_i + \frac{\varrho_i}{\varrho_e}\left(\frac{1}{2}I + K_{\kappa_e}\right) u_i\right] \\ &= \left[V_{\kappa_i} + \frac{\varrho_e}{\varrho_i} V_{\kappa_e}\right] t_i - [K_{\kappa_i} + K_{\kappa_e}] u_i = 0 \end{aligned}$$

and

$$\begin{aligned} \frac{\partial}{\partial n_i} U_i - \frac{\partial}{\partial n_i} U_e &= \left(\frac{1}{2}I + K'_{\kappa_i}\right) t_i + D_{\kappa_i} u_i - \left[\left(\frac{1}{2}I - K'_{\kappa_e}\right) t_i - \frac{\varrho_i}{\varrho_e} D_{\kappa_e} u_i\right] \\ &= \left[K'_{\kappa_i} + K'_{\kappa_e}\right] t_i + \left[D_{\kappa_i} + \frac{\varrho_i}{\varrho_e} D_{\kappa_e}\right] u_i = 0, \end{aligned}$$

i.e. (U_i, U_e) is a solution of the homogeneous transmission problem

$$-\Delta U_i - \kappa_i^2 U_i = 0 \quad \text{in } \Omega, \quad -\Delta U_e - \kappa_e^2 U_e = 0 \quad \text{in } \Omega^c$$

with the transmission conditions

$$\varrho_i U_i = \varrho_e U_e, \quad \frac{\partial}{\partial n_i} U_i = \frac{\partial}{\partial n_i} U_e \quad \text{on } \Gamma.$$

Since U_e also satisfies the radiation condition (8.3), we can apply Theorem 8.26 to conclude $U_i \equiv 0$ in Ω and $U_e \equiv 0$ in Ω^c. From this,

$$V_{\kappa_i} t_i + \left(\frac{1}{2}I - K_{\kappa_i}\right) u_i = 0, \quad \left(\frac{1}{2}I + K'_{\kappa_i}\right) t_i + D_{\kappa_i} u_i = 0 \quad \text{on } \Gamma \tag{8.58}$$

follows. Next we define

$$\widetilde{U}_e(x) = -\int_\Gamma U^*_{\kappa_i}(x,y) t_i(y) ds_y + \int_\Gamma \frac{\partial}{\partial n_y} U^*_{\kappa_i}(x,y) u_i(y) ds_y \quad \text{for } x \in \Omega^c$$

which is a solution of the exterior Helmholtz equation

$$-\Delta\widetilde{U}_e(x) - \kappa_i^2\widetilde{U}_e(x) = 0 \quad \text{for } x \in \Omega^c,$$

and with the Cauchy data for $x \in \Gamma$, by using (8.58),

$$\widetilde{U}_e(x) = -(V_{\kappa_i}t_i)(x) + \left(\frac{1}{2}I + K_{\kappa_i}\right)u_i(x) = u_i(x),$$

$$\frac{\partial}{\partial n_i}\widetilde{U}_e(x) = \left(\frac{1}{2}I - K'_{\kappa_i}\right)t_i(x) - (D_{\kappa_i}u_i)(x) = t_i(x).$$

Hence, Green's formula for the exterior Helmholtz equation reads

$$\int_{\Omega^c} |\nabla\widetilde{U}_e(x)|^2 dx - \kappa_i^2 \int_{\Omega^c} |\widetilde{U}_e(x)|^2 dx = -\int_\Gamma t_i(x)\overline{u_i(x)}ds_x,$$

From the radiation condition (8.3) we conclude

$$\begin{aligned} 0 &= \lim_{r\to\infty} \int_{|x|=r} \left|\frac{\partial}{\partial n_x}\widetilde{U}_e(x) - i\kappa_i\widetilde{U}_e(x)\right|^2 ds_x \\ &= \lim_{r\to\infty}\left\{\int_{|x|=r}\left|\frac{\partial}{\partial n_x}\widetilde{U}_e(x)\right|^2 ds_x + \kappa_i^2\int_{|x|=r}\left|\widetilde{U}_e(x)\right|^2 ds_x \right. \\ &\qquad \left. -2\kappa_i\,\Im\left(\int_{|x|=r}\frac{\partial}{\partial n_x}\widetilde{U}_e(x)\overline{\widetilde{U}_e(x)}ds_x\right)\right\}, \end{aligned}$$

and using Green's formula for the bounded exterior domain $\Omega^c \cap B_r$ we further obtain

$$\Im\left(\int_\Gamma t_i(x)\overline{u_i(x)}ds_x\right) = \Im\left(\int_{|x|=r}\frac{\partial}{\partial n_x}\widetilde{U}_e(x)\overline{\widetilde{U}_e(x)}ds_x\right) = 0.$$

Hence we have

$$\lim_{r\to\infty}\int_{|x|=r}|\widetilde{U}_e(x)|^2 ds_x = 0,$$

and by the Rellich lemma, e.g. [8, Lemma 3.11], $\widetilde{U}_e \equiv 0$ in Ω^c follows. From this we further conclude $u_i = \gamma_0^{\text{ext}}\widetilde{U}_e = 0$, and therefore, by using (8.58),

$$V_{\kappa_i}t_i = 0, \quad \left(\frac{1}{2}I + K'_{\kappa_i}\right)t_i = 0.$$

Since the first relation implies $(\frac{1}{2}I - K'_{\kappa_i})t_i = 0$, also $t_i = 0$ follows. □

Recall that coercivity and injectivity of the boundary integral operator system as considered in Lemma 8.29 ensure unique solvability of the boundary integral equation system (8.56) and (8.57), in particular we find $(t_i, u_i) \in H^{-1/2}(\Gamma) \times H^{1/2}(\Gamma)$ as the unique solution of the variational problem

$$\left\langle \left(V_{\kappa_i} + \frac{\varrho_e}{\varrho_i} V_{\kappa_e}\right) t_i, \tau \right\rangle_\Gamma - \langle (K_{\kappa_i} + K_{\kappa_e}) u_i, \tau \rangle_\Gamma = \left\langle \frac{\varrho_e}{\varrho_i} V_{\kappa_e} g + \frac{1}{\varrho_i}\left(\frac{1}{2} I - K_{\kappa_e}\right) f, \tau \right\rangle_\Gamma$$

$$\left\langle \left(K'_{\kappa_i} + K'_{\kappa_e}\right) t_i, v \right\rangle_\Gamma + \left\langle \left(D_{\kappa_i} + \frac{\varrho_i}{\varrho_e} D_{\kappa_e}\right) u_i, v \right\rangle_\Gamma = \left\langle \left(\frac{1}{2} I + K'_{\kappa_e}\right) g + \frac{1}{\varrho_e} D_{\kappa_e} f, v \right\rangle_\Gamma$$

for all $(\tau, v) \in H^{-1/2}(\Gamma) \times H^{1/2}(\Gamma)$. In fact, (8.56) and (8.57) are considered as boundary integral equations of the first kind.

Alternatively, one may consider the formulation of second-kind boundary integral equations, where the single-layer boundary integral operators V_{κ_i} and V_{κ_e} as well as the hypersingular boundary integral operators D_{κ_i} and D_{κ_e} appear in the coupling terms. For the particular choice

$$\alpha_e = 1, \quad \alpha_i = -1, \quad \beta_i = -1, \quad \beta_e = \frac{\varrho_e}{\varrho_i}$$

we conclude the system of boundary integral equations

$$\begin{aligned} &[V_{\kappa_e} - V_{\kappa_i}] t_i + \frac{1}{2}\left(\frac{\varrho_i}{\varrho_e} + 1\right) u_i - \left[\frac{\varrho_i}{\varrho_e} K_{\kappa_e} - K_{\kappa_i}\right] u_i \\ &= \frac{1}{\varrho_e}\left(-\frac{1}{2} I + K_{\kappa_e}\right) f - V_{\kappa_e} g, \end{aligned} \tag{8.59}$$

$$\begin{aligned} &[D_{\kappa_e} - D_{\kappa_i}] u_i + \frac{1}{2}\left(\frac{\varrho_e}{\varrho_i} + 1\right) t_i + \left[\frac{\varrho_e}{\varrho_i} K'_{\kappa_e} - K'_{\kappa_i}\right] t_i \\ &= \frac{\varrho_e}{\varrho_i}\left[\left(-\frac{1}{2} I - K'_{\kappa_e}\right) g - \frac{1}{\varrho_e} D_{\kappa_e} f\right]. \end{aligned} \tag{8.60}$$

A closer look on the difference $V_{\kappa_e} - V_{\kappa_i}$ of the single-layer boundary integral operators indicates a reduced order of the singularity involved, i.e.

$$\begin{aligned} (V_{\kappa_e} t_i)(x) - (V_{\kappa_i} t_i)(x) &= \frac{1}{4\pi} \int_\Gamma \frac{e^{i\kappa_e|x-y|} - e^{i\kappa_i|x-y|}}{|x-y|} t_i(y) ds_y \\ &= \frac{i}{2\pi} \int_\Gamma \frac{e^{i\frac{\kappa_e+\kappa_i}{2}|x-y|} \sin \frac{\kappa_e-\kappa_i}{2}|x-y|}{|x-y|} t_i(y) ds_y. \end{aligned}$$

Since also the difference $D_{\kappa_e} - D_{\kappa_i}$ has a reduced order in its singularity, the system of boundary integral equations (8.59) and (8.60) is known as *minimal coupling formulation*, see [19], i.e. *the minimal coupling formulation involves lower order singularities in the kernels of the coupling terms than does any other formulation generated from* the combined boundary integral equations.

5 Conclusions

In this work, several boundary integral formulations for the solution of the exterior Dirichlet boundary value problem, and for transmission problems of the Helmholtz equation with piecewise constant wave numbers were presented. Although the focus of this contribution was on the existence and uniqueness of solutions, and the equivalence with the solution of the underlying boundary value and transmission problems, almost all formulations allow for the use of standard arguments, see, e.g. [25, 27], within a stability and error analysis of related Galerkin boundary element methods. Note that it is almost impossible to compare and to rate different boundary integral formulations and different boundary element implementations. The method of choice always depends on several aspects, e.g. the application in mind, the mathematical foundation, the required accuracy, the availability of fast and accurate boundary element implementations including parallel and preconditioned iterative solution strategies, the use of a posteriori error estimators and adaptivity, just to name a few. Although there are already some numerical studies around for a comparison of different boundary element approaches, in particular when considering exterior boundary value problems, it seems that more work is required to state a fair comparison of different boundary integral formulations and related boundary element implementations.

The focus of future work will be on the numerical analysis of stable, robust, and efficient boundary element domain decomposition methods for the solution of transmission problems, including multiple trace formulations [7] and tearing and interconnecting methods [31, 32]. In addition to local and global preconditioning strategies, which are required to be robust with respect to local wave numbers, the handling of adaptive and non-matching discretisations becomes more challenging. The latter also includes the coupling with conforming and non-standard finite element methods, including mixed finite elements, discontinuous Galerkin, and Mortar finite elements.

While the current considerations of this contribution were restricted to the case of acoustic scattering problems, almost all approaches and methodologies as presented here can be carried over to the case of electromagnetic scattering problems, see, e.g. the discussion in [7], in [30, 34], and the references given therein.

References

[1] W. Benthien and A. Schenck, Nonexistence and nonuniqueness problems associated with integral equation methods in acoustics. *Comput. & Structure* **65** (1997), 295–305.

[2] H. Brakhage and P. Werner, Über das Dirichletsche Aussenraumproblem für die Hemholtzsche Schwingungsgleichung. *Arch. Math.* **16**, (1965) 325–329.

[3] A. Buffa and R. Hiptmair, Coercive combined field integral equations. *J. Numer. Math.* **11** (2003), 115–133.

[4] A. Buffa and R. Hiptmair, Regularized combined field integral equations. *Numer. Math.* **100** (2005), 1–19.

[5] A. J. Burton and G. F. Miller, The application of integral equation methods to the numerical solution of some exterior boundary value problems. *Proc. R. Soc. Lond. Ser. A* **323** (1971), 201–210.

[6] S. N. Chandler–Wilde, I. G. Graham, S. Langdon, and E. A. Spence, Numerical–asymptotic boundary integral methods in high–frequency acoustic scattering. *Acta Numer.* **21** (2012), 89–305.

[7] X. Claeys, R. Hiptmair, and C. Jerez–Hanckes, Multi–trace boundary integral equations. In this volume, 51–100.

[8] D. Colton and R. Kress, Integral Equation Methods in Scattering Theory. John Wiley & Sons, Chichester, 1983.

[9] M. Costabel, Boundary integral operators on Lipschitz domains: Elementary results. *SIAM J. Math. Anal.* **19** (1988), 613–626.

[10] S. Engleder and O. Steinbach, Modified boundary integral formulations for the Helmholtz equation. *J. Math. Anal. Appl.* **331** (2007), 396–407.

[11] S. Engleder and O. Steinbach, Stabilized boundary element methods for exterior Helmholtz problems. *Numer. Math.* **110** (2008), 145–160.

[12] S. Esterhazy and J. M. Melenk, On stability of discretizations of the Helmholtz equation. In: Numerical Analysis of Multiscale Problems (I. G. Graham, T. Y. Hou, O. Lakkis, R. Scheichl eds.). Lecture Notes in Computational Science and Engineering, vol. 83, Springer, Heidelberg, pp. 285–324, 2012.

[13] R. Hiptmair, C. Jerez–Hanckes, Multiple traces boundary integral formulation for Helmholtz transmission problems. *Adv. Comput. Math.* **37** (2012), 39–91.

[14] G. C. Hsiao and W. L. Wendland, Boundary Integral Equations. Applied Mathematical Sciences, vol. 164. Springer, Berlin, 2008.

[15] R. Kress, G. F. Roach, Transmission problems for the Helmholtz equation. *J. Math. Phys.* **19** (1978), 1433–1437.

[16] T.–C. Lin, A proof for the Burton and Miller integral equation approach for the Helmholtz equation. *J. Math. Anal. Appl.* **103** (1984), 565–574.

[17] A. W. Maue, Zur Formulierung eines allgemeinen Beugungsproblems durch eine Integralgleichung. *Z. Phys.* **126** (1949), 601–618.

[18] W. McLean, Strongly Elliptic Systems and Boundary Integral Equations. Cambridge University Press, 2000.

[19] K. M. Mitzner, Acoustic scattering from an interface between media of greatly different density. *J. Math. Phys.* **7** (1966), 2053–2060.

[20] J.–C. Nédélec, Integral equations with nonintegrable kernels. *Integral Equations Operator Theory* **5** (1982), 562–572.

[21] J.–C. Nédélec, Acoustic and electromagnetic equations. Applied Mathematical Sciences, vol. 144, Springer, New York, 2001.

[22] G. Of and O. Steinbach, A fast multipole boundary element method for a modified hypersingular boundary integral equation. In: Analysis and Simulation of Multifield Problems (W. L. Wendland, M. Efendiev eds.). Lecture Notes in Applied and Computational Mechanics, vol. 12, Springer, Heidelberg, pp. 163–169, 2003.

[23] O. I. Panič, On the solvability of exterior boundary-value problems for the wave equation and for a system of Maxwell's equations. *Uspehi Mat. Nauk* **20** (1965), 221–226.

[24] T. von Petersdorff, Boundary integral equations for mixed Dirichlet, Neumann and transmission problems. *Math. Meth. Appl. Sci.* **11** (1989), 185–213.

[25] S. A. Sauter and C. Schwab, Boundary Element Methods. Springer Series in Computational Mathematics, vol. 39, Springer, Berlin, 2011.

[26] H. A. Schenck, Improved integral formulation for acoustic radiation problems. *J. Acoust. Soc. Am.* **44** (1968), 41–58.

[27] O. Steinbach, Numerical Approximation Methods for Elliptic Boundary Value Problems. Finite and Boundary Elements. Springer, New York, 2008.

[28] O. Steinbach and G. Unger, Convergence analysis of a Galerkin boundary element method for the Dirichlet Laplacian eigenvalue problem. SIAM J. Numer. Anal. **50** (2012), 710–728.

[29] O. Steinbach and W. L. Wendland, On C. Neumann's method for second–order elliptic systems in domains with non–smooth boundaries. *J. Math. Anal. Appl.* **262** (2001), 733–748.

[30] O. Steinbach and M. Windisch, Modified combined field integral equations for electromagnetic scattering. *SIAM J. Numer. Anal.* **47** (2009), 1149–1167.

[31] O. Steinbach and M. Windisch, Robust boundary element domain decomposition solvers in acoustics. In: Domain Decomposition Methods in Science and Engineering XIX (Y. Huang, R. Kornhuber, O. B. Widlund, J. Xu eds.). Lecture Notes in Computational Science and Engineering, vol. 78, Springer, Heidelberg, pp. 277–284, 2011.

[32] O. Steinbach and M. Windisch, Stable boundary element domain decomposition methods for the Helmholtz equation. *Numer. Math.* **118** (2011), 171–195.

[33] G. Unger, Analysis of boundary element methods for Laplacian eigenvalue problems. PhD thesis, Monographic Series TU Graz, Computation in Engineering and Science, vol. 6, 2009.

[34] M. Windisch, Boundary element tearing and interconnecting methods for acoustic and electromagnetic scattering. PhD thesis, Monographic Series TU Graz, Computation in Engineering and Science, vol. 11, 2011.

Color plates

Color plates for the chapter by Ammari

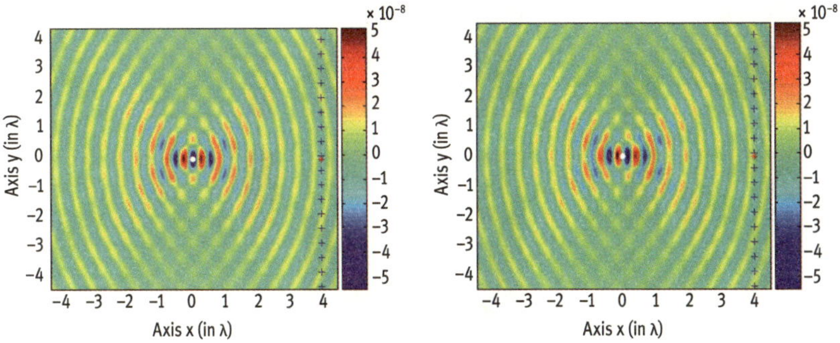

Figure 1.3: Real and imaginary part of the back-propagation. Here "*" and "+" respectively show the transceiver and the receiver positions.

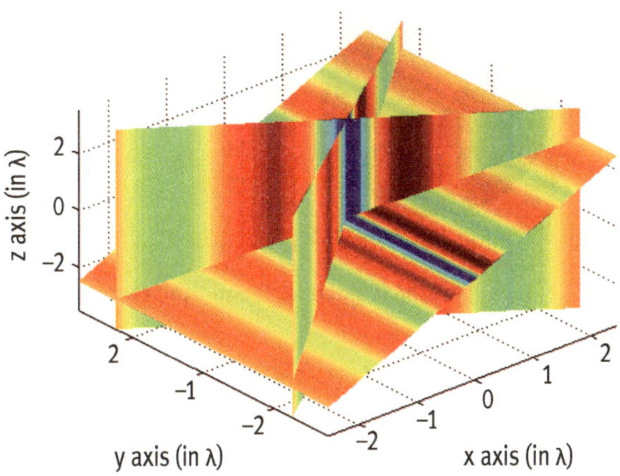

Figure 1.4: Intersection of the three planes where the real parts of the Kirchhoff functionals attain their maximum for three different receivers.

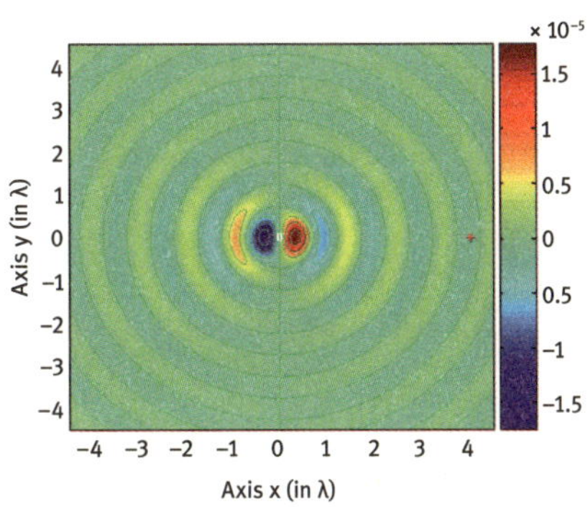

Figure 1.5: Detection result with time-reversal technique, "*" indicates transceiver location.

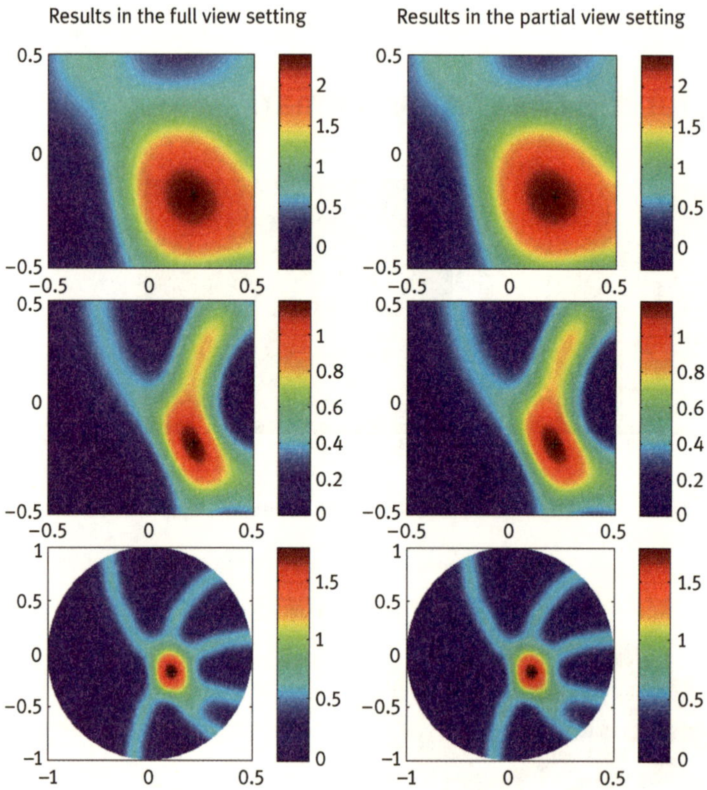

Figure 1.6: Kirchhoff results. The (black/white) × denotes the (numerical/theoretical) center of the target.

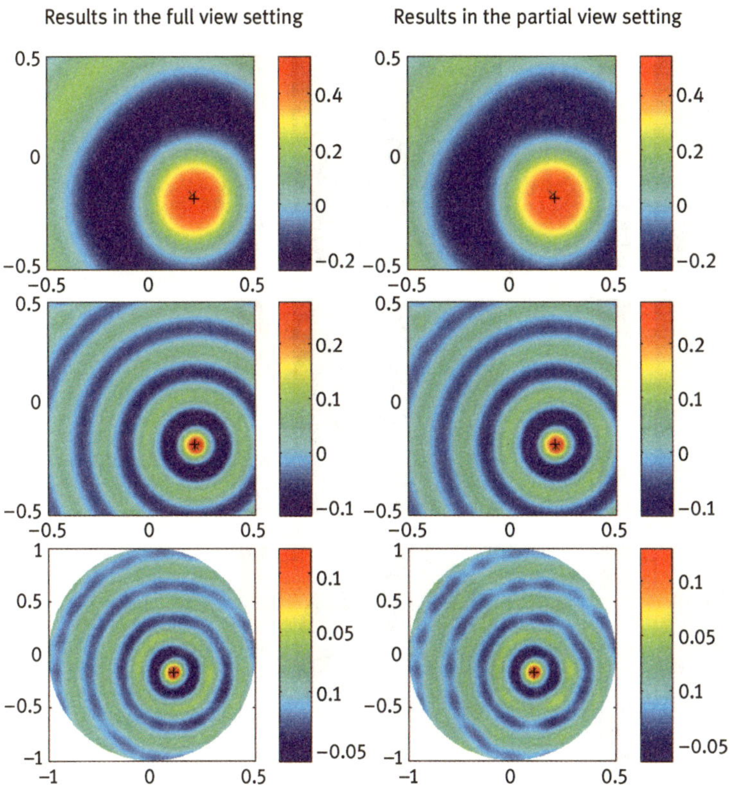

Figure 1.7: Back-propagation results. The (black/white) × denotes the (numerical/theoretical) center of the source.

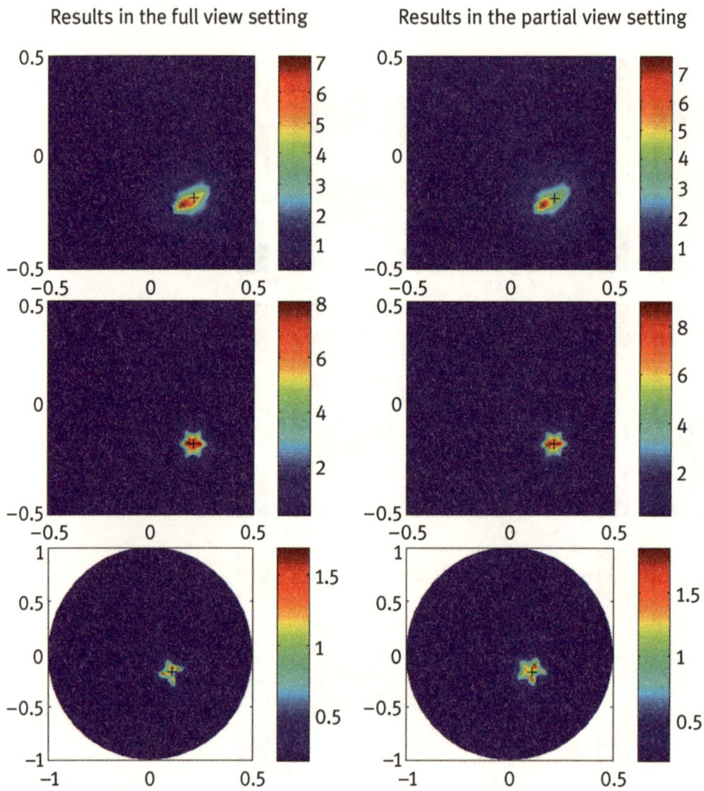

Figure 1.8: MUSIC results. The (black/white) × denotes the (numerical/theoretical) center of the source.

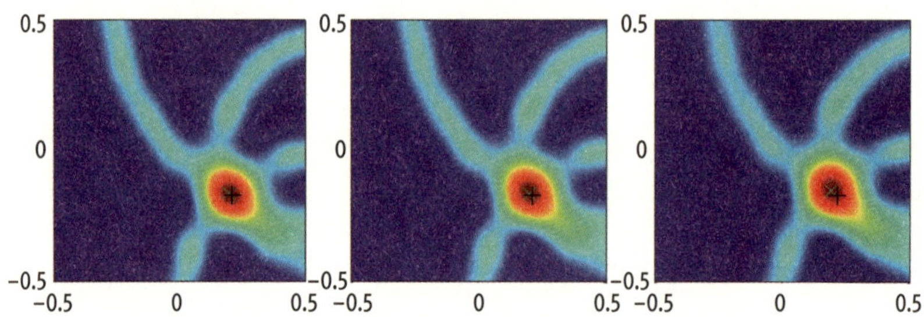

Figure 1.9: Kirchhoff results with perturbed boundary (from left-to-right, 0.01, 0.025 and 0.05). The (black/white) × denotes the (numerical/theoretical) center of the source.

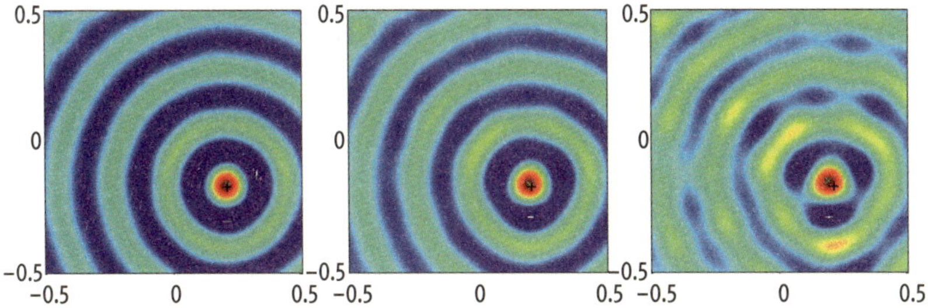

Figure 1.10: Back-propagation results with perturbed boundary (from left-to-right, 0.01, 0.025 and 0.05). The (black/white) $\times$ denotes the (numerical/theoretical) center of the source.

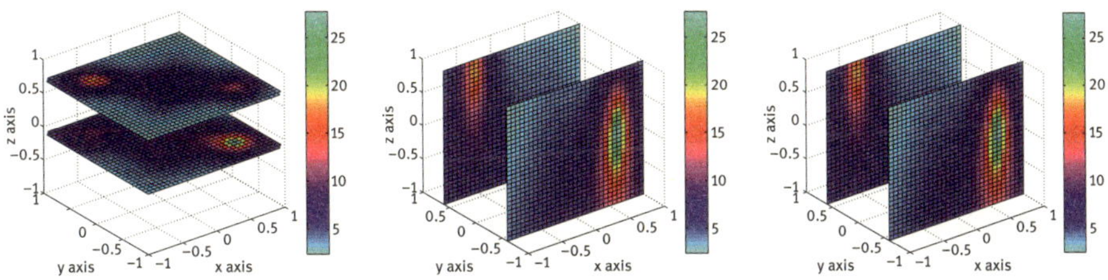

Figure 1.13: (Dielectric and permeability contrasts): color maps of W_{MU} in cross-sectional planes $z = \{-0.1, 0.7\}$, $x = \{-0.6, 0.6\}$, and $y = \{-0.6, 0.6\}$ in the case of noisy data with 20dB signal-to-noise ratio.

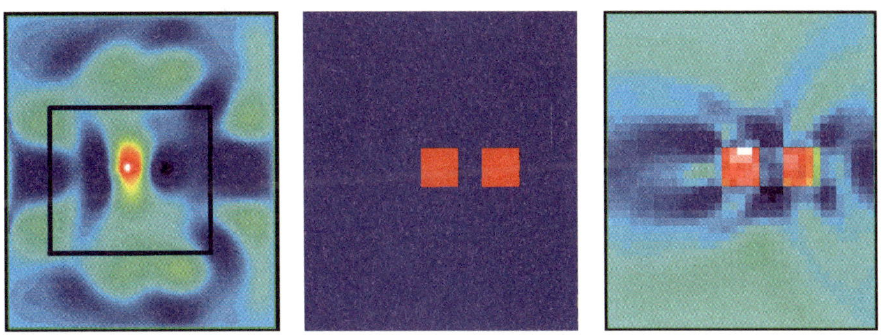

Figure 1.23: Reconstruction using the data on the whole domain on the left, a zoom on the target in the middle, and on the right the reconstruction limited on the subregion defined by the boxed region on the left.

Color plates for the chapter by Elschner & Hu

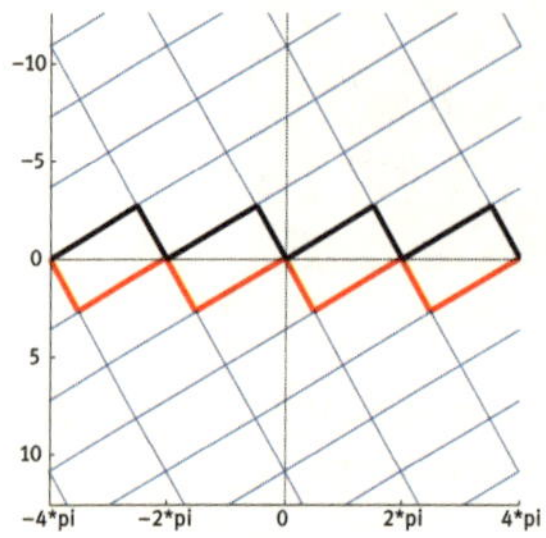

Figure 3.1: $\mathcal{D}_2(\theta, k_p)$ with $\theta = -\frac{\pi}{6}, k_p = 2$.

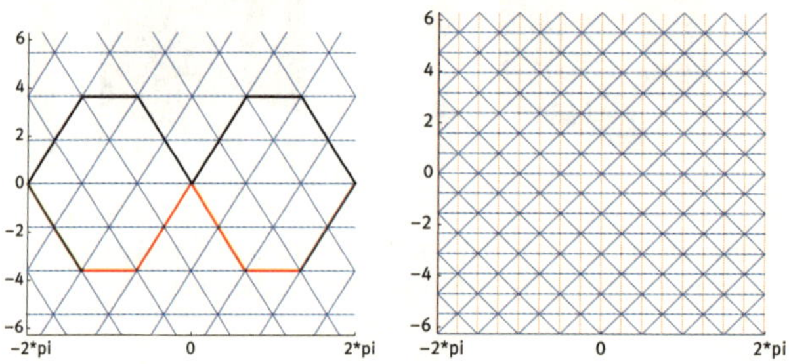

Figure 3.2: Left: $\mathcal{N}_3(\theta, k_p)$ with $\theta = -\frac{\pi}{6}$, $k_p = 2$. Right: $\mathcal{N}_4(0, k_p)$ with $k_p = 4$.

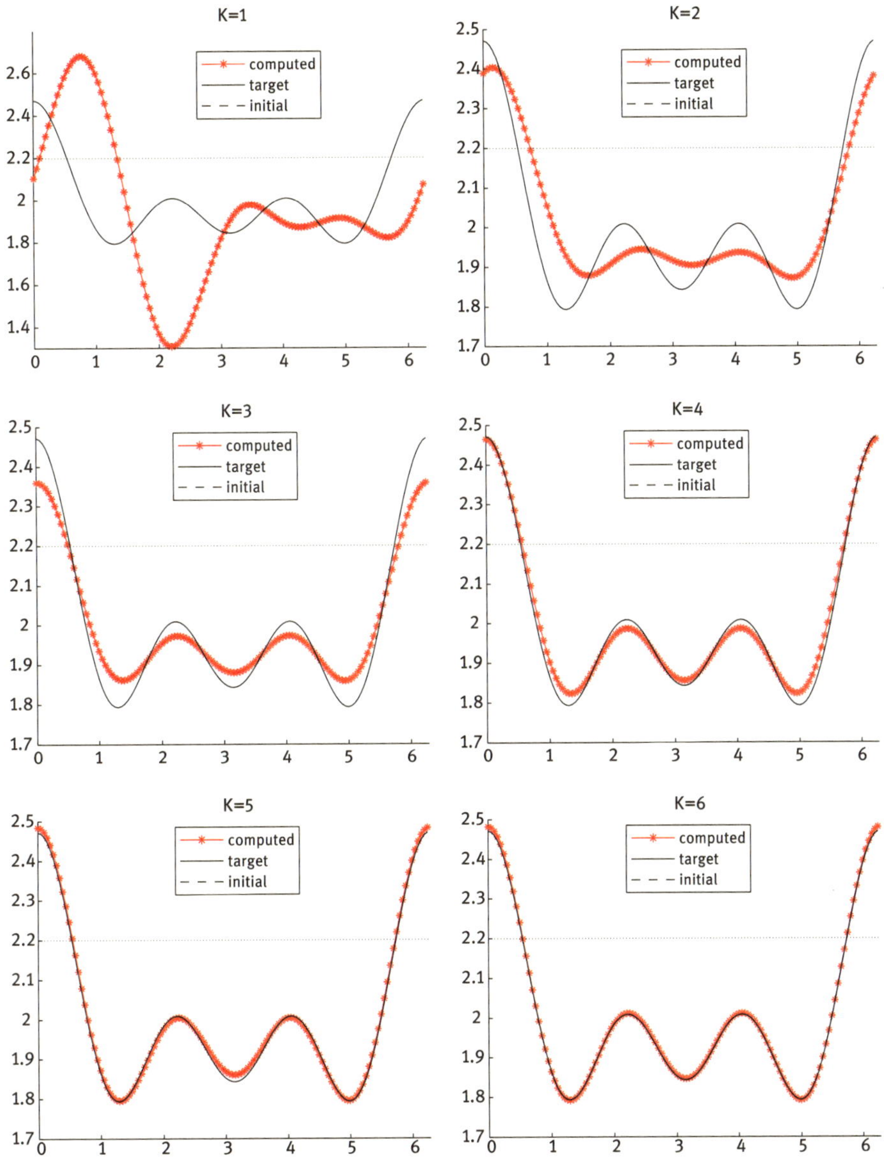

Figure 3.3: Sensitivity of the method to the parameter K in Example 3.19.

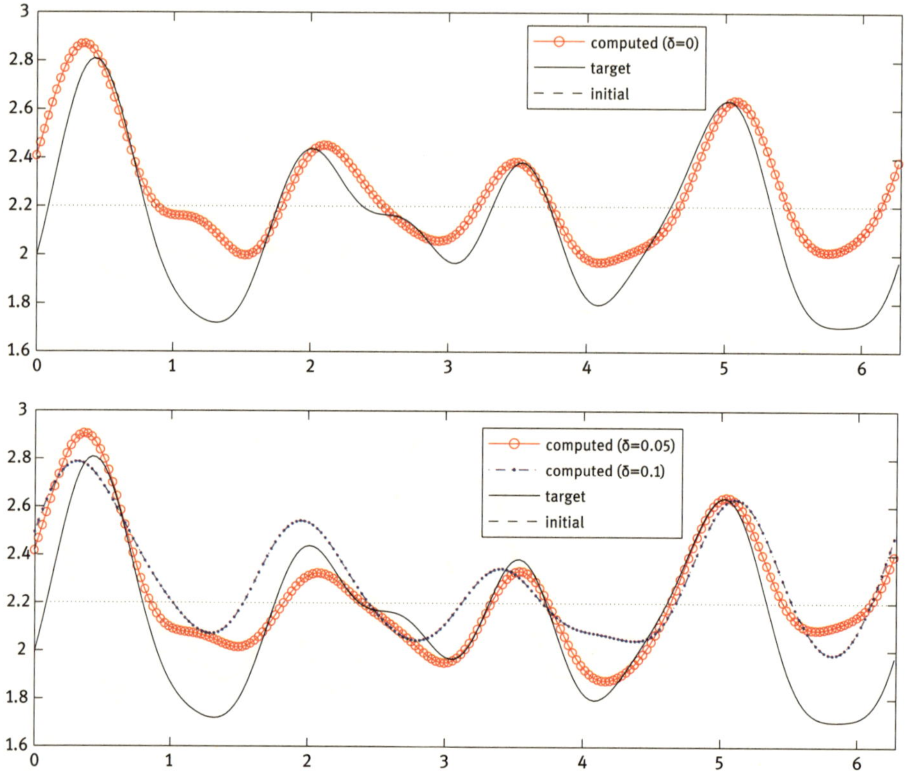

Figure 3.4: Example 3.20. $K = 7, M = 8$.

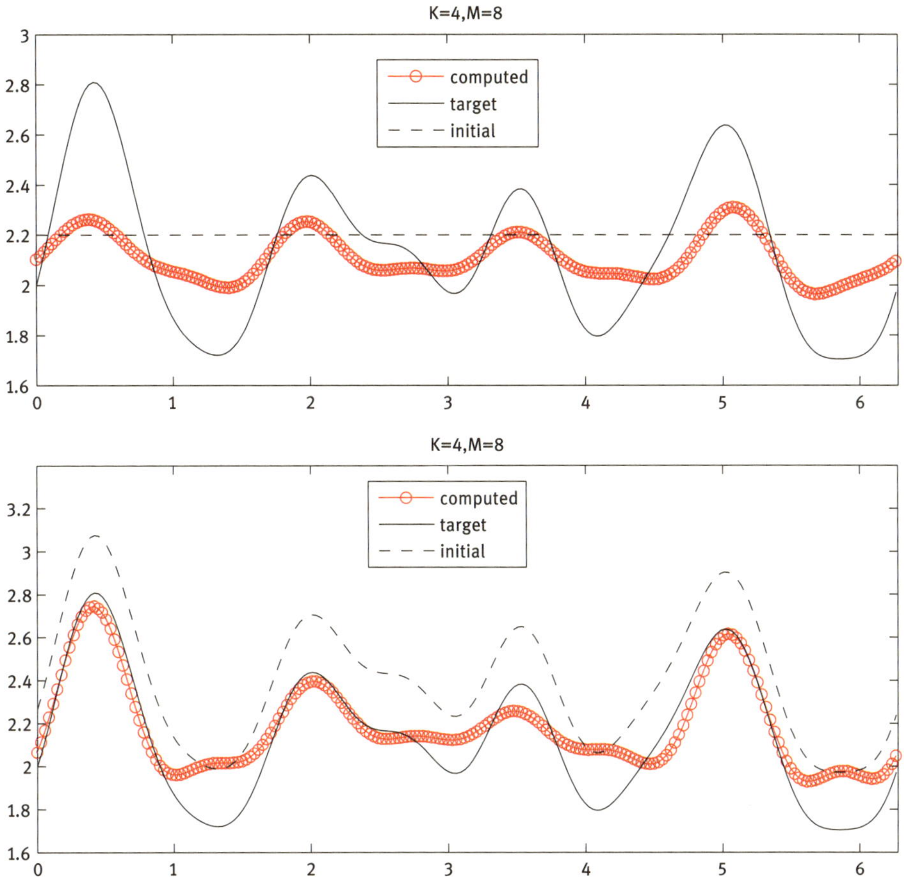

Figure 3.5: Example 3.20. $K = 4, M = 8, \delta = 0$.

Color plates for the chapter by Ernst & Gander

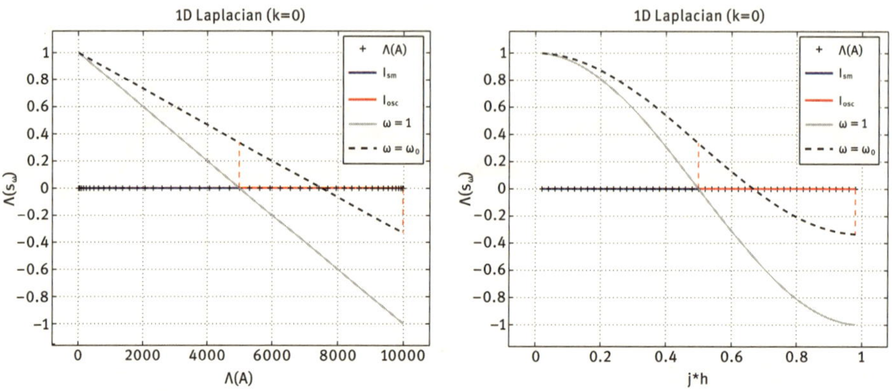

Figure 4.2: Left: eigenvalues of the undamped and optimally damped Jacobi smoother plotted against those of the associated 1D Laplacian $-\Delta_h$, $h = 1/50$, divided into smooth and oscillatory parts I_{sm} and I_{osc}. The dashed red lines indicate the spectral radius of $\mathbf{S}_\omega$ restricted to the space of oscillatory eigenfunctions. Right: same results, but with smoother eigenvalues σ_j plotted against jh.

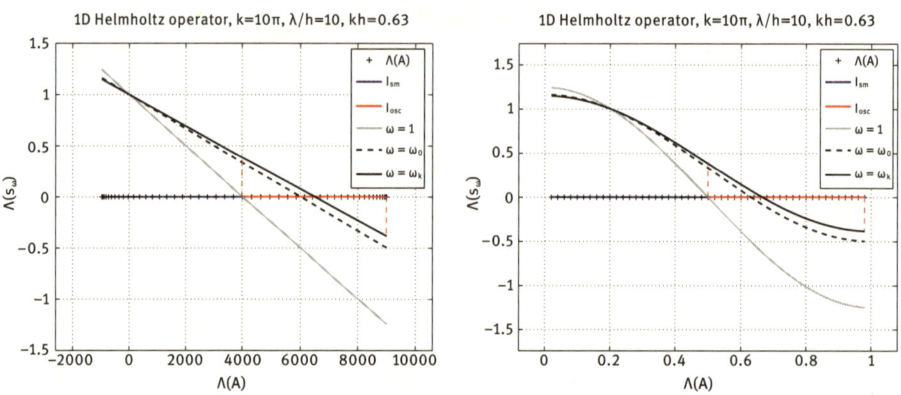

Figure 4.3: Left: eigenvalues of the Jacobi smoother plotted with $\omega = \omega_k$, $\omega = 1$ and $\omega = \omega_0$ against those of the associated 1D Helmholtz operator $-\Delta_h - k^2$, $h = 1/50$ with wavelength-to-mesh ratio $\lambda/h = 10$. Right: same result in equispaced representation.

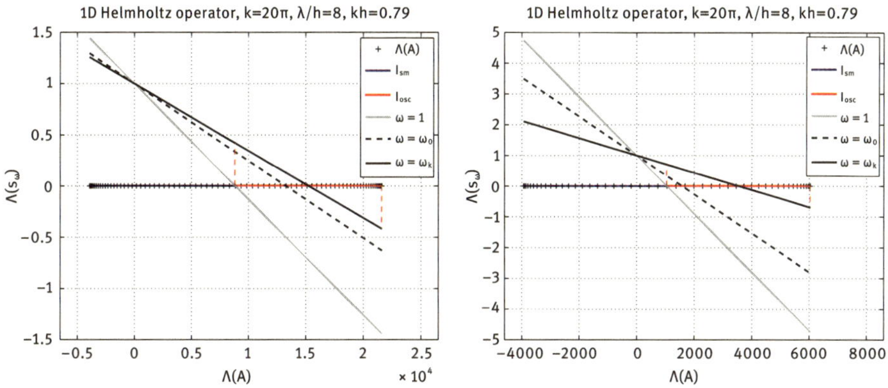

Figure 4.4: Eigenvalues of the damped Jacobi smoother with larger wavelength-to-mesh ratio $\lambda/h = 8$ and $\lambda/h = 5$. The damping on I_{osc} is becoming less while smooth modes at the lower end of I_{sm} are amplified significantly.

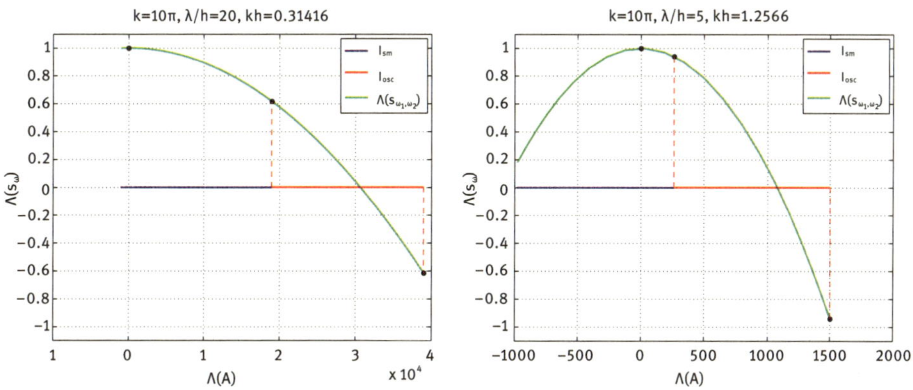

Figure 4.5: Two-step Jacobi smoothing with parameter choice (2.14).

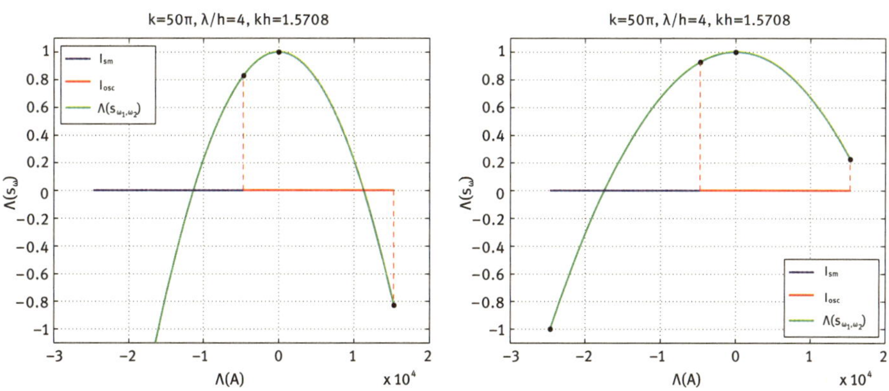

Figure 4.6: Left: two-step Jacobi smoother with parameter choice (2.14) from Case (1) at resolution $\lambda/h = 4$, for which $\lambda_{\text{mid}} < 0 < \lambda_N$. The smooth modes are seen to be strongly amplified. Right: two-step Jacobi smoother with parameter choice (2.16) from Case (2) at resolution $\lambda/h = 4$.

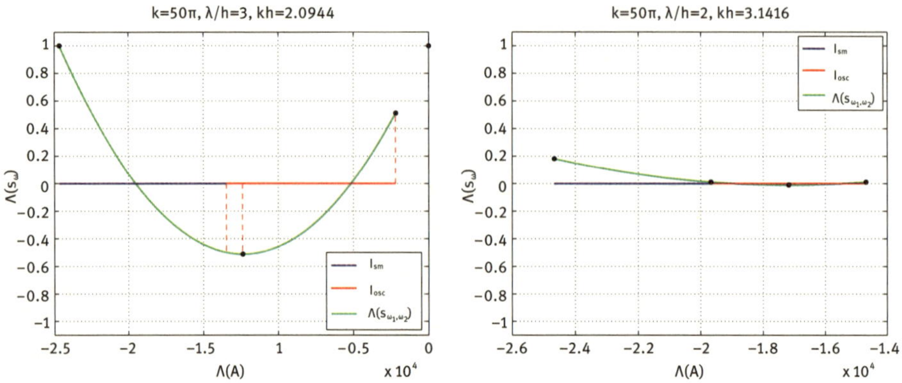

Figure 4.7: Two-step Jacobi smoothing with parameter choices (2.19) (left) and (2.20) (right), respectively.

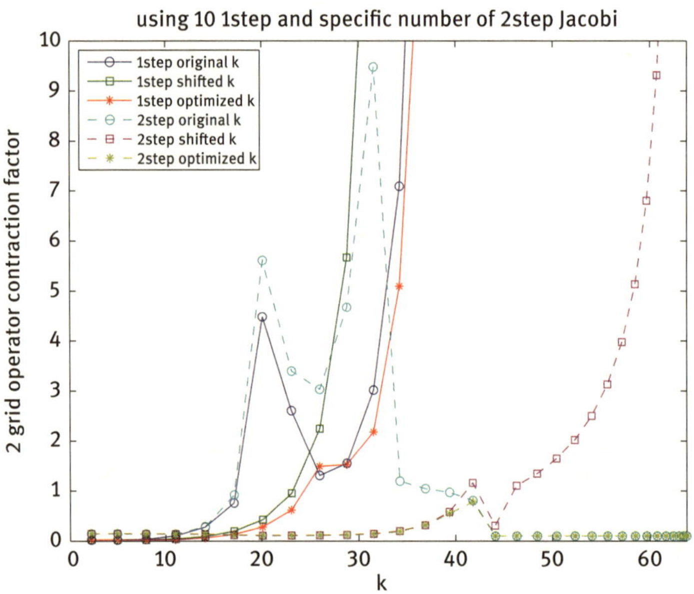

Figure 4.15: Spectral radii of the two grid algorithm using various components, for a fixed mesh size h, and increasing wave number k.

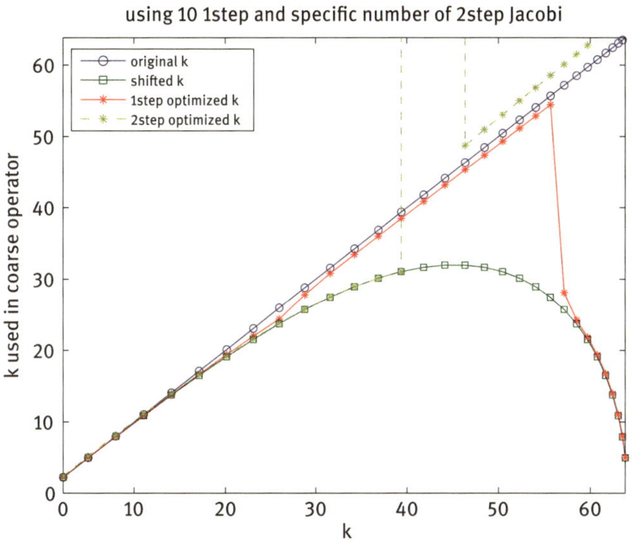

Figure 4.16: Shifts used in the two grid algorithm of Figure 4.15.

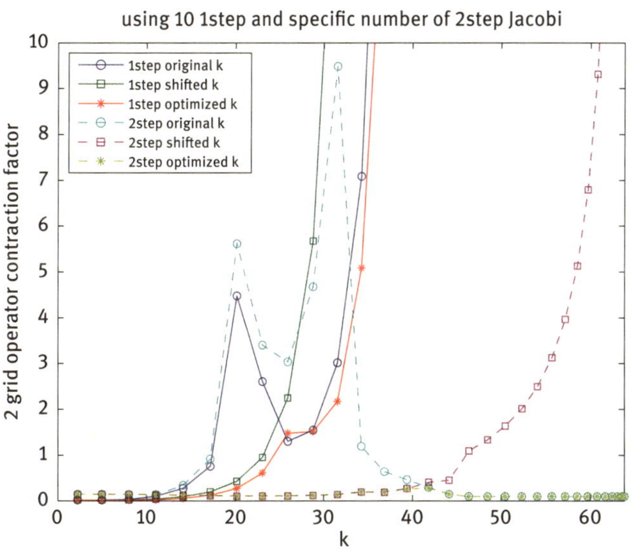

Figure 4.18: Contraction factors of the two grid algorithm using various components, for a fixed mesh size h, and increasing wave number k, now using the $O(k^2)$ smoothing steps already a bit before the resonance level.

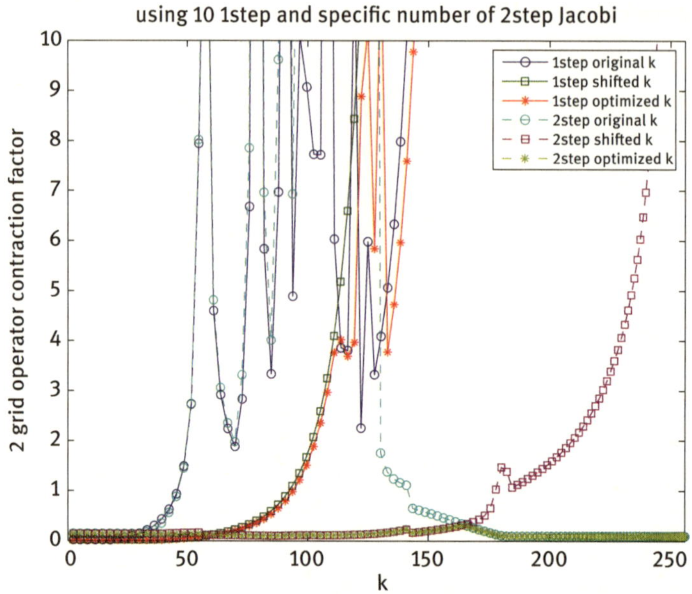

Figure 4.20: Contraction factors of the two grid algorithm using various components, for a fixed mesh size $h = 1/128$, and increasing wave number k.

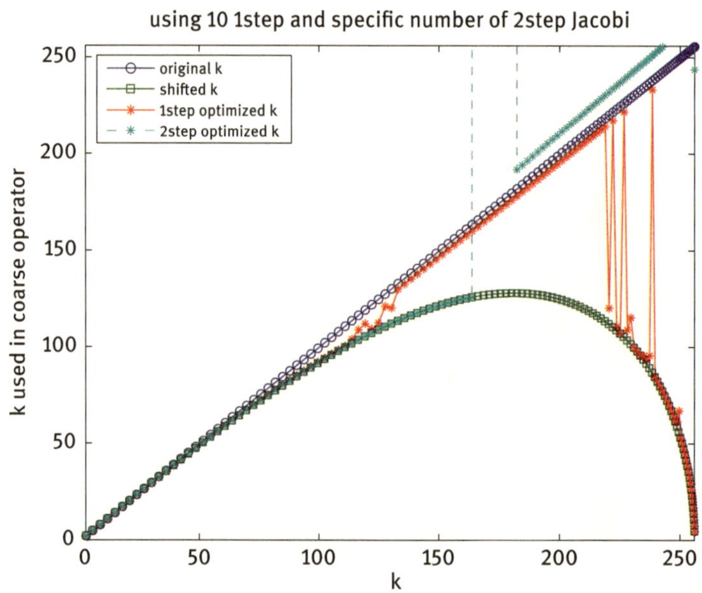

Figure 4.21: Shifts used in the two grid algorithm of Figure 4.20.

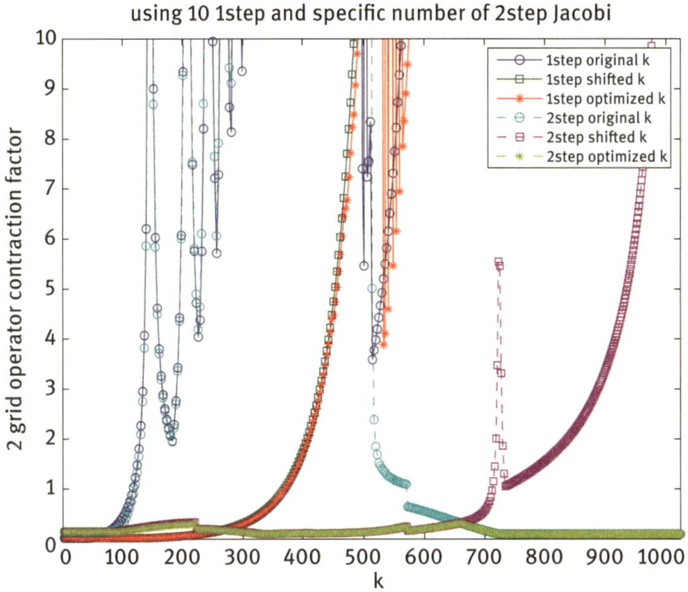

Figure 4.22: Contraction factors of the two grid algorithm using various components, for a fixed mesh size $h = 1/512$, and increasing wave number k.

Color plates for the chapter by Grote & Mitkova

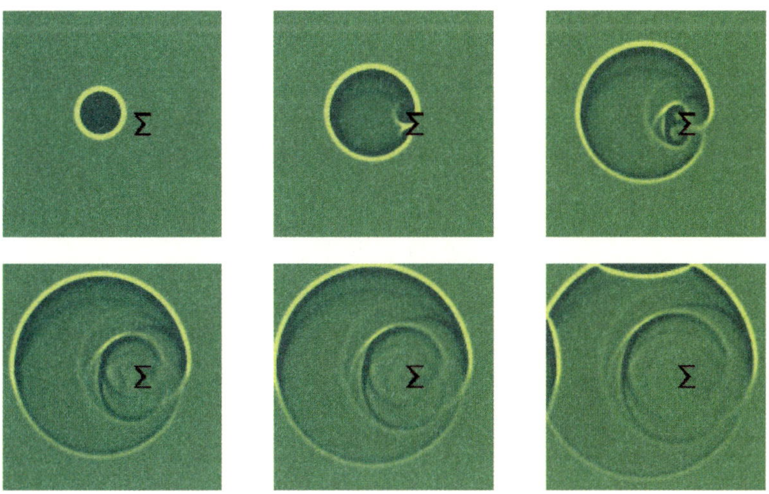

Figure 5.4: Gaussian pulse penetrating a cavity with a small sigma-shaped hole. The solution is shown at times $t = 0.1, 0.2, 0.3, 0.4, 0.44$ and 0.5.

Index

Radon Series on Computational and Applied Mathematics

Volume 13
Mike Cullen, Melina A Freitag, Stefan Kindermann, Robert Scheichl (Eds.), 2013
Large Scale Inverse Problems.
Computational Methods and Applications in the Earth Sciences
ISBN: 978-3-11-028223-8, eISBN: 978-3-11-028228-3, ISSN: 1865-3707

Volume 12
Peter Bastian, Johannes Kraus, Robert Scheichl, Mary Wheeler (Eds.), 2013
Simulation of Flow in Porous Media. Applications in Energy and Environment
ISBN 978-3-11-028221-4, e-ISBN 978-3-11-028224-5, Set-ISBN 978-3-11-028225-2

Volume 11
Pascale Charpin, Alexander Pott, Arne Winterhof (Eds.), 2013
Finite Fields and Their Applications
ISBN 978-3-11-028240-5, e-ISBN 978-3-11-028360-0, Set-ISBN 978-3-11-028361-7

Volume 10
Thomas Schuster, Barbara Kaltenbacher, Bernd Hofmann, Kamil S. Kazimierski, 2012
Regularization Methods in Banach Spaces
ISBN 978-3-11-025524-9, e-ISBN 978-3-11-025572-0, Set-ISBN 978-3-11-220450-4

Volume 9
Massimo Fornasier (Ed.), 2010
Theoretical Foundations and Numerical Methods for Sparse Recovery
ISBN 978-3-11-022614-0, e-ISBN 978-3-11-022615-7, Set-ISBN 978-3-11-174177-2

Volume 8
Hansjörg Albrecher, Wolfgang J. Runggaldier, Walter Schachermayer (Eds.), 2009
Advanced Financial Modelling
ISBN 978-3-11-021313-3, e-ISBN 978-3-11-021314-0, Set-ISBN 978-3-11-173185-8

Volume 7
Jan H. Maruhn, 2009
Robust Static Super-Replication of Barrier Options
ISBN 978-3-11-020468-1, e-ISBN 978-3-11-020851-1, Set-ISBN 978-3-11-916585-3

www.degruyter.com